Fluorine-Containing Free Radicals

Kinetics and Dynamics of Reactions

John W. Root, EDITOR
University of California, Davis

Based on a symposium sponsored by the Division of Physical Chemistry at the 169th Meeting of the American Chemical Society, Philadelphia, Pennsylvania, April 7–11, 1975.

ACS SYMPOSIUM SERIES 66

AMERICAN CHEMICAL SOCIETY
WASHINGTON, D. C. 1978

Library of Congress CIP Data

Main entry under title:
Fluorine-containing free radicals.
(ACS symposium series; 66 ISSN 0097-6156)

Includes bibliographical references and index.

1. Radicals (Chemistry)—Congresses. 2. Fluorides—Congresses.
I. Root, John W., 1935– . II. American Chemical Society. Division of Physical Chemistry. III. Series: American Chemical Society. ACS symposium series; 66.

QD471.F67 546'.731 77-26667
ISBN 0-8412-0399-7 ACSMC 8 66 1-423

PRINTED IN THE UNITED STATES OF AMERICA

ACS Symposium Series

Robert F. Gould, *Editor*

FOREWORD

The ACS SYMPOSIUM SERIES was founded in 1974 to provide a medium for publishing symposia quickly in book form. The format of the SERIES parallels that of the continuing ADVANCES IN CHEMISTRY SERIES except that in order to save time the papers are not typeset but are reproduced as they are submitted by the authors in camera-ready form. As a further means of saving time, the papers are not edited or reviewed except by the symposium chairman, who becomes editor of the book. Papers published in the ACS SYMPOSIUM SERIES are original contributions not published elsewhere in whole or major part and include reports of research as well as reviews since symposia may embrace both types of presentation.

CONTENTS

PREFACE

This volume was conceived originally after a symposium that I organized for the Physical Chemistry Division. The symposium consisted of two half-day invited-lecture sessions presented in Philadelphia in April, 1975.

Chemical kinetics and dynamics research with fluorine-rich radicals (including atomic fluorine) has become increasingly popular during the seventies. Since the Philadelphia symposium was not comprehensive, this subject received further discussion one year later during three half-day invited-lecture sessions that were given in New York City. This second symposium, which was co-sponsored by the Divisions of Physical Chemistry and Fluorine Chemistry and Technology, included topics on the kinetics, reactivities, and structures of fluorine-containing radicals.

The present volume was intended initially to provide the first comprehensive treatment of the kinetic and dynamic characteristics of these interesting chemical species. Following acceptance of this proposal by the A.C.S. Symposium Series Editorial Advisory Board, invitations to participate were sent to many individuals throughout the English-speaking scientific community. All of the contributors in kinetics and dynamics at both American Chemical Society symposia were included, but multiple coverage of individual topics was strongly discouraged. Participants were requested specifically to aim for comprehensive state of the art reviews and to prepare theoretical presentations addressed mainly to experimentally oriented readers.

Partially because of last minute cancellations, several important topics have not been included in the present work. Papers on fluorine atom chemical lasers, high-energy fluorine atom reactions, and classical trajectory simulations of hydrogen transfer reactions by atomic fluorine are conspicuously absent. Other more specialized topics not included were olefin addition reactions by fluorine-rich carbenes, H_2/F_2 explosions, unimolecular reaction dynamics of fluorine-containing aliphatic radicals, and high-power fluorine atom lasers in energy research.

Hopefully these omissions will not detract too severely from the usefulness of this work. The above unpublished topics and the rapid pace of ongoing research in this exciting field clearly suggest the need for another project of this type in the near future.

I wish to thank the participants for their excellent work, the many individuals with whom this project has been discussed during the past two years, Mr. Robert Gould of the A.C.S. Books Department for guidance and assistance, Ms. Cecelia Damian who performed the bulk of the minor editorial revisions, and Dr. James Muckerman of Brookhaven National Laboratory for the materials used in the design of the dust jacket.

University of California, Davis
Davis, California
November 1, 1977

JOHN W. ROOT

Classical Kinetics: Atomic Fluorine

1

Elementary Reaction Kinetics of Fluorine Atoms, FO, and NF Free Radicals

E. H. APPELMAN
Argonne National Laboratory, Argonne, IL 60439

M. A. A. CLYNE
Department of Chemistry, Queen Mary College, London E1 4NS, England

The study of reactions of free fluorine atoms in their ground $2p^5\ {}^2P_{3/2,1/2}$ states has progressed rapidly in the last several years. Major technique difficulties have been gradually overcome, and several methods of determining F atom concentrations are now available. Elementary reaction kinetics in low pressure flow systems have received particular attention, probably because of the relative ease of interfacing flow systems to F-atom detection devices such as mass spectrometers and vacuum-uv atomic resonance spectrometers. Photolytic production of F atoms is less easy - F_2, for instance, absorbs only very weakly in the near ultraviolet or visible regions. Similarly, detection of F ^2P_J atoms in static systems presents problems that have not been fully solved at the present time. Therefore, the work described in this article has mostly been carried out in flow systems.

Current interest in fluorine-based chemical lasers extends from HF and DF lasers based on F atom reactions into the area of elementary processes involving electronically excited radicals such as NF. Also, the chemistry of NF and OF radicals is of considerable interest, and we summarize here selected aspects of the kinetic behaviour of these species.

Production of F ^2P_J atoms

Ground state ^2P_J halogen atoms may be produced by direct dissociation of the molecular halogens, F_2, Cl_2, Br_2 and I_2. A convenient and commonly-used technique is to pass a mixture of molecular halogen diluted with argon or helium through a microwave discharge at a total pressure near 100-200 N m^{-2}. This is the simplest technique for forming F ^2P_J atoms (1-4); the typical degree of dissociation of F_2 using an uncoated silica discharge tube is 70-80%(1). Mixtures of fluorine diluted with inert gas can be handled in a conventional glass flow vacuum system. A

© 0-8412-0399-7/78/47-066-003$10.00/0

problem with microwave-excited dissociation of F_2 is that SiF_4 and molecular and atomic oxygen are produced as a result of attack on the SiO_2 material of the discharge and flow tubes (1, 2, 5). Rosner and Allendorf (3) have replaced the silica discharge tube by one fabricated from pure fused alumina, which is inert to attack by fluorine atoms. However, this reduces the efficiency of dissociation somewhat. The pyrex or silica flow tube may be protected from attack by F atoms by application of a thin coating of teflon or Kel-F fluorocarbon polymer (2). A technique for coating with a thin fused layer of teflon has been described (6). The presence of undissociated F_2 in the discharge products causes complications in kinetic studies involving hydrogen atoms or alkyl radicals, which react rapidly with F_2. However, F_2 appears to be unreactive towards such other common atoms as O ^3P, N ^4S, Cl ^2P_J and Br ^2P_J (2,7).

Production of F ^2P_J atoms can also be achieved by microwave dissociation of fluorides such as SF_6 and CF_4 (8). These sources are satisfactory for spectroscopic studies such as the epr spectroscopy of F $^2P_{3/2}$ and F $^2P_{1/2}$ (8), or for the formation of excited BrF and IF from the recombination of Br + F or I + F in the presence of singlet oxygen (9). However, they are in some degree suspect for systematic quantitative kinetic studies, in as much as the nature and reactivity of the discharge products other than F ^2P_J are very incompletely known.

In principle, a cleaner source of F ^2P_J atoms than any of the preceding ones is the rapid reaction of N ^4S atoms with NF_2 radicals produced by thermolysis of N_2F_4 (10,11). This reaction occurs either directly to give F atoms $N + NF_2 \rightarrow 2F + N_2$ (10) or via the formation of NF radicals; $N + NF_2 \rightarrow 2NF$ (11), followed by $NF + NF \rightarrow N_2 + 2F$ or $N + NF \rightarrow N_2 + F$. In practice, the instability of N_2F_4 and its expense have precluded extensive use of the $N + NF_2$ method for forming F atoms.

The formation of F ^2P_J atoms from the bimolecular reaction of NO with F_2 is extensively used in HF, DF chemical lasers: $NO + F_2 \rightarrow FNO + F$. The kinetics of this reaction have been studied (12,13).

Measurement of Atom Concentrations: absolute concentrations.

The range of methods that have been employed to measure atom concentrations is wide, and includes thermal and diffusion methods which are not considered here. Most of the physical methods available are suitable only for relative concentration measurements, and require calibration by chemical means in order to yield absolute concentrations. This need not always be a disadvantage; for instance, in kinetic studies of simple atom reactions under pseudo-first order conditions it is sometimes only necessary to monitor relative changes in atom concentrations.

Titration reactions based on simple stoichiometry, and proceeding extremely rapidly, have been developed for the measurement of absolute concentrations of ground state atoms in flow systems. Well-known examples for N, O and H atoms include the reactions, $N + NO \rightarrow N_2 + O$, $O + NO_2 \rightarrow NO + O$, $H + ClNO \rightarrow HCl + NO$ and $H + NO_2 \rightarrow OH + NO$. These reactions permit the determination of absolute atom concentrations by the addition of known quantities of a stable reagent. Several similar titration reactions have been characterized and used to determine F atom concentrations, namely

$$F + ClNO \rightarrow NO + ClF \qquad (1),$$

$$F + Cl_2 \rightarrow ClF + Cl \qquad (2),$$

$$F + Br_2 \rightarrow BrF + Br \qquad (3),$$

$$F + H_2 \rightarrow HF + H \qquad (4).$$

The reaction of ClNO with F has been used to determine F atom concentrations by mass spectrometric measurement of the ClF produced after the addition of excess ClNO. The ClF^+ ion peak was calibrated by the addition of measured quantities of ClF from a cylinder (14). To eliminate the need to work with the reactive ClF gas, a variant of this method was developed in which the F atoms were titrated with ClNO to an endpoint identified as the amount of added ClNO beyond which the ClF^+ peak no longer incr- eased (1). The ClNO methods are less well established than some other, inasmuch as the rate constant of reaction (1) has not been determined. Also, at high concentrations, the reaction $NO + F_2 \rightarrow FNO + F$, which has a rate constant of 8×10^{-15} cm^3 $molecule^{-1}$ s^{-1}at 300 K, may occur to an appreciable extent, leading to an erroneous value for [F].

The reaction of F atoms with Cl_2 appears to be an excellent means of determining absolute F atom concentrations. Reaction (2) has a forward rate constant around 1×10^{-10} cm^3 $molecule^{-1}$ s^{-1} at 300 K and an equilibrium constant greater than 5 (1, 7,15). Hence the reverse reaction is unlikely to be important under the correct conditions. Furthermore, the possible secondary reaction

$$Cl + F_2 \rightarrow ClF + F$$

has a rate constant less than 5×10^{-14} cm^3 $molecule^{-1}$ s^{-1} at 300 K (1) and may be safely neglected. Reaction (2) has been used to determine F atom concentrations mass spectrometrically by addition of a measured excess concentration of Cl_2 and subsequent determination of either the amount of Cl_2 consumed (7) or the amount of ClF produced (4). In the latter case the ClF^+ ion current was calibrated by measuring its intensity when an excess of F atoms was added to a known amount of Cl_2. Measurement of ClF production is more sensitive than measurement of Cl_2 consumption and is also free from errors that might result from recombination of Cl atoms or reaction of Cl_2 with other species such as H or O. Reaction (2) has also been utilized as a titration of F atoms, with the endpoint estimated to be the point of critical extinction of the F atom fluorescence (7). In an alternative titration procedure, the chemiluminescence accompanying the recombination of the Cl atoms formed in reaction (2) is monitored, and Cl_2 is added until the intensity of this emission reaches a plateau. At low pressures, this method yields a sharp endpoint (16).

Concentrations of F atoms have also been determined by titration with Br_2 to the point of extinction of the F atom fluorescence (7). Reaction(3) has a rate constant of about 2×10^{-10} cm^3 $molecule^{-1}$ sec^{-1} at 298 K, and an equilibrium constant in excess of 10^9 (7). Hence its back reaction is of no concern and in this respect it should be superior to reaction (2) as a means of determining F atom concentrations. On the other hand, the rate of the reaction (5),

$$Br + F_2 \rightarrow BrF + F \qquad (5)$$

does not appear to be known. In practice reactions (2) and (3) appear to be equally satisfactory for the determination of F atom concentrations.

Reaction (4) has a rate constant of about 2×10^{-11} cm^3 $molecule^{-1}$ s^{-1} at 300 K (10) and has been used to determine F atom concentration. However, the secondary reaction

$$F_2 + H \rightarrow HF + F \qquad (6)$$

has a rate constant of about 3×10^{-12} cm^3 $molecule^{-1}$ s^{-1} at 300 K. (17). Inasmuch as reaction (6) regenerates the F atoms consumed in reaction (4), when a substantial amount of F_2 is present the added H_2 will first consume the F_2 and only subsequently will it consume the F. Advantage has been taken of this effect to titrate both F_2 and F with hydrogen mass spectrometrically, using an inlet system that incorporates an inhomogeneous magnetic field to permit monitoring of both the F and F_2 (2).

Measurement of Atom Concentrations: Atomic Resonance

The method of atomic resonance spectrometry in the vacuum ultraviolet, with detection of either absorption or fluorescence, has become one of the most useful direct methods for the measurements of reaction rates of ground (18)(and metastable excited (19,20)) state atoms. The sensitivity and scope of atomic resonance in this respect rivals, and possibly surpasses, that of other methods such as epr and mass spectrometry. Recently, it has been shown that atomic resonance may be used to measure fluorine atom concentrations in a flow system (7,21). The sensitivity for F is not yet as high as for atoms such as Cl, whose resonance transitions lie in the Schumann region of the vacuum ultraviolet.

The essence of the method is as follows. The source of atomic resonance radiation, usually a microwave-excited discharge in a low-pressure flowing gas, is incident on the reaction vessel, either static or flowing, in which ground state atoms are present.

On account of the coincidence of the emission line source and the absorption line of ground state atoms, specific and very intense absorption of the exciting radiation occurs. Either the fractional absorption intensity, $A=I_{abs}/I_o$, or the corresponding fluorescence intensity, I_F, may be measured and related to the concentration of absorbing atoms, N. Figure 1 shows a schematic diagram of the atomic resonance experiment.

The central problem of relating light intensity to concentration of absorbers, N, has been discussed (22-24). The fluorescence intensity varies in direct proportion to N at sufficiently low values of N (typically $\leqslant 10^{12}$ cm^{-3} (24)). This proportionality clearly facilitates kinetic studies. However, absorption intensities at such low atom concentrations cannot normally be measured, and the dependence of I_{abs}/I_o upon N at measureable absorbances is complex (22,25). On the other hand, the magnitude of detected fluorescence intensity, I_F, is always several orders of magnitude less than that of I_{abs}, which can lead to low signal-to-noise ratios when short integrating times are employed for the measurement of light intensity. In such cases, e.g. in flash photolysis experiments (26), it is not generally possible to use both time resolution and wavelength resolution when atomic resonance fluorescence is being used to measure the kinetics of elementary reactions. On account of the long integrating times possible in discharge-flow systems, both time resolution (using linear displacement) and wavelength resolution may routinely be used. There is then no difficulty in working in the region where I_F varies proportionally with N (27).

The most usual approach in resonance absorption and fluorescence work has been to use microwave lamps, and to calibrate resonance absorption or fluorescence intensity by chemical titration (see above). Extensive studies of this type on I and Br $^2P_{3/2,1/2}$ atoms (19, 20, 27,28) and on Cl $^2P_{3/2,1/2}$ atoms (26 - 30) have been described. Detection of F atoms by resonance absorption using a conventional microwave lamp has been successful (7,21). In this case, the fluorine atom lamp developed was based on a dilute mixture of F_2 with He. (Argon lamps gave extremely low intensities of F atom lines). Figures 2 and 3 show the relevant energy levels of F and the spectrum of the F_2 + He microwave-excited lamp. Because the first resonance transition of $F(3s-2p^5)$ lies well below the lithium fluoride absorption cut-off ($\sim$ 105 nm), the usual windows were replaced by collimated hole structures in order to separate the lamp from the absorbing atoms in the flow tube and from the spectrometer. The arrangement used (7) is shown in Figure 4.

For reasons already given, resonance fluorescence would be a preferred technique for kinetic studies of F atom reactions. Bemand and Clyne (7) detected resonance fluorescence in a number of lines of the $3s-2p^5$ multiplet, using the apparatus of figure 4.

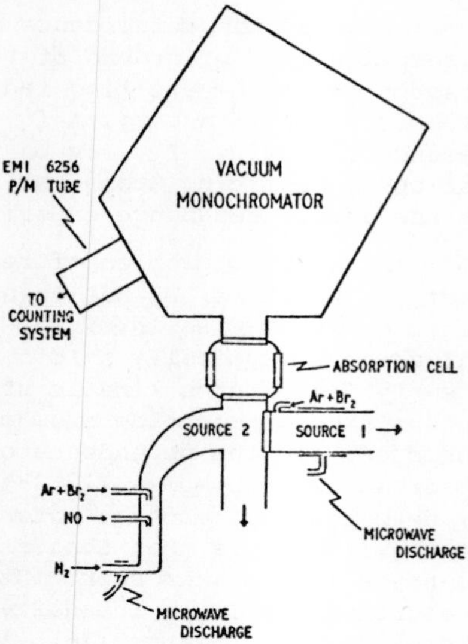

Figure 1. Schematic of experimental system for atomic resonance

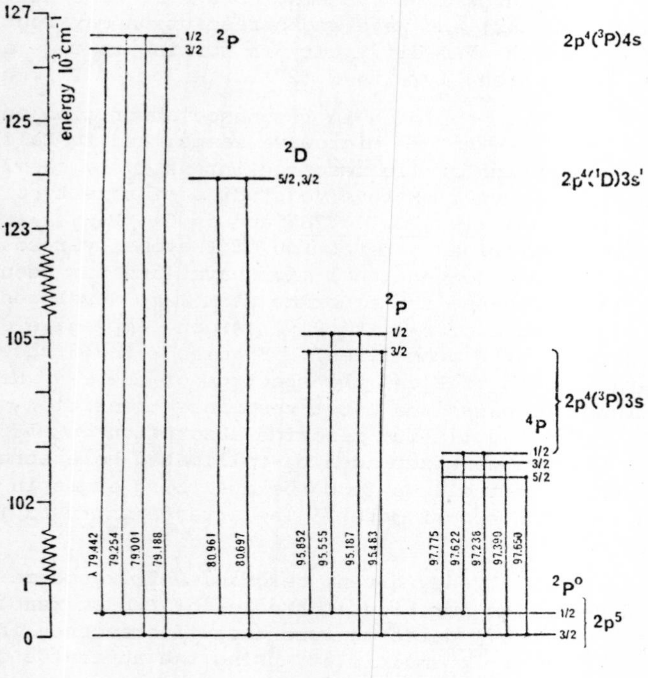

Figure 2. Atomic resonance transitions for F(I). Wavelengths and wave numbers of the lower resonance transitions of F(I).

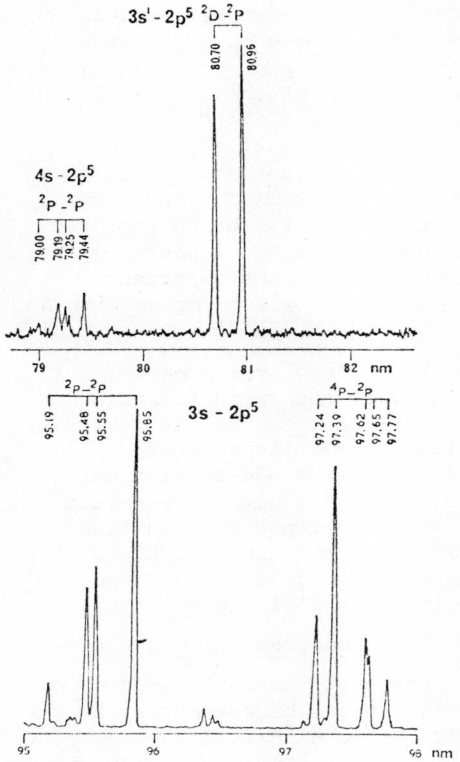

Figure 3. Emission spectrum of F atom resonance lamp. F_2 (1 mol %) + He, 25-W incident microwave power. Lower spectrum: $^{2,4}P_J - {}^2P_j$ multiplets of the 3s − 2p^5 transition using 5μm × 10 mm slits. Detected photon flux rate at 95.85 nm = 1.8 kHz. Upper spectrum shows the much weaker 3s′ ^2D_J − 2p^5 ^2P_J and 4s ^2P_J − 2p^5 ^2P_J multiplets, using 13μm × 10 mm slits. Detected photon count at 80.70 nm = 50 Hz. Note that the lamp spectrum shows negligible continuum background. Noise (photomultiplier dark count + scattered light) count rate = 2.5 Hz.

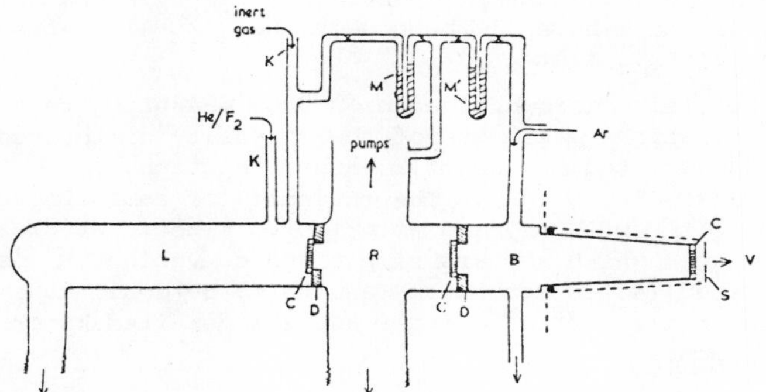

Figure 4. Windowless system for resonance spectrometry of F ^2P_J atoms. B, buffer chamber; C, collimated hole structures, two of which were mounted on specially fabricated glass discs, D; K, inlet to lamp L; M, M^1, differential manometers; R, section of flow tube; S, spectrometer slit sealed via an O-ring seal to the silica apparatus; V, vacuum spectrometer.

Although weak fluorescence in the $^4P-^2P$ lines was observed, from a practical standpoint only the $^2P-^2P$ lines were found intense enough for resonance fluorescence work at low atom concentrations. The fluorescence count rates using the whole of the fully-allowed $^2P-^2P$ transition of F were typically 2 counts s^{-1} at $[F] = 1 \times 10^{12}$ cm^{-3}. These data set a lower concentration limit of 1×10^{11} cm^{-3} for the smallest detectable concentration of F $^2P_{3/2}$ atoms (7). The corresponding lower limits for Cl, Br and I atoms (27,28) are appreciably less, and are continually being improved by better attention to collimation and detection. Because of the low count rates observed in the F atom resonance fluorescence studies, a better approach is to use resonance absorption with a non-reversed line source (67).

The sensitivity of resonance fluorescence by F atoms was, however, sufficient for kinetic studies of the rapid titration reaction of F with Br_2: $F + Br_2 \rightarrow BrF + Br$ (7). It was necessary to make corrections both for non-linearity of fluorescence intensity with $[F]$, and for deviations from pseudo-first-order kinetic conditions, both due to the relatively high F $^2P_{3/2}$ atom concentrations used ($> 1 \times 10^{11}$ cm^{-3}). The final rate constant for the $F + Br_2$ reaction was $(1.4 \pm 0.5) \times 10^{-10}$ at 298 K, in fair agreement with results from the comparably difficult mass spectrometric study of Appelman and Clyne (5) ($k = (3.1 \pm 0.9) \times 10^{-10}$).

Measurement of atom concentrations: EPR and chemiluminescence

Although electron paramagnetic resonance has been used as a very successful technique for kinetic studies of elementary reactions, including those of ground state O ^3P, H ^2S, and N ^4S atoms and of OH $^2\Pi$ radicals (31-35), little effort appears to have been expended on the use of this technique to study reactions of halogen atoms. However, the epr spectra of the $^2P_{3/2}$ and $^2P_{1/2}$ states of F, Cl, Br and I have all been described (36-41), with the exception of the I $^2P_{1/2}$ state.

A completely different technique for measuring relative F atom concentrations makes use of the intensity of the red chemiluminescence emitted in the third-order reaction, $F + NO + M \rightarrow FNO^* + M$ (42). The intensity of emission is reported to vary directly with $[F]$ and with $[NO]$. Kinetic studies using this technique, which appears simple and convenient to use, have been described (43). More work remains to be done, however, to clarify the nature of the emitter and the detailed kinetics of the emission process.

Measurement of atom concentrations: Mass spectrometry.

In principle, mass spectrometry provides a specific and highly sensitive method for the detection of gaseous free radicals. Foner (41) has reviewed the techniques available for sampling free radicals, and for their mass spectrometric detection. Unless the radical source is at pressures comparable to atmospheric pressure, sampling of radicals through orifices usually occurs under conditions approximating to effusive flow, or intermediate between effusive and supersonic flow, thus minimizing collisions within the sampling system. Most of the sampling systems in which radicals were produced in a discharge-flow system (near 100 N m^{-2} total pressure) have been designed utilizing the principles laid down by Foner (44). In a typical system the mass spectrometer is separated from the flow tube by two in-line orifices of 0.7 mm diameter spaced 1 cm apart (45). The space between the orifices is maintained at a pressure of 0.2 N m^{-2} or less by differential pumping with a large diffusion pump. Ion source pressures are of the order of 10^{-3} N m^{-2}. A typical system is shown in Figure 5.

Although it is sometimes possible to detect radicals in the presence of large concentrations of a parent molecule by the use of low energy ionizing electrons, the ready dissociation of the F_2 molecule makes it impractical to detect F atoms mass spectrometrically in the presence of excess F_2. Hence, direct mass spectrometric detection of F atoms is limited to those situations in which little or no F_2 is present. This is the case when F atoms are generated by a microwave discharge in a very dilute mixture of F_2 with an inert carrier gas (< 0.1 mol-% F_2). It is also the case when F atoms are produced by a discharge in compounds such as CF_4, or by such reactions as

$$NF_2 + N \rightarrow N_2 + 2F$$

$$NF_2 + O \rightarrow NO + 2F$$

or $$NO(excess) + F_2 \rightarrow FNO + F.$$

However, formation of F^+ in the ion source by fragmentation of CF_4, NF_2, or FNO must be considered. The formation of F^+ by fragmentation of spin-paired molecules can be avoided by placing a barrier between the molecular beam inlet and the ionizer and then using an inhomogeneous magnetic field to bend paramagnetic molecules around the barrier. In this way spin-paired molecules are largely prevented from reaching the ionizer, and F may be detected in the presence of F_2. Removal of the barrier allows F_2 to be monitored (46). When possible, direct monitoring of F as F^+ ions represents one of the most successful approaches to quantitative kinetic studies of F atom reactions, as demonstrated by the work of Warnatz, Wagner and Zetzsch (10,14).

Mass spectrometry may also be used for the indirect detection of F atoms by the attenuation of the F_2^+ peak when a microwave discharge is struck in a mixture of F_2 with inert carrier gas. This method obviously makes no allowance for reactions that consume F atoms. Indirect detection of F atoms in the mass spectrometer may also be effected by monitoring either the consumption of reagents that react rapidly with F atoms or the formation of products in such reactions. These techniques have already been discussed in the section on measurement of absolute F atom concentrations.

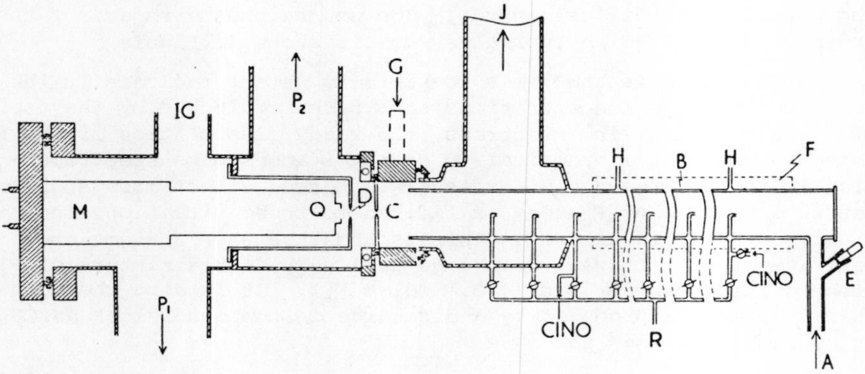

Figure 5. Schematic of flow system for on-line mass spectrometric studies of F atom reactions. A, flow of F atoms in He carrier from discharge; B, silica flow tube; C, first pinhole (0.7 mm) separating flow tube from differentially-pumped chamber; D, second pinhole (0.7 mm) separating differentially-pumped chamber from quadrupole mass spectrometer Q(E.A.I. 150/A); E, access to discharge tube; F, furnace; G, 50 mm gate valve; H, connections to total pressure manometer; IG, ion gage; J, connection to cold traps and large pump (60 ls^{-1}); M, multiplier; P_1, P_2, pumping lines for quadrupole chamber and for differentially-pumped chamber; R, reagent inlets.

Addition of Fluorine Atoms to Spin-Paired Inorganic Molecules.

There has been very little study of reactions in which fluorine atoms add to spin-paired inorganic molecules with the ultimate formation of stable products (addition of F atoms to olefins is discussed elsewhere in this monograph). We may expect such reactions to be rather slow termolecular processes and hence to be rather difficult to investigate. The reaction of F with excess CO has been studied in a discharge-flow system (4). Both COF_2 and CO_2 were identified as products by mass spectrometry, and the F atom concentration was measured by mass spectrometric monitoring of the ClF formed when an excess of Cl_2 was added at the end of the reaction zone. The results were consistent with a rate determining step

$$CO + F + M \xrightarrow{k} COF + M, \qquad (7)$$

although the dependence of the rate constant on [M] was not uniquely determined. Values of 3.4 ± 1 and $5.7 \pm 2 \times 10^{-32}$ cm^6 $molecule^{-2}$ s^{-1}, respectively, were obtained for k with helium and argon as third bodies at 298 K. A more recent study of reaction (7) has used epr monitoring of the F atoms to give values of 6.5 ± 1 and $7 \pm 1 \times 10^{-32}$ cm^6 $molecule^{-2}$ sec^{-2} in helium and nitrogen, respectively at 293 K (47).

A study of reaction (7) by flash photolysis of mixtures of N_2F_4, CO, and N_2 resulted in the observation of a transient ultraviolet spectrum that was attributed to COF (48). The decay of the transient produced the spectrum of $(COF)_2$ (48). (COF_2 did not absorb in the spectral region that was examined).

Formation of FCO has been observed to result from the reaction of CO with photolytically formed F atoms in a solid matrix at liquid helium temperature (49). The authors of this article conclude that the reaction of CO with F has little or no activation energy. The COF reacts further in the matrix to form both COF_2 and $(COF)_2$.

The reaction of F atoms with excess xenon was studied in the same discharge-flow apparatus used for the F + CO reaction, and XeF_2 was identified as a product (4). The rate determining step was assumed to be

$$Xe + F + M \xrightarrow{k} XeF + M, \qquad (8)$$

and on the basis of a single measurement k was estimated to be about 2.3×10^{-33} cm^6 $molecule^{-2}$ s^{-1} at 298 K with argon as the third body.

In the same discharge-flow system no evidence was found for the reaction of F atoms with krypton in helium at 298 K, and an upper limit of 2×10^{-34} cm^6 $molecule^{-2}$ s^{-1} was set for the rate constant of the reaction

$$Kr + F + M \rightarrow KrF + M \qquad (4)$$

These differences in the rate of reaction of F atoms with Kr, Xe and CO correlate with the fact that the fluorine atoms become progressively more strongly bound as one goes from KrF_2 to XeF_2 to COF_2. Both XeF and KrF have been produced by the reaction of photolytically produced F atoms with Xe and Kr in low temperature matrices. (50,51).

It is not known what subsequent steps produce the final products from the reactions of F atoms with CO and Xe. The most plausible process appears to be disproportionation, e.g.,

$$2COF \rightarrow CO + COF_2.$$

Also possible are additions of a second F atom by a more rapid termolecular process, e.g.

$$COF + F + M \rightarrow COF_2 + M,$$

or reaction with F_2 and F in schemes such as

$$COF + F_2 \rightarrow COF_2 + F$$

$$COF + F \rightarrow CO + F_2.$$

In the case of CO, the reaction $2COF \rightarrow (COF)_2$ is apparently also involved.

We may note that COF_2 and XeF_2 can be prepared by the thermal reactions of F_2 with CO and Xe respectively. (52,53). These reactions very probably proceed via the formation of F atoms from the readily dissociated F_2 molecule, followed by reactions (7) and (8) to form the Xe-F and C-F bonds.

Warnatz (54) looked at the addition of F to HCN and $(CN)_2$, obtaining bimolecular rate constants of 1.5×10^{-10} and 5×10^{-13} cm^3 $molecule^{-1}$ s^{-1} at 300 K respectively for the reactions

$$F + HCN \rightarrow HFCN$$

and

$$F + (CN)_2 \rightarrow F(CN)_2.$$

These reactions were measured mass spectrometrically in a discharge flow system at pressures of 470 N m^{-2}, and they thus appear to be true bimolecular processes. Both reactions have apparent negative activation energies (-5 kJ/mole for F + HCN and -2 kJ/mole for F + $(CN)_2$). The reaction of F with HCN is followed by

$$F + HFCN \rightarrow HF + FCN \quad (54).$$

Bimolecular Reactions of F Atoms.

a) Hydrogen Abstraction

These reactions are discussed elsewhere in this monograph and have been reviewed by Foon and Kaufman (55) and by Jones and Skolnik (43). The reactions are generally very fast, with rate constants that are within a factor of 10 of the hard-sphere bimolecular collision frequency. Notable exceptions are

F + NH_3 ($k_{300\ K}$ = 1 x 10^{-12} cm^3 $molecule^{-1}$ s^{-1}) and F + trihalogenated methanes, such as CHF_3, $CHCl_2F$ and $CHClF_2$ (k_{300} ≃ 2 x 10^{-12} cm^3 $molecule^{-1}$ s^{-1}), and $CHCl_3$ (k_{300} = 5 x 10^{-12} cm^3 $molecule^{-1}$ s^{-1}). (42, 1, 54, 68).

We have already discussed the reaction of F with H_2 as a means of determining F atom concentrations. A novel hydrogen abstraction reaction is the reaction of F atoms with the recently prepared hypofluorous acid:

$$F + HOF \rightarrow HF + OF.$$

A lower limit of 2 x 10^{-10} cm^3 $molecule^{-1}$ s^{-1}has been set for the rate constant of this reaction at 300 K on the basis of a mass-spectrometric discharge-flow study (4). It is one of the two reactions known to produce OF radicals. Inasmuch as the latter decompose to reform F atoms (45):

$$2OF \rightarrow O_2 + 2F,$$

we have, in effect, an F-atom catalyzed decomposition of HOF. These reactions may contribute to the spontaneous decomposition of HOF at room temperature, a decomposition that is enchanced by F_2 and inhibited by such F-atom scavengers as H_2 and CO (56).

b) Oxygen Abstraction

Abstraction of oxygen by fluorine atoms requires a reagent molecule containing a relatively weakly bonded oxygen atom, inasmuch as the OF product is only bound by 213 kJ/mole (57). The only well established case is the reaction with ozone, which is important as a source of OF radicals:

$$F + O_3 \rightarrow OF + O_2.$$

A mass spectrometric discharge flow study gives for the rate constant k = 2.8 x 10^{-11} exp(-450 ± 400 cal mol^{-1}/RT) cm^3 $molecule^{-1}$ s^{-1}, corresponding to a value of 1.3 x 10^{-11} cm^3 $molecule^{-1}$ s^{-1} at 300 K. In this study, [O], [F] and [OF] were all monitored, and conditions ranged from excess O to excess F. The F atom concentration tends to remain constant even with excess O, because of the regeneration of F atoms in the subsequent decomposition of OF. Hence a pseudo-first order kinetic analysis can be applied (58).

Inasmuch as the energy of a xenon-oxygen bond is only about 88 kJ/mole (53), F atoms should be able to extract oxygen from such molecules as XeO_4 and $XeOF_4$. These reactions do not, however, appear to have been investigated.

c) Halogen Abstraction.

The reaction of F atoms with Cl_2 and Br_2 have already been discussed as means for determining F atom concentrations. The reaction

$$F + Cl_2 \rightarrow ClF + Cl$$

has been the subject of three mass spectrometric discharge-flow studies, one with excess Cl_2 and the others with excess F. (1,4, 14). The studies using excess F monitored the disappearance of Cl_2; the study with excess Cl_2 monitored the disappearance of F and the formation of ClF and Cl. The reported rate constants range from 0.86 to 1.6×10^{-10} cm^3 $molecule^{-1}$ s^{-1} at 300 K. A mean value of 1.2×10^{-10} may be as good a choice as any. The reaction

$$F + Br_2 \rightarrow BrF + Br,$$

has been studied in discharge-flow systems with mass spectrometric monitoring of Br_2 in the presence of excess F, (4) and by monitoring F atom resonance fluorescence in the presence of excess Br_2 (7). The two studies yielded respective 300 K rate constants of 3.1 ± 0.9 and $1.4 \pm 0.5 \times 10^{-10}$ cm^3 $molecule^{-1}$ s^{-1}. The higher value may be a little more reliable inasmuch as the fluorescence study was subject to uncertain attenuation of F atoms on surfaces and also involved a significant correction for self-reversal of the fluorescence. Warnatz reported a preliminary value of 0.8×10^{-10} for this constant (54).

The reactions of F atoms with Cl_2 and Br_2 can constitute convenient sources of Cl and Br atoms, respectively. It is often possible to generate higher concentrations of such atoms in this way than by the passage of a discharge through chlorine or bromine (4).

The reactions

$$F + I_2 \rightarrow IF + F \qquad (9),$$

$$F + ICl \rightarrow IF + Cl \qquad (10),$$

$$F + ICl \rightarrow ClF + F \qquad (11),$$

have been studied by mass spectrometric monitoring of the consumption of I_2 and ICl and the formation of ClF in a discharge flow system in the presence of excess F (4). Respective rate constants of 4.3 ± 1.1, 3.8 ± 1.5, and $1.2 \pm 0.7 \times 10^{-10}$ cm^3 $molecule^{-1}$ s^{-1} were obtained at 300 K. The rate constant for reaction (10) sets a lower limit to the binding energy of IF and allows a choice to be made among the several estimates of this quantity that have been put forth.

The reactions of F atoms with Br_2, I_2 and ICl are complicated by the fact that BrF and IF are unstable to disproportionation, with the subsequent formation of higher fluorides and Br_2 and I_2. As a result, the reaction of an excess of F atoms with Br_2 forms BrF rapidly and then, on a longer time scale, goes on to form BrF_5. Surprisingly, there is no evidence for the formation of substantial amounts of the stable intermediate fluoride BrF_3. When an excess of F atoms reacts with I_2 or ICl, formation of IF_5 appears to be nearly competitive with IF formation, particularly at higher concentrations. It is difficult to imagine a homogeneous process that could form IF_5 so rapidly, and it seems that surface reactions are involved. A mechanism has been proposed in which the IF polymerizes reversibly on the walls of the reaction tube and IF_5 is formed by the subsequent heterogeneous reaction of F atoms with the polymer (4). It seems likely that the formation of BrF_5 takes place by a similar process.

Inasmuch as the F_2 molecule is only bound by about 155 kJ/mole, fluorine abstraction by F atoms requires a reagent with a very weakly bound fluorine. One possibility is the reaction

$$F + OF_2 \rightarrow F_2 + OF.$$

A kinetic analysis of the thermal decomposition of OF_2 yields for this reaction a rate constant $k = 8 \times 10^{-14} \exp(-57.3 \pm 4\ kJ/mole^{-1}/RT)\ cm^3\ molecule^{-1}\ s^{-1}$ and implies a negligibly slow reaction rate at room temperature (39).

Xenon-fluorine bonds have energies around 125 kJ/mole, (53) and we might expect F atoms to abstract fluorine from the xenon fluorides. However, upper limits of about $7 \times 10^{-16}\ cm^3\ molecule^{-1}\ s^{-1}$ have been set for the reaction of F with XeF_2 and XeF_4 at 300 K on the basis of mass spectrometric studies in a discharge-flow system (4). It is possible that removal of the first fluorine from either of these compounds may require considerably more than 125 kJ/mole. Johnston and Woolfolk (60) found that NO and NO_2 reacted much more slowly with XeF_2 and XeF_4 than with F_2. By correlating reaction rates with bond energies, they concluded that the reactions

$$XeF_2 \rightarrow XeF + F$$

and $$XeF_4 \rightarrow XeF_3 + F$$

were endoergic by 226 and 201 kJ/mole, respectively.

The reactions of F atoms with perhalogenated alkanes have been reviewed by Foon and Kaufman (55) and by Jones and Skolnik (43). As might be expected, those reactions that are exoergic, such as

$$CCl_3Br + F \rightarrow BrF + CCl_3$$

and $$CF_3I + F \rightarrow IF + CF_3,$$

proceed at rates close to the bimolecular collision frequency. Endoergic abstraction reactions are very much slower, and in some cases they may have to compete with fluorine substitution. There are wide discrepancies in the rates and mechanisms reported for these latter reactions. Thus the reported rate constants for the reaction of F with CCl_4 span a range of 3×10^7!

Production and reactions of FO radicals.

The gaseous ClO, BrO and IO radicals have all been observed by uv absorption spectroscopy in both static and flow systems. However, no optical or epr spectrum of the gaseous FO radical has been reported, even though definite evidence for this species was obtained in Ar and N_2 matrices at 4° K using infrared spectrometry (61). This is surprising in view of the ready identification of FO radicals by mass spectrometry after their production in a discharge-flow system by the rapid reaction, $F + O_3 \rightarrow FO + O_2$, (57,58). It is possible that the FO analogue of the well-known $^2\Pi$-$X^2\Pi$ uv band absorption systems of ClO, BrO and IO (62) is diffuse and therefore difficult to identify. Franck-Condon factors for discrete vibronic transitions from v" = 0 of FO $X^2\Pi$ may be low, with the main fraction of the oscillator strength of the v" = 0 transitions occurring in the continuum region. It is also possible to rationalize the failure to observe the epr spectrum of FO. According to unpublished experiments of Clyne and Sales, any epr spectrum of FO must be at least a factor of 20 weaker than that of BrO, for the same concentrations of FO and BrO. This is most likely due to the FO radical having a very small dipole moment because of the similar electronegativities of O and F.

Figure 6 shows the flow apparatus, inlet system, and mass spectrometer used by Clyne and Watson for kinetic studies of FO radical reactions (45). The source of FO was the rapid reaction $F + O_3 \rightarrow FO + O_2$. Absolute concentrations of FO were determined mass spectrometrically by addition of excess NO and monitoring of the NO_2 produced in the rapid reaction:

$$FO + NO \rightarrow F + NO_2,$$

which has a rate constant in excess of 5×10^{-12} cm^3 $molecule^{-1}$ s^{-1} at 298 K. The concentration of FO at any time is equated to the concentration of NO_2 produced by reaction with excess NO at that time, the FO + NO reaction being essentially instantaneous under the conditions used. The FO^+ ion current can thus be related via the NO_2^+ ion current to the absolute [FO] concentration. This method is also applicable to the determination of ClO and BrO concentrations (45).

Production of FO radicals suitable for quantitative kinetic studies required careful attention to the F^2P atom source employed. For example, use of the N + NF_2 reaction to generate F led to the formation of significant quantities of NO, and hence of NO_2 when the F was subsequently converted to FO by means of the F + O_3 reaction. This NO_2 background complicated the measurement of absolute FO concentrations. The source of NO was either impurity $NONF_2$ in the N_2F_4 used to generate NF_2, or the reaction of NF_2 with O ^3P atom impurities in the N ^4S atom stream, or possibly both. The use of a discharge in dilute F_2 + He mixtures to produce the F atoms led to an acceptably low concentration of NO_2.

O ^3P atoms are known to be formed to the extent of $\lesssim 0.1$[F] when F_2 is dissociated in a silica microwave discharge tube. These atoms arise from dissociation of O_2 liberated by attack of F on SiO_2. O ^3P atoms should not greatly affect the measurement of absolute concentrations of FO by the NO_2 method inasmuch as most

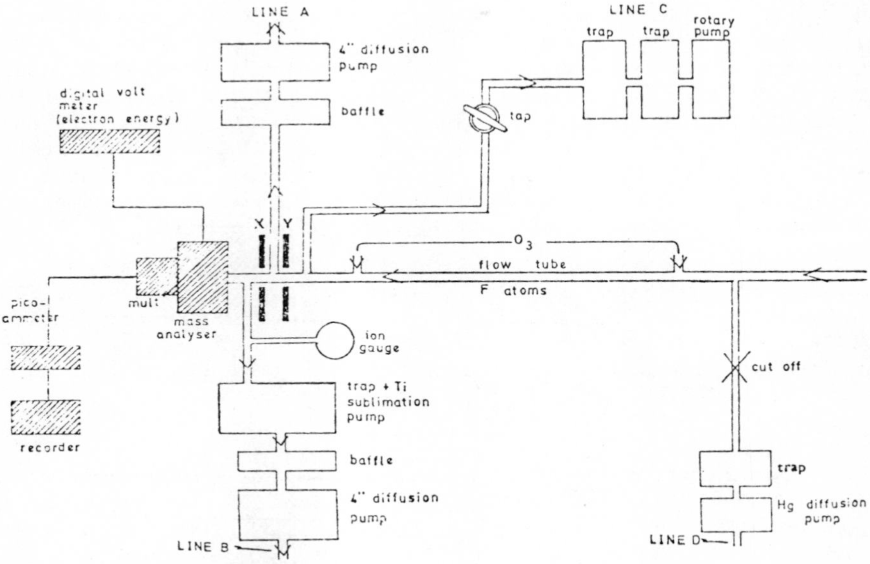

Figure 6a. Flow apparatus, inlet system, and mass spectrometer used by Clyne and Watson for kinetic studies of FO radical reactions. Block diagram of mass spectrometric system. Pumping lines as follows: A, differential chamber (between pinholes X and Y); B, mass spectrometer ion source; C, flow tube; D, flow tube diffusion pump (used only for overnight clean-up of flow tube); X and Y, sampling pinholes. System shown is for study of FO radical reactions.

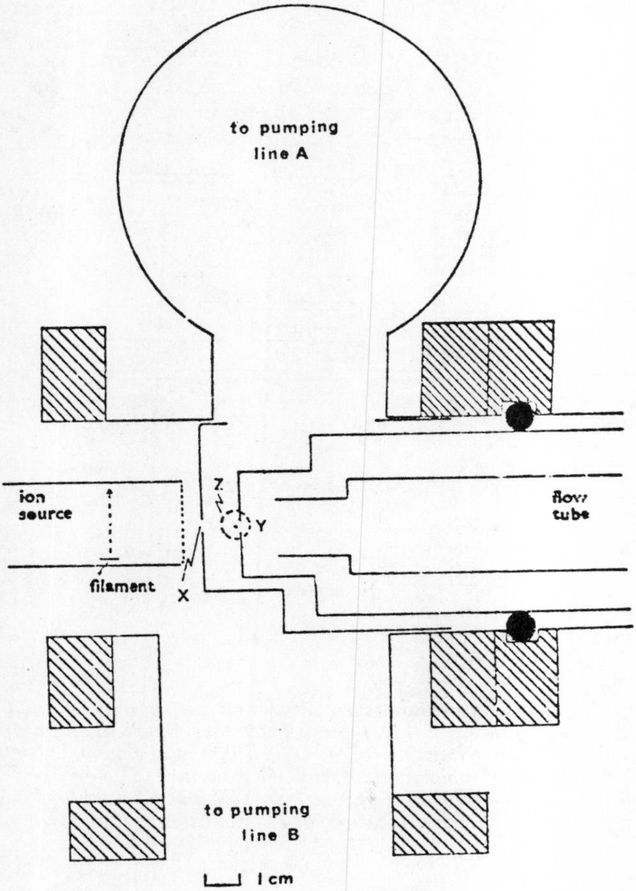

Figure 6b. Flow apparatus, inlet system, and mass spectrometer used by Clyne and Watson for kinetic studies of FO radical reactions. Scale diagram of sampling system, showing pinholes X and Y, crossed-beam inlet Z, and connections to flow tube and pumping lines A and B.

of the O ^3P would be expected to be removed by excess FO radicals before reaching the point of addition of NO:

$$O + FO \rightarrow O_2 + F.$$

Preliminary work on the FO + FO reaction, indicating this reaction to be kinetically second-order in [FO] (57) was confirmed by later work of Clyne and Watson (45). The second order rate constant was determined to be $(8.5 \pm 2.8) \times 10^{-12}$ cm^3 $molecule^{-1}$ s^{-1} at 298 K. The mechanism of this reaction is believed to involve regeneration of F atoms and to be analogous to the similar second-order disproportionation reaction of ClO + ClO and BrO + BrO, i.e.:

$$FO + FO \rightarrow 2F + O_2.$$

Although the reactions of FO with NO, H atoms and O atoms are known to be fast ($k \gtrsim 5 \times 10^{-12}$ cm^3 $molecule^{-1}$ s^{-1} at 298 K), no quantitative kinetic data on these reactions appear to be available. Since the F-O bond energy is only (215 ± 17) kJ $mole^{-1}$ (57), there are many exoergic elementary reactions of FO radicals possible. Further investigation of such reactions should be rewarding.

Production and reactions of NF

The NF radical is isoelectronic in structure with O_2. Its electronic states are closely similar to those of O_2. The ground state is NF $X^3\Sigma^-$, whilst the low-lying $a^1\Delta$ and $b^1\Sigma^+$ metastable states, closely analogous to the singlet states of O_2, are expected and known. The singlet states of NF are somewhat more energetic than those of O_2. Douglas and Jones (62) were the first to report NF. They observed the forbidden b-X and a-X band systems in emission, using a low-pressure discharge in NF_3 as the source of excited NF radicals. Subsequently, Clyne and White (63) reported strong NF b-X and a-X emission from the reactions H + NF_2 and O + NF_2. The 0-0 band of NF a-X near 874.2 nm, and the 0-0 (and 0-1) bands of the b-X system near 528.8 nm, were reported (63). These workers (63) also reported intense $B^3\Pi_g - A^3\Sigma_u^+$ First Positive emission bands of N_2 in the H + NF_2 reaction zone. Since the observed N_2 B-A bands had the same vibrational energy distribution to that from the N + N + M recombination, Clyne and White (63) proposed that N ^4S ground state atoms were a product of the H + NF_2 reaction. A likely reaction mechanism would comprise the following steps (1) and (2):-

$$H + NF_2 \rightarrow NF + HF \quad (1),$$

$$H + NF \rightarrow HF + N \quad (2).$$

Reaction (2) would be the only known example of a chemical reaction where N atoms could be produced exoergically.

The details of reaction (2) are of interest. The reacting NF radical can be in either the $X^3\Sigma^-$ ground state, or the $a^1\Delta$ or $b^1\Sigma^+$ metastable excited states. H is the 1s ^2S ground state, and

HF is $X^1\Sigma^+(v \geq 0)$. If the reacting NF is $X^3\Sigma^-$, spin and orbital angular momentum conservation between the reactants and products determine that the N atom formed in reaction (2) shall be ^4S or ^2S. The latter is energetically impossible; so we conclude that H+ NF $X^3\Sigma^-$ gives N ^4S ground state atoms. On the other hand, H ^2S + NF $a^1\Delta$ must give N ^2D, i.e. metastable excited atoms. It is interesting that H ^2S+ NF $b^1\Sigma^+$ must give N ^2S which is energetically impossible. Hence we would expect that H ^2S does not react with NF $b^1\Sigma^+$ at a significant rate. In addition, formation of N ^2D atoms in the H + NF_2 system would be a diagnostic for the occurrence of reaction between H and NF $a^1\Delta$. Experimental work along these lines to investigate the relative reactivity of NF $X^3\Sigma^-$,$a^1\Delta$ and $b^1\Sigma^+$, is currently being carried out by Clyne and Whitefield.

The relative reactivity of the electronic states of NF is important in considering proposals for an NF electronic transition laser based on the $a^1\Delta$ or $b^1\Sigma^+$ states. According to Herbelin and coworkers (64), their experimental results are consistent with a product yield of 100% NF $a^1\Delta$ in the H + NF_2 reaction. The $b^1\Sigma^+$ state is believed to be formed by pumping of the a state by vibrationally-excited HF:- NF $a^1\Delta$ + HF(v) $\rightarrow$ NF $b^1\Sigma^+$ + HF(v-1). It is not certain whether the apparent high yields of metastable singlet NF in the H + NF_2 reaction might in fact be due to the superior reactivity of the triplet NF $X^3\Sigma^-$.

Herbelin (65) has suggested that NF $a^1\Delta$ should be less reactive towards atoms and radicals than NF $X^3\Sigma^-$. He assumed that NF can be considered as analogous to an N atom - i.e., the F atom in NF behaves as an inert spectator. The NF a and b states would then both correlate with F + N ^2D excited atoms, whilst NF X would correlate to F + N ^4S. The bimolecular NF + NF reaction (3) was considered in light of this model. If both the NF radicals are $a^1\Delta$ or $b^1\Sigma^+$, N_2 would be formed in the $a^1\Sigma_u^-$ excited state. For reaction between two NF $a^1\Delta$ radicals, formation of N_2 $a^1\Sigma_u^-$ is endoergic. Hence, this reaction was argued to be very slow at 298 K. On the other hand, NF $X^3\Sigma^-$ + NF $X^3\Sigma^-$ could give N_2 $X^1\Sigma_g^+$ + 2F by exoergic reaction, consistent with the assumption of a high rate constant for reaction between two ground state NF radicals.

The failure so far to unequivocally detect NF $X^3\Sigma^-$in H + NF_2 systems using epr, even though NF $^1\Delta$ was detected in the same systems (66), could be interpreted as support for the greater reactivity of NF $X^3\Sigma^-$ which has been suggested above. Further work on this question is required, and it appears that detection mass spectrometrically of NF, with atomic resonance for detection of atoms, will be a promising approach.

Acknowledgement: We wish to thank Dr M Kaufman and Dr W E Jones for making available to us preprints of their reviews. The preparation of this article was partially supported by the U.S. Energy Research and Development Adminstration.

References

1. M.A.A. Clyne, D.J. McKenney and R.F. Walker, Canad.J.Chem., (1973), 51 3596.

2. C.E. Kolb and M. Kaufman, J.Phys.Chem., (1972), 76, 947.

3. D.E. Rosner and H.D. Allendorf, J.Phys.Chem., (1971), 75, 308.

4. B.R. Zegarski, T.J. Cook and T.A. Miller, J.Chem.Phys., (1975), 62, 2952.

5. E.H. Appelman and M.A.A. Clyne, J.C.S. Faraday I, (1975), 71, 2072.

6. H.C. Berg and D. Kleppner, J.Sci.Instr., (1962), 33, 248.

7. P.P. Bemand and M.A.A. Clyne, J.C.S.Faraday II,(1976), 72, 191.

8. A. Carrington and D.H. Levy, J.Chem.Phys.,(1966), 44, 1298; ibid., (1970), 52, 309.

9. M.A.A. Clyne, J.A. Coxon and L.W. Townsend, J.C.S.Faraday II, (1972), 68, 2134.

10. K.H. Homann, W.C. Solomon, J. Warnatz, H.Gg. Wagner and C.Zetzsch, Ber.Bunsenges.Phys.Chem., (1970), 74, 585.

11. M.A.A. Clyne and I.F. White, Chem.Phys.Letters,(1970), 6, 465.

12. H.S. Johnston and D. Rapp, J.Chem.Phys.,(1960), 33, 695.

13. C.E. Kolb, J.Chem.Phys.,(1976), 64, 3087.

14. J. Warnatz, H.Gg. Wagner and C. Zetzsch,Ber.Bunsenges.Phys. Chem., (1971), 75, 119.

15. P.Kim, D.I. MacLean and W.G. Valence, NTIS AD 751456, Abstracts of 164th Am.Chem.Soc.Meeting (1972).

16. P.C. Nordine, J.Chem.Phys., (1974), 61, 224.

17. R.G. Albright, A.F. Dodonov, G.K. Lavrovskaya, I.I.Morosov and V.L. Tal'roze, J.Chem.Phys., (1969), 50, 3632.

18. M.A.A. Clyne in Physical Chemistry of Fast Reactions (Ed. B.P. Levitt), (1973), 1, 245 (Plenum Press, N.Y.).

19. R.J. Donovan and H.M. Gillespie, Spec.Per.Repts., Reaction Kinetics, Vol.1 (The Chemical Society, London 1975), Ch.2.

20. D. Husain and R.J. Donovan, Adv.Photochem.,(1971), 8,1.

21. P.P. Bemand and M.A.A. Clyne, Chem.Phys.Letts.,(1973), 21,555.

22. W Braun and T. Carrington, J.Quant.Spectr.Rad.Trans.,(1969), 9, 1133.

23. F. Kaufman and D.A. Parkes, Trans.Faraday Soc.,(1970),66,1579.

24. P.P. Bemand and M.A.A. Clyne, J.C.S.Faraday II, (1973),69, 1643.

25. A.C.G. Mitchell and M.W. Zemansky, Resonance Radiation and Excited Atoms, Cambridge Univ.Press, (1971).

26. for example, W. Braun, A.M. Bass and D.D.Davis, Int.J.Chem. Kin., (1970), 2, 101.

27. M.A.A.Clyne and H.W. Cruse, J.C.S.Faraday II,(1972),68,1281.

28. M.A.A.Clyne and H.W. Cruse, J.C.S.Faraday II,(1972),68,1377.

29. R.J. Donovan, D. Husain, W. Braun, A.M. Bass and D.D. Davis, J.Chem.Phys., (1969), 50, 4115.

30. M.A.A. Clyne and H.W. Cruse, Trans.Faraday.Soc.,(1971),67, 2869.

31. A.A. Westenberg, Ann.Rev.Phys.Chem., (1973), 24, 77.

32. A.A. Westenberg, Prog.Reaction Kinetics,(1973), 7, 23.

33. P.Kim, R.B. Timmons, Int.J.Chem.Kin., (1975), 7,77.

34. C.N. Wei and R.B. Timmons, J.Chem.Phys.,(1975), 62, 3240.

35. G.A. Takacs and G.P. Glass, J.Phys.Chem.,(1973),77, 1182.

36. H.E. Radford, V.W. Hughes and V. Beltran-Lopez, Phys.Rev., (1961), 123, 153.

37. V. Beltran-Lopez and H.G. Robinson, Phys.Rev.,(1961),123,161.

38. J.S.M. Harvey, R.A. Kamper and K.R. Lea, Proc.Phys.Soc., (1960), B76, 979.

39. K.D. Bowers, R.A. Kamper and C.D. Lustig, Proc.Phys.Soc., (1957), B70, 1176.

40. A. Carrington, D.H. Levy and T.A. Miller,J.Chem.Soc., (1966), 45, 4093.

41. P.B. Davies, B.A. Thrush, A.J. Stone and F.D. Wayne, Chem.Phys.Letters, (1972), 17, 19.

42. T.L. Pollock and W.E. Jones, Canad.J.Chem.,(1973),51, 2041.

43. W.E. Jones and E.G. Skolnik, Chem.Revs.,(1976) (in press).

44. S.N. Foner, Adv.Atomic Mol.Phys.,(1966),2,385.

45. M.A.A. Clyne and R.T. Watson,J.C.S.Faraday I,(1974),70, 1109.

46. M. Kaufman and C.E. Kolb, Chem.Instrumentation,(1971),3,175.

47. V.S. Arutyunov, S.N. Buben and A.M. Chaikin, React.Kinet. Catal.Lett.,(1975), 3, 205.

48. D.K.W. Wang and W.E. Jones, J.Photochem., (1972),1,147.

49. D.E. Milligan, M.E. Jacox, A.M. Bass, J.J. Comeford and D.E. Mann, J.Chem.Phys., (1965), 42, 3187.

50. J.J. Turner and G.C. Pimentel, Science (1963), 140, 974.

51. J.J. Turner and G.C. Pimentel, in "Noble Gas Compounds", H.H. Hyman, Ed. Univ. of Chicago Press, Chicago (1963), pp 101-105.

52. O. Ruff and Shih-Chang Li, Z. Anorg.Chem. (1939), 242, 272.

53. J.G. Malm and E.H. Appelman, Atomic Energy Review, (1969), VII, p.3-48.

54. J. Warnatz, Ph.D. Dissertation, Georg-August Univ., Göttingen (1971). (Discussed in references 43 and 55).

55. R. Foon and M. Kaufman, Progress in Reaction Kinetics, Vol.8, Pt. 2 (1975).

56. E.H. Appelman, Accounts of Chem.Research, (1973), 6, 113.

57. M.A.A. Clyne and R.T. Watson, Chem.Phys.Letts., (1971), 12, 344.

58. H. Gg.Wagner, C.Zetzsch and J. Warnatz, Ber.Bunsenges. Physik.Chem., (1972), 76, 526. see also, Angew.Chem.Intl.Ed. (1971), 10, 564.

59. J. Czarnowski and H.J. Schumacher, Chem.Phys.Letts., (1972), 17, 235.

60. H.S. Johnston and R. Woolfolk, J.Chem.Phys., (1964), 41, 269.

61. A. Arkell, R.R. Reinhard and L.P. Larson, J.Am.Chem.Soc., (1965), 87, 1016.

62. A.E. Douglas and W.E. Jones, Canad.J.Phys., (1966), 44, 2251; W.E. Jones, Canad.J.Phys., (1967), 45, 21.

63. M.A.A. Clyne and I.F. White, Chem.Phys.Letts., (1970), 6, 465.

64. J.M. Herbelin, D. Spencer and M. Kwok, Report TR-0076(6603)-2 The Aerospace Corporation, (1976), (cited in ref. 65).

65. J.M. Herbelin, Chem.Phys.Letts., (1976), 42, 367.

66. A.H. Curran, R.G. McDonald, A.J. Stone and B.A. Thrush, Proc.Roy.Soc.A, (1973), 332, 355.

67. M.A.A. Clyne and W.S. Nip, (1976) to be published.

68. Recent results according to ref. 67 for k_{300} (cm^3 $molecule^{-1}$ s^{-1}) are as follows:- F + $CHCl_2F$, (1.04 ± 0.22) x 10^{-12}; F + $CHClF_2$, (5.3 ± 1.1) x 10^{-13}; F + $CHCl_3$, (6.3 ± 1.4) x 10^{-12}. Other values for k_{300} are:- F + CH_4, (7.5 ± 1.1) x 10^{-11}; F + HCl, (1.6 ± 0.4) x 10^{-11}. Atomic resonance absorption at 95.48 nm with a non-reversed resonance lamp was used to determine F ^2P_J atom concentrations.

2

Reactions of Radioactive ^{18}F with Alkenes, Alkynes, and Other Substrates

F. S. ROWLAND, FLEET RUST, and JOAN P. FRANK

Department of Chemistry, University of California, Irvine, CA 92717

Radioactive ^{18}F atoms at tracer levels offer several special advantages, and some problems, in comparison with stable ^{19}F atoms for the study of gaseous chemical reactions. Thermal fluorine atoms are exceedingly reactive with a wide variety of substrates (and surfaces), leading to a variety of experimental difficulties: (1-4) (a) the reactive, often corrosive, nature of many F atom sources; (b) the rapid abstraction of H from most hydrogenous substrates, with the formation of HF; (c) the exothermicity of many reactions, especially the formation of HF, with corollary problems of rapid, uncontrolled temperature rises; and (d) the high chemical reactivity of many product molecules, again including HF. The use of ^{18}F atoms at tracer levels avoids several of these difficulties, permitting the systematic study of the reactions of atomic fluorine with many types of substrate molecules, e.g., alkenes and alkynes.

Since ^{18}F can be readily detected at mole fractions of 10^{-11}-10^{-14}, many of the macroscopic problems of ^{19}F chemistry simply do not occur. For example, glass containers never show visible etching; heat increases are always negligible; reactive products do not cause macroscopic corrosion, etc. At these concentration levels certain corollary limitations are also observed. Two ^{18}F atoms do not react with one another, nor are two ^{18}F atoms detectable in the same molecule. No macroscopic property of the molecule (e.g. infrared, NMR, etc.) can be used as an aid in identification of the qualitative or quantitative aspects of any of the products. The 110 minute half-life of ^{18}F limits experiments to those that can be carried to completion in a few hours. Further, the ^{18}F atoms must be freshly produced for each experiment, and the ^{18}F atoms emerging from nuclear reactions ordinarily have kinetic energies vastly in excess of thermal energies and this excess energy must be removed both to retain the atoms within the physical boundaries of the experiment and to avoid confusion between thermal and hot atom reactions in each particular system. Overall, then, the radioactive tracer ^{18}F approach has a separate set of experimental advantages and disadvantages, and studies with ^{18}F can furnish information complementary to that obtained with stable ^{19}F. In many instances, the ^{18}F studies can readily furnish details not accessible to the macroscopic studies, and this review emphasizes some of the systems in which the ^{18}F techniques have shown such utility.

Experimental Aspects of Radioactive ^{18}F Studies.

While several nuclear reactions are potentially available for the formation of ^{18}F, most of the chemical studies described here have utilized ^{18}F formed by the $^{19}F(n, 2n)^{18}F$ nuclear reaction with fast neutrons, usually from a special fast neutron generator. Some studies have utilized the $^{19}F(\gamma,n)^{18}F$ reaction, with results which are comparable to those from the (n, 2n) reaction, since the many inelastic collisions made prior to reaching the chemical energy range completely erase the source and energy history of the ^{18}F atoms at the time of chemical reaction. However, the accompanying radiation damage from the various nuclear methods for producing ^{18}F is not necessarily comparable, and different distributions of ^{18}F among various possible products can arise from the presence of such radiation effects.

A discussion of the experimental aspects of ^{18}F recoil chemistry in the gas phase has been presented elsewhere (5-10), and only a brief summary follows. The $^{19}F(n, 2n)^{18}F$ reaction is initiated by 14 MeV fast neutrons from a neutron generator (Kaman A 711). In the usual target arrangement an irradiation of twenty minutes forms approximately 1 microcurie of ^{18}F from 0.3 grams of ^{19}F source gas. The total absolute yield for ^{18}F formation is routinely monitored through the measured ^{18}F production in a standard (a sleeve of Teflon) surrounding the glass sample ampoule during irradiation. Comparison of absolute yields between duplicate samples shows that the reproducibility of irradiation is approximately 10%. The analysis of relative yields of products within a given sample can be carried out much more accurately, with 1% reproducibility often attained. However, it is possible to detect compounds of which as few as 10^4 molecules have been made, and random statistical fluctuations become an important limitation in a manner not found with the 10^{16}-10^{18} molecules formed in typical macroscopic experiments.

The ^{18}F atoms formed in nuclear reactions can possess 10^6 electron volts of kinetic energy per atom (2 x 10^7 kilocalories/mole) and approximately 1-2 atmospheres of gas is necessary to remove this energy in a 1 cm path length. Consequently, either high pressures or large containers are required, and the geometry of the fast neutron source itself limits glass ampoules to about 2 cm diameter and 4 cm length for most of our experiments. Loss of ^{18}F from the gas phase by recoil into the ampoule walls is unimportant (<10%) at about 1000 Torr and progressively less important at higher pressures. In most studies to date, the samples are maintained at about $10^{\circ}C$ (the cooling temperature of the generator ion target) and are not thermally affected by the neutron irradiation process itself. Irradiation at other temperatures is not a difficult experimental problem, but would involve redesign of the target area with some loss of neutron intensity.

Since macroscopic measurements are not applicable, the analytical techniques used in such tracer systems involve the qualitative identification of the nature of the compound through some property such as gas chromatographic retention time, and the quantitative assay through detection of the radioactive decay of ^{18}F atoms. Numerous experiments have demonstrated that small macroscopic peaks and tracer peaks unaccompanied by observable macroscopic species both have identical gas chromatographic retention times for ^{19}F and ^{18}F compounds. The standard product identification procedure consists of the sequential measurement of thermal conductivity and radioactivity for the gases contained in the effluent helium stream from the gas chromatographic columns (11). Molecular identification of the ^{18}F-labelled species is normally made from coincidence in retention times with macroscopic amounts of the corresponding non-radioactive molecule on several different chromatographic columns. This procedure is especially satisfactory when the number of likely products is rather limited, e.g. two-carbon compounds containing hydrogen and only one fluorine atom, etc.

Almost all of the excess kinetic recoil energy of the ^{18}F atoms is lost in various collision processes before the kinetic energy is low enough to permit the formation of a stable bond with any substrate molecule. The substrate alkene or alkyne is normally present as a minor component mixed with the source gas; other scavenger gases, such as HI, are often present as well.

The energy of the reactive ^{18}F atom cannot in general be precisely controlled. However, it has been demonstrated that the addition of an inert moderator species can greatly reduce the fraction of ^{18}F atoms reacting while possessing excess kinetic energy. Under conditions in which the moderator species is present in mole fraction >0.9 virtually all ^{18}F reactions are attributable to thermal mechanisms with hot reactions suppressed to about 1% or less (5-10). The thermalization of the ^{18}F atoms can be accomplished in two ways: (a) through collision with noble gases which are inert moderators (9, 10); and (b) through collision with an excess of perfluorinated source gases which are inert toward reaction with thermal ^{18}F atoms. The necessity for ^{19}F as a target for the $^{19}F(n, 2n)^{18}F$ reaction requires that at least one major component of each system must be a fluorinated molecule, but some of these are chemically quite inert. Typical source gases which can also serve as moderators are NF_3, CF_4, SF_6 and C_2F_6 (12-15).

^{18}F Reactions with Alkenes and Alkynes.

The most widely studied ^{18}F atom reaction system is that of ^{18}F plus ethylene, which was initially studied for the purpose of using ethylene as a scavenger in CF_4 systems, and more recently with interest in the excited fluoroethyl radical intermediate. Studies using widely different sample preparation procedures and irradiation conditions are mutually consistent, (9, 10, 13) and have demonstrated that the kinetics of the ^{18}F atom reactions under study are independent of their irradiation history.

In the first study, valuable information concerning the elementary processes involved in the reactions of atomic fluorine with ethylene was obtained from CF_4-C_2H_4 mixtures with ^{18}F atoms formed by the $^{19}F(\gamma, n)^{18}F$ reaction in CF_4 (9, 10). The ^{18}F atoms which failed to react with CF_4 were effectively removed by addition to ethylene, as in Equation 1. The

$$^{18}F + CH_2{=}CH_2 \longrightarrow CH_2{}^{18}FCH_2^* \qquad (1)$$

subsequent fate of these $CH_2{}^{18}FCH_2$ radicals was then determined by a pressure-dependent competition between stabilization, (equation 2) and decomposition, (equation 3). The stabilized

$$CH_2{}^{18}FCH_2^* + M \longrightarrow CH_2{}^{18}FCH_2 + M \qquad (2)$$

$$CH_2{}^{18}FCH_2^* \longrightarrow CH^{18}F{=}CH_2 + H \qquad (3)$$

$CH_2{}^{18}FCH_2$ radical was then detected after reaction with scavenger molecular I_2 as $CH_2{}^{18}FCH_2I$ (equation 4).

$$CH_2{}^{18}FCH_2 + I_2 \longrightarrow CH_2{}^{18}FCH_2I + I \qquad (4)$$

The linear pressure dependence of the $CH^{18}FCH_2$ to $CH_2{}^{18}FCH_2I$ ratio in excess CF_4 further showed that the $CH_2{}^{18}F$-CH_2^* radicals were almost uniformly monoenergetic indicating that the radicals received negligible additional excitation from extra translational energy of the ^{18}F atom (see below). A similar F atom addition-plus-decomposition mechanism has been invoked to explain the presence of C_2H_3F as a product from the photolysis of ONF in the presence of C_2H_4 (16).

More recently, Rowland and coworkers reported the results of some experiments on fluorine atom additions to olefins and acetylene using tracer levels of ^{18}F generated by the $^{19}F(n, 2n)$-^{18}F reaction (5-8, 12-15). These experiments were carried out under more nearly ideal conditions: (a) the ampoule contents suffered negligible radiation damage because of the low level of radioactivity per sample; (b) hot reactions of ^{18}F were reduced to <1% of the total by maintaining the mole fraction of moderator (source gas) above 0.90; and (c) improved radio gas chromatographic methodology, i.e., column specificity, more advanced detection techniques and better separations. One other substantial advantage involved the use of HI as the scavenger (equation 5), providing much greater ease of radio gas chroma-

$$CH_2{}^{18}FCH_2 + HI \longrightarrow CH_2{}^{18}FCH_3 + I \qquad (5)$$

tographic determination of the scavenged radical product, CH_3-$CH_2{}^{18}F$, than of the iodo-compound formed in (4). The radioassay of $CH_2{=}CH^{18}F$ and $CH_3CH_2{}^{18}F$ is illustrated in Figure 1. No $CH_3CH_2{}^{18}F$ is observed in the absence of HI, but a major peak appears when it is present. In the illustrated assay the product $CH_3{}^{18}F$ is also found when HI is in the sample. This product is formed by reaction (6) which is about 50 kcal/mole endothermic, and can only occur when initiated by energetic

$$^{18}F + CH_2{=}CH_2 \rightarrow CH_2{}^{18}FCH_2^{*} \rightarrow CH_2{}^{18}F + CH_2 \qquad (6)$$

^{18}F atoms. The system in Figure 1 contained about 20% C_2H_4 and 80% SF_6, permitting many collisions between "hot" ^{18}F atoms and C_2H_4. When similar experiments are carried out with only 1% C_2H_4 and 99% SF_6, the yield of $CH_3{}^{18}F$ is reduced by a factor of 20, and is negligible overall.

If all of the radicals in (2) and (3) are essentially monoenergetic(i. e. radicals formed by thermal ^{18}F atoms, excited only by the bond-formation energy), then the relative rates of formation of $C_2H_3{}^{18}F$ and $C_2H_5{}^{18}F$ are given by (7), with a

$$\frac{d(CH_2CH^{18}F)}{d(CH_3CH_2{}^{18}F)} = \frac{(\text{Decomposition})}{(\text{Stabilization})} = \frac{D}{S} = \frac{k_3}{k_2M} = \frac{k}{P}$$

straight line expected in a D/S plot against the inverse pressure.

The agreement in data from experiments in two laboratories with widely varying techniques is shown in the D/S plot for CF_4-C_2H_4 mixtures shown in Figure 2. The slope of the lines is a measure of the ratio of k_3/k_2. At D/S = 1, the half-stabilization point, the lifetime of the fluoroethyl radical is about 1×10^{-9} sec (13). The derivation of equation 7 has implicitly assumed that one collision is always sufficient to stabilize the radical in equation 2--the "strong collision" assumption in the kinetic derivation. In many mixtures further research has shown that this assumption is not completely valid, as discussed later. In fact, the CF_4-C_2H_4 system is one in which the collisions are not "strong" (i. e. do not remove large quantities of energy per collision), as demonstrated by isotopic effects in comparisons of CF_4-C_2H_4 and CF_4-C_2D_4 systems. These isotope effects are discussed in the final section. When D/S >1, the observation of concave curvature upward instead of the straight lines of Figure 2 serves as a diagnostic test for "weak" collisions.

The reactions of thermal ^{18}F atoms with alkynes are analogous to the alkene reactions just described. The ^{18}F atoms add rapidly (17) to acetylenic bonds, as illustrated in (8) and (9) (7).

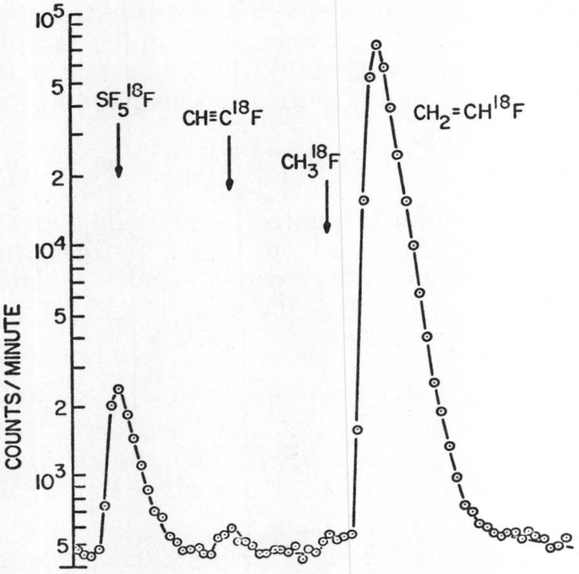

Figure 1. Radio gas chromatographic separation of CH_2═$CH^{18}F$ from $^{19}F(n, 2n)^{18}F$ reaction in gaseous SF_6–C_2H_2–HI mixtures (27)

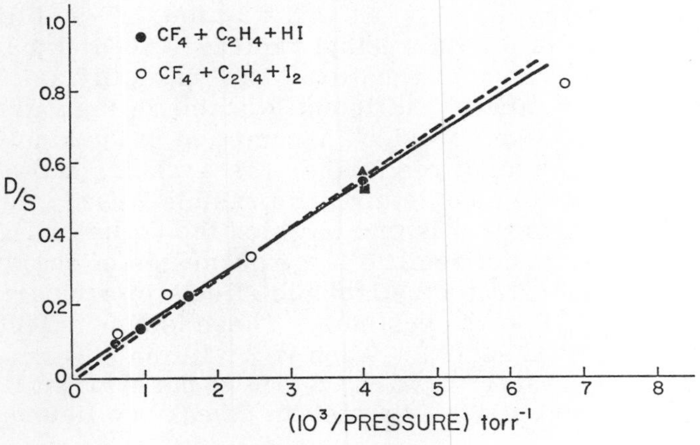

Figure 2. Ratio of decomposition stabilization vs. pressure^{-1} for ^{18}F-labeled products from ^{18}F reactions with C_2H_4 in excess CF_4 (all samples contained CF_4:C_2H_4 in the approximate ratio 18/1). ○ = $C_2H_3{}^{18}F/CH_2ICH_2{}^{18}F$ with I_2 as scavenger at 25°C (10).

$$^{18}F + HC{\equiv}CH \longrightarrow HC^{18}F{=}CH \qquad (8)$$

$$CH{=}CH^{18}F + HI \longrightarrow CH_2{=}CH^{18}F + I \qquad (9)$$

Addition to ethylene is 0.83±0.02 as fast as addition to acetylene, as determined by experiments in which both ethylene and acetylene are simultaneously present and directly competing with one another (18).

Product Analyses and Identification of Reaction Channels.

All of the interpretations which follow are based on the assumption that the reacting ^{18}F species is in the neutral, ground electronic state, designated as hot if it possesses excess kinetic energy and thermal if it does not. The possibility that the resulting species is occasionally either charged or electronically excited cannot be rigorously excluded. However, the general reaction pattern of ^{18}F from nuclear recoil, especially the yield behavior in the presence of inert moderators of varying ionization potential, has not yet furnished any evidence inconsistent with the hypothesis of the neutral, ground state as the reacting entity (9, 10, 19-23). The very high energies for the first excited electronic state (12.7 eV above the ground state) and the ionization limit (17.4 eV) make appreciable fractions of these species very unlikely at energies comparable to chemical bond energies, while the hot reactions of ^{18}F are almost entirely suppressed by the use of moderators. The observation of normal D/S behavior (equation 7) for radicals such as $CH_2{}^{18}F$-CH_2^* is good evidence for the absence of severe complications from the reacting species (13).

Typical data for the systems, $C_2H_4/CF_4/HI$ and $C_2H_4/SF_6/HI$ are contained in Table I (13). Additional data for the SF_6/C_2H_4 (18:1) system are included in the D/S vs $(P)^{-1}$ plot of Figure 3 (24). The extrapolated pressure for half-stabilization obtained from Figure 3 is 80 Torr. Flores and Darwent (16) also measured the stabilization of $CH_2FCH_2^*$ and found that pressures of about 100 Torr were required. In their photolysis system, however, product analyses were difficult and accurate pressures for half-stabilization were not obtained.

As seen from Table I, yields of about 60% are found for the addition of thermal ^{18}F to C_2H_4 at high pressures. The experiments at pressures less than 1000 Torr clearly show the loss of

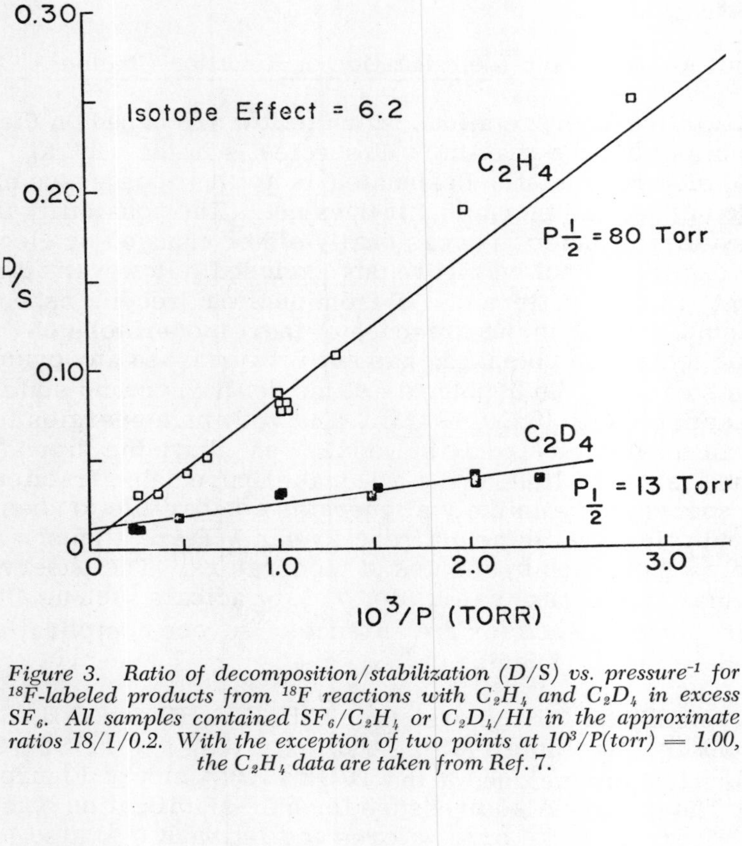

Figure 3. Ratio of decomposition/stabilization (D/S) vs. pressure^{-1} for ^{18}F-labeled products from ^{18}F reactions with C_2H_4 and C_2D_4 in excess SF_6. All samples contained SF_6/C_2H_4 or C_2D_4/HI in the approximate ratios 18/1/0.2. With the exception of two points at $10^3/P(torr) = 1.00$, the C_2H_4 data are taken from Ref. 7.

energetic ^{18}F into the ampoule walls because of insufficient available path length for removing the 10^6eV of initial kinetic energy from the (n, 2n) reaction. The minor yield of $CH_3{}^{18}F$ is caused by the direct hot substitution reaction (6). Most of the "missing" ^{18}F in the high pressure samples has formed $H^{18}F$ by abstraction of H from C_2H_4 (equation 10). The $H^{18}F$ so formed reacts on the ampoule wall and is not detected by normal radio gas chromatographic techniques.

$$^{18}F + CH_2{=}CH_2 \longrightarrow H^{18}F + CH_2{=}CH \qquad (10)$$

In general, at least half of the thermal ^{18}F atoms can be expected to add to olefinic or acetylenic substrates with the formation of the corresponding radicals. Stabilization of these excited radicals is favored by high pressures (or by condensed phase experiments) although some radicals undergo even more decomposition at low pressures than observed with $CH_2{}^{18}FCH_2$ (17). In most systems the radicals stabilized by collision with SF_6 can usually be converted to the corresponding hydrocarbon by reactions with scavengers in analogy with equation (5).

Table I. Volatile Radioactive Products from ^{18}F Atom Reactions in Typical Mixtures of C_2H_4 with CF_4 or SF_6 (13).

	Pressure, Torr						
CF_4	1480	570	232				
SF_6				3840	1570	287	94
C_2H_4	83	31	13	209	87	16	5
HI	17	6	2	42	17	3	1
	Product Yields, % Absolute ^{18}F						
$CF_3{}^{18}F$	2.24	1.90	2.33				
$CHF_2{}^{18}F$	0.58 (combined with $CH_3{}^{18}F$)	0.21	0.22				
$CH_3{}^{18}F$		0.32	0.52	0.21	0.20	0.35	
$CH^{18}F{=}CH_2$	5.61	10.6	17.9	1.67	3.0	11.2	16.9
$CH_3CH_2{}^{18}F$	58.7	47.9	32.9	53.2	59.0	45.2	23.0
$SF_5{}^{18}F$				1.29	1.06	1.09	0.3

Absolute Rate and Cross Section Measurements.

A. Absolute Rate Measurements: The direct abstraction of H atoms from saturated hydrocarbons by fluorine atoms occurs very rapidly, with the formation of HF. The reaction of ^{18}F with CH_4 by Equation 11 leads to a product ($H^{18}F$) which is

$$^{18}F + CH_4 \longrightarrow H^{18}F + CH_3 \qquad (11)$$

not readily analyzed by gas chromatography because of its high chemical reactivity. Direct chemical detection of such $H^{18}F$ has been carried out by Root et. al.(25). In our own experiments, the rate of reactions such as Equation 11 has been estimated by its competitive diminution of the yields of reactions with ethylene or acetylene (26, 27). The absolute rate constant of the corresponding ^{19}F reaction with CH_4 was determined independently in a flow system using mass spectrometric detection and accurately measured flows of very small amounts of the reactant gases (28). Fluorine atoms, free of F_2, were generated by the reaction of N atoms with NF_2 (29). The rate constant with ^{19}F was determined to be 7.1×10^{-11} $cm^3 molecule^{-1} sec^{-1}$ at 283^0K, or reaction in about 1 of every 5 collisions with CH_4. Since the ratio of H atom abstraction from CH_4 versus addition to HC≡CH was measured to be 0.41, (26, 27) the absolute value for the rate of thermal F atom addition to HC≡CH is about $1.7 \times 10^{-10} cm^3 molecule^{-1} sec^{-1}$. Similarly, the absolute rate constant for thermal F atom addition to ethylene can be estimated from the relative rates of addition to HC≡CH versus CH_2=CH_2 to be $1.4 \times 10^{-10} cm^3 molecule^{-1} sec^{-1}$. Addition to both C_2H_2 and C_2H_4 thus requires three collisions or less on the average.

The relative rate constant ratio for addition of thermal ^{18}F to perfluoropropylene to abstraction by thermal ^{18}F of H from H_2, has been measured by Grant (25) to be $k_{C_3F_6}/k_{H_2} = 0.056 \pm 0.002 \exp(1502 \pm 17 \text{ cal/mol})/RT$. Rowland and Williams (27) have measured the rate constants of thermal ^{18}F atom abstraction of H from H_2 and CH_4 relative to addition of ^{18}F to HC≡CH. This ratio, measured to be 0.14: 0.41:1.0, may be combined with the results of Wagner et. al. (28) to yield an absolute value of $2.0 \times 10^{-11} cm^3 molecule^{-1} sec^{-1}$ for $k_{C_3F_6}$. Rowland and Milstein (30) found that the initial ^{18}F addition step is approximately equal in rate for C_2H_2 and CHCl=CHCl and about six times slower for CFCl=CFCl. With both CF_3CF=CF_2 and CFCl=CFCl addition is considerably slower than with olefins containing fewer F and more H atoms. While such reactions are all very rapid, and are reasonably consistent with one another, Grant and Root (31) have pointed out that certain inconsistencies do exist in the calculation of Arrhenius parameters for the reactions of F with CF_3H, CH_4, H_2 and C_3F_6, based on the measurements of Wagner et. al.(28). Nevertheless, Foon and Kaufman (32) believe that Wagner's result is the preferable one at this time. Work in the area of absolute rate constant determinations for fluorine atoms is still generally in the preliminary stage. The nature of tracer ^{18}F experiments makes direct absolute rate determinations very difficult, while facilitating relative rate measurement. Measurements with ^{18}F at various temperatures will eventually be important, and Root et al. have already constructed a target that can be used with cyclotron-produced fast neutrons at temperatures as high as 475°K (33).

B. Cross Section Measurements: In recent years crossed molecular beam techniques have been employed by Lee and coworkers (34) to study extensively the reactions of fluorine atoms with several different olefins. Very high cross-sections were observed, consistent with the rapid thermal rate constants measured above. In these experiments, multichannel decomposition is observed (i.e., loss of H or CH_3 after ^{19}F addition to the π-bond). Since the C-F bond is stronger than any other single bond with a carbon atom, C-F bond scission back to the original reactants does not compete with the exothermic processes involving breakage of C-C and C-H bonds. A tabulation of ΔH values for addition reactions of thermal ^{18}F to various olefins is included (Table II) (6). Some highly endothermic pathways are listed for the purposes of comparison. The experimental measurements of D/S in this table confirm that the only radicals undergoing appreciable C-C bond rupture with carbene formation after thermal ^{18}F addition are those that form CF_2. The extraordinary stability of the CF_2 radical makes C-C bond rupture exothermic for these radicals.

A summary of some average product translational energies and reactive cross sections is presented in Table III (34, 35). According to Lee and coworkers, the higher average translational energies typical of CH_3 release reactions can be attributed to a higher potential energy barrier in the exit channel, under the assumption that some of the excess potential energy over that necessary for dissociation is converted into product translational energy. In all of the cases listed in Table III the cross

Table II. Energetics and Yields for Decomposition Pathways for Fluoroethyl Radicals Formed by Addition of ^{18}F to Olefins (6).

Olefin	Radical	Decomposition products	ΔH, kcal/mole	D/S: Olefin alone (a)	D/S: SF_6/olefin (b)
$CF_2{=}CF_2$	$CF_2{}^{18}FCF_2$	$CF_2{}^{18}F + CF_2$	-17	0.42	0.28
		$CF^{18}F{=}CF_2 + F$	0	<0.02	<0.01
$CHF{=}CF_2$	$CHF^{18}FCF_2$	$CHF^{18}F + CF_2$	+ 2	0.34	0.12
		$CH^{18}F{=}CF_2 + F$	0	<0.1	<0.02
		$CF^{18}F{=}CF_2 + H$	- 5	<0.02	<0.01
	$CF_2{}^{18}FCHF$	$CF_2{}^{18}F + CHF$	+ 9	0.07	<0.2
		$CF^{18}F{=}CHF + F$	0	<0.2	<0.05
$CH_2{=}CF_2$	$CH_2{}^{18}FCF_2$	$CH_2{}^{18}F + CF_2$	+18	0.29	0.01
		$CH^{18}F{=}CF_2 + H$	- 3	0.05	<0.01
	$CF_2{}^{18}FCH_2$	$CF_2{}^{18}F + CH_2$	+38	0.04	<0.01
		$CF^{18}F{=}CH_2 + F$	0	<0.2	<0.01
$CHF{=}CHF$ (trans)	$CHF^{18}FCHF$	$CHF^{18}F + CHF$	+13	0.05	~0.05
		$CH^{18}F{=}CHF + F$	0	<0.15	<0.03
		$CF^{18}F{=}CHF + H$	-20	<0.12	<0.02
$CH_2{=}CHF$	$CH_2{}^{18}FCHF$	$CH_2{}^{18}F + CHF$	+27	0.06	<0.1
		$CH^{18}F{=}CHF + H$	- 5	<0.05	<0.01
	$CHF^{18}FCH_2$	$CHF^{18}F + CH_2$	+40	0.03	<0.01
		$CH^{18}F{=}CH_2 + F$	0	<0.1	<0.03
		$CF^{18}F{=}CH_2 + H$	-21	<0.05	<0.01
$CH_2{=}CH_2$	$CH_2{}^{18}FCH_2$	$CH_2{}^{18}F + CH_2$	+52	No ^{18}F source	~0.02
		$CH^{18}F{=}CH_2 + H$	- 7		~0.05

(a) Reactions of hot and thermal ^{18}F mixed together; (b) Reactions of thermal ^{18}F alone.

Table III. Some Average Product Translational Energies, Reactive Cross Sections and the Ratios between CH_3 Emission and H Emission Channels (34, 35).

Reactant	Products	<E'> (kcal/mole)	σ_r ($Å^2$)	σ_{CH_3}/σ_H
ethylene	$C_2H_3F + H$	6.7	2.2	--
propylene	$C_3H_5F + H$	5.8		
cis 2-butene	$C_4H_7F + H$	4.6	0.30	22
	$C_3H_5F + CH_3$	10.1	6.6	
trans 2-butene	$C_4H_7F + H$	4.6	0.23	40
	$C_3H_5F + CH_3$	10.1	9.1	
isobutene	$C_4H_7F + H$	4.6	0.24	30
	$C_3H_5F + CH_3$	10.9	7.3	
1-butene	$C_4H_7F + H$	4.6	0.026	115
	$C_3H_5F + CH_3$	9.3	3.0	
Tetramethylethylene	$C_6H_{11}F + H$	5.4	0.48	32
	$C_5H_9F + CH_3$	8.3	15.4	
$(H_3C)_2C{=}CH(CH_3)$	$C_5H_9F + H$	5.3	0.19	15
	$C_4H_7F + CH_3$	9.3	2.9	

sections for CH_3 release are much larger than those for H release, reflecting the greater exothermicity of CH_3 loss.

Addition to Asymmetric Olefins.

Quantitive studies of fluorine atom addition to simple asymmetric alkenes have not been numerous because of experimental difficulties in the handling of fluorine atoms and the eventual reaction products. However, two investigations have been undertaken which permit estimates of the ratio of terminal to central attack on propylene in the gas phase. A value of 3.0 for this ratio was estimated from the respective yields of $CH_3CH(NF_2)CH_2F$ and $CH_3CHFCH_2NF_2$ in the gas phase photolysis of N_2F_4-propylene mixtures (36). The mechanism of this photodifluoramination reaction involves the direct photolysis of NF_2 (in equilibrium with N_2F_4) to F + NF; the addition of F to the olefin; and the further reaction of the adduct radical with NF_2 (or N_2F_4) to give an observable stable product. Terminal F atom attack thus leads to $CH_3CH(NF_2)CH_2F$, with the other isomer being formed by initial central attack by F.

In contrast, a set of experiments based on standard tracer ^{18}F techniques (12) has yielded a value of 1.4 (37) for the ratio of terminal to central attack (Equations 12, 13). These studies

$$^{18}F + CH_3CH{=}CH_2 \rightarrow CH_3CHCH_2{}^{18}F^* \qquad (12)$$

$$^{18}F + CH_3CH{=}CH_2 \rightarrow CH_3CH^{18}FCH_2^* \qquad (13)$$

were performed over a greater pressure range (400-4000 Torr) than the N_2F_4 experiments (all done at 150 Torr) (36). The results of these ^{18}F experiments demonstrate that the radical formed from central addition readily undergoes unimolecular decomposition at low pressures (Equation 14). Such decompo-

$$CH_3CH^{18}FCH_2^* \longrightarrow CH_3 + CH^{18}F{=}CH_2 \qquad (14)$$

sition is in agreement with the molecular beam results of Lee and coworkers (34). Consequently, quantitative interpretation of relative ratios of attack by fluorine atoms on asymmetric olefins requires experimentation at pressures high enough to avoid such decomposition problems (Table IV). The yields of each of the ^{18}F-fluoropropylenes were negligibly small at all pressures, indicating no appreciable excitation decomposition of

Table IV. Yields of ^{18}F-Labeled Products from HI-Scavenged Addition of ^{18}F Atoms to Propylene in Excess SF_6 (12).

Sample composition Pressure, Torr			Absolute Product Yield, Per Cent			
			Radical: $CH_3CH^{18}FCH_2*$		$CH_3CHCH_2{}^{18}F*$	
SF_6	Propylene	HI	Product[b]: $CH_2{=}CH^{18}F$	$CH_3CH^{18}FCH_3$	$CH_3CH_2CH_2{}^{18}F$	Ratio[a]
3800	191	19	0.77	18.4	25.9	1.35
3780	101	100	0.47	14.0	19.7	1.36
1920	48	48	0.62	13.7	19.3	1.34
1920	48	48	0.72	14.2	19.9	1.33
1060	26	26	1.02	14.2	20.7	1.36
360	18	19	2.29	10.5	17.4	1.36
360	18	18	2.22	10.0	16.7	1.36

[a] $(CH_3CH_2CH^{18}F)/\{(CH_2{=}CH^{18}F) + (CH_3CH^{18}FCH_3)\}$. [b] Also observed (from hot reactions of energetic ^{18}F atoms), yields in absolute per cent: $SF_5{}^{18}F$, 0.8-1.3%; $CH_3{}^{18}F$, 0.2%; traces of ^{18}F-fluoropropenes, <0.3% total.

$CH_3CHCH_2{}^{18}F^*$ radicals resulting from terminal addition. Extrapolation of the decomposition/stabilization ratio of Table IV ($CH_2{=}CH^{18}F$ vs. $CH_3CH^{18}FCH_3$) to lower pressures indicates that perhaps 50% of the CH_3CHFCH_2 radicals might be stabilized at about 150 Torr. A precise comparison cannot be made because the main component gases are not the same. The observed terminal/central ratio for surviving fluoropropyl radicals would then be 1.4/0.5 or about 3.0, as reported for the N_2F_4 system (36).

The measured terminal to central ratio of 1.4 contrasts sharply with the ratio of about 15 found for the thermal hydrogen atom addition to propylene (38). The value of 1.4 indicates only a moderate preference for reaction at the CH_2 end of propylene, indicating that atomic fluorine is a rather indiscriminate, highly reactive species. Both the ^{18}F recoil and N_2F_4 photolysis experiments show substantial H atom abstraction from propylene by atomic F, with about 50% of the ^{18}F going to $H^{18}F$. The complementary nature of tracer ^{18}F and macroscopic ^{19}F experiments can be illustrated by a further comparison of these two experimental systems. In the ^{18}F experiments, the residual allyl radicals presumably left by the abstraction process (equation 15) cannot be detected since they have no radioactive label,

$$^{18}F + CH_3CH{=}CH_2 \longrightarrow H^{18}F + CH{\because}CH{\because}CH_2 \qquad (15)$$

and the abstraction process can be directly followed only by observing $H^{18}F$. This was not done in reference 11 and the existence of reaction (15) was determined only indirectly through the absence of 50% of the ^{18}F. In the $N_2F_4/CH_3CH{=}CH_2$ system, however, the abstraction reaction leaves major yields of allyl radicals, traceable after reaction with NF_2. The high yield of $CH_2{=}CHCH_2NF_2$ found in the N_2F_4 system results from this combination of allyl radicals with NF_2 (36).

The addition of fluorine atoms to butene has been investigated as an extension of the experiments outlined above. Isobutylene was also studied with the N_2F_4 technique and behaved very similarly to propylene (36). The ratio of terminal to central addition was 3.1/1.0, again with no correction for decomposition of $(CH_3)_2CFCH_2{}^*$ formed at relatively low pressures. Standard ^{18}F experiments with isobutylene and HI are not possible because they react with each other thermally. Tracer ^{18}F studies with 1-butene have shown that it resembles ^{18}F-

propylene in its behavior (39), with a terminal/central ratio of addition of 1.4. Again, when the isobutylene photolysis system calculations are corrected for the unimolecular decomposition of $(CH_3)_2CFCH_2{}^*$, the results of both experiments are mutually consistent.

Since the results of both the propylene and butene experiments are essentially the same, the conclusion that the F atom is a rather indiscriminate reactive entity little influenced by the presence of a methyl group is reinforced. In contrast, studies of thermal H atom addition to 1-butene, as with propylene, have shown a much more pronounced directional effect with a terminal to central addition ratio of about 20 (40).

The addition of ^{18}F has also been studied with ethylenes (6). In intramolecular competition with fluorohydroolefins, the ^{18}F atoms react preferentially with less fluorinated positions. Intermolecularly, ^{18}F addition is also preferred at the less fluorinated positions. Among multiply-bonded molecules, no extreme preference is found for reaction with one molecule or the other. For example, ^{18}F atoms add about four times as rapidly to $CH_2{=}CH_2$ as to $CHF{=}CF_2$ (6).

The competitive addition reactions available with halogenated olefins are illustrated in equations 16 and 17 for vinyl fluoride.

$$^{18}F + CH_2{=}CHF \longrightarrow CH_2{}^{18}FCHF^* \qquad (16)$$

$$^{18}F + CH_2{=}CHF \longrightarrow CH_2CHF^{18}F^* \qquad (17)$$

The two difluoroethyl radicals can be stabilized by collision and observed as $CH_2FCH_2{}^{18}F$ and $CH_3CHF^{18}F$, respectively, after reaction with HI.

The relative yields for various molecules are summarized in Table V. For convenience, the reactivity per methylene group in ethylene has been given the value of 1.0 and both the intermolecular and intramolecular competitions have been expressed in comparison to this standard. The addition of ^{18}F to ethylenes containing C-H or C-F bonds consistently favors, on both intermolecular and intramolecular bases, addition to the less fluorinated positions. The CH_2 end of $CH_2{=}CF_2$ is about five times as reactive toward addition of thermal ^{18}F as is the

Table V. Intermolecular and Intramolecular Selectivity in ^{18}F Atom Addition to Olefins in Excess SF_6 (6).

	Normalized yield per olefinic carbon atom [(a)]		
Olefin	CH_2	CHF	CF_2
$CH_2=CH_2$	1.0		
$CHF=CH_2$	0.7(0.8)	0.6	
$CF_2=CH_2$	0.8(1.1)		0.2
trans-CHF=CHF		0.3	
$CHF=CF_2$		0.4	0.1
$CF_2=CF_2$			0.14(0.2)

(a)
Yield per carbon atom of the stabilized radical formed by addition of ^{18}F to the listed group, relative to CH_2 in ethylene as 1.0. The numbers in parentheses represent the approximate yields after correction for decomposition of excited radicals.

CF_2 end (17), but no examples have yet been found in which attack is negligible at either end of the multiple bond. Similar behavior has been observed for the addition of CCl_3, CF_3 and C_3F_7 radicals to fluorinated ethylenes (41).

The preference for attacking the more hydrogenated and less fluorinated end of the double bond is consistent with electrophilic (^{18}F) attack on the molecular end with the greater π-electron density, i.e., usually the end with fewer highly electronegative substituents (42). However, the extended Hückel molecular orbital calculations of Libit and Hoffmann (43) have shown similar π-electron densities on the terminal (0.945) and central (0.936) carbon atoms in propylene. Satisfactory rationalization of these preferential orientation effects probably will require additional experiments with still other asymmetric alkenes. Some experimental complications enter through the reactivity of HI with some substituted olefins, preventing the use of the standard experimental mixtures. Experiments in this laboratory suggest that H_2Se can frequently be used as a replacement scavenger for HI (44).

When ^{18}F atoms are added to unsaturated carbon atoms carrying chlorine or bromine atom substituents, the heavier halogen atom is lost very rapidly ($<10^{-10}$sec) (17). In such cases substitution formally occurs as in equation 18. Further

$$^{18}F + RCH{=}CHX \rightarrow RCHCHX^{18}F \rightarrow RCH{=}CH^{18}F + X \qquad (18)$$
$$(X{=}Cl, Br)$$

experiments with (a) cis- and trans-CHCl=CHCl; and (b) CFCl=CFCl (mixture of isomers) have been reported (30). The decomposition rates of the resulting $CHCl^{18}FCHCl$ and $CFCl^{18}FCFCl$ radicals have been studied through the pressure dependence of the relative yields of the decomposition and stabilization products. The stabilized radicals were once again identified as the corresponding saturated molecule after H abstraction from HI. The $CHCl^{18}FCHCl$ radicals decompose very rapidly (98% decomposition at 2000 Torr), about 50 times faster than $CFCl^{18}FCFCl$ radicals (50% decomposition at 2000 Torr). The trans to cis ratios of the $CH^{18}F{=}CHCl$ formed by $CHCl^{18}FCHCl$ decomposition are different (0.6 from cis-CHCl=CHCl, 0.9 from trans) for each isomeric parent indicating that decomposition by Cl atom loss is kinetically competitive with rotation about the C-C bond in the excited $CHCl^{18}FCHCl$ radical.

Atomic fluorine addition to allene had at first been assumed to proceed entirely through attack at the central position, since only the product corresponding to initial F atom attack (and NF_2 termination) has been isolated from N_2F_4-allene photolysis experiments (45, 46). No evidence was found for the presence of either $CH_2{=}C(NF_2)CH_2F$ or its expected isomerization product $(CH_2F)_2C{=}NF$ (45). However, Rowland and coworkers (47, 48) found both terminal and central addition with relative yields in the approximate ratio of 2:1 (Table VI), indicating a moderate selective preference toward reaction with the terminal end of the π-bond.

Table VI. Absolute Percentage Yields of Fluoropropene-^{18}F molecules from ^{18}F Reaction with Allene in Excess SF_6 (47).

Pressure(torr)			Absolute Yield, Per cent		
SF_6	Allene	HI	$CH_3C^{18}F{=}CH_2$	$CH_2{}^{18}FCH{=}CH_2$	Ratio
3360	158	16	20.4	34.3	1.7
3350	112	56	19.2	32.3	1.7
3330	106	85	15.9	29.8	1.9
3360	46	126	8.3	17.9	2.2
827	27	14	10.8	18.9	1.8

The most important reactions with alkynes correspond to those found with alkenes and include:

(a) substitution of ^{18}F for H

$$^{18}F + HC{\equiv}CH \longrightarrow CH{\equiv}C^{18}F + H \qquad (19)$$

(b) abstraction of H

$$^{18}F + HC{\equiv}CH \longrightarrow H^{18}F + C{\equiv}CH \qquad (20)$$

(c) addition to the π-bond to form $CH{=}CH^{18}F$ (Equation 8).

The $SF_6/HC{\equiv}CH/HI$ system has been thoroughly investigated by Williams and Rowland (7). Typical results are presented in

Table VII.

Table VII[a]. Pressure Dependence of Product Yields from SF_6—Moderated ^{18}F Atom Reactions with HI-Scavenged Acetylene (7)

Total pressure, Torr	3800	690	300	250	100
Absolute yield, %					
$SF_5{}^{18}F$	1.16	1.05	0.77	0.70	~1
$CH{\equiv}C^{18}F$	<0.1	<0.2	<0.4	<0.3	<1
$CH_2{=}CH^{18}F$	59.3	67.0	57	58	48

[a] Molar composition ratio, $SF_6/C_2H_2/HI = 20/1.0/1.0$.

The yields indicate that about 60% of the ^{18}F is recovered as the stabilized product, $CH_2{=}CH^{18}F$. When the HI/C_2H_2 ratio is reduced below the 1.0 of Table VII, the yield of $CH_2{=}CH^{18}F$ increases, leading to an estimated 85% at low HI concentration.

The extrapolated half-stabilization pressure of ≤ 200 Torr implies a lifetime for $CH{=}CH^{18}F^*$ of $\geq 5\times10^{-10}$ sec. At 100 Torr pressure, the observed limit on the loss of an H atom from $CH^{18}F{=}CH^*$ (< 1% to form $HC{\equiv}C^{18}F$) indicates that the lifetime against H atom loss is $>10^{-8}$ sec. Moreover, crossed molecular beam studies of low energy F atom reactions with acetylene failed to show any evidence for the F for H replacement reaction (7), and H atom loss is at most a minor path for the decomposition of $CH^{18}F{=}CH$ radicals under collision-free conditions. Since the C-H bond in C_2H_2 is very much stronger then in C_2H_4, the substitution of ^{18}F for H in C_2H_2 is probably endothermic.

The ^{18}F atoms which would normally react with acetylene to form $CH^{18}F{=}CH^*$ can be deflected from this route by prior reaction with some other substrate molecule, such as by abstraction of H from HI or from CH_4. The inclusion of an amount of CH_4 1.4 times that of acetylene is sufficient to reduce the $CH_2{=}CH^{18}F$ yield almost by half, as shown in Table VIII (7).

Table VIII. Pressure Dependence of Yields of ^{18}F Products in C_2H_2-CH_4 Competition (7).

Total pressure,[a] Torr	3840	2490	690	380	300	250
Absolute yield, %						
$SF_5{}^{18}F$	1.12	1.06	0.9	0.9	0.8	0.8
$CH{\equiv}C^{18}F$	0.06	0.04	<0.2	<0.2	<0.4	<0.4
$CH_2{=}CH^{18}F$	43.3	42.2	39.9	38	40	33
$CH_3{}^{18}F$	0.59	0.59	0.5	0.6	~ 0.7	~ 0.4

[a] Molar composition ratio, $SF_6/C_2H_2/CH_4/HI$ = 20/1.0/1.4/1.0.

The observed competition among addition and the various abstraction reactions shows that more detailed studies of the variation in $H^{18}F$ yield, or in diminution of the $CH_2{=}CH^{18}F$ yield can provide relative rate constants for these reactions. The substitution of other RH molecules for methane can then permit the extension of such relative rate constant measurements to many molecules. For example, the relative rate constants for hydrogen atom abstraction by thermal ^{18}F atoms from acetylene and methane are 0.06 and 0.11 per hydrogen atom, respectively (26), a factor of nearly two.

The abstraction reaction must always be considered as a likely path if hydrogen-containing molecules are present in an irradiation mixture. However, some hydrogen-containing compounds with especially strong C-H bonds do not favor $H^{18}F$ formation as readily as most hydrocarbons and do not appreciably diminish the possible yields for other reaction paths. Even so, the formation of HF by H atom abstraction is still an exothermic reaction (17).

Milstein et al. (18) have investigated the competition between ethylene and acetylene for thermal ^{18}F atoms. Several samples containing both compounds were irradiated and analyzed to establish a relative rate constant ratio between them. Experiments without acetylene present were required so that the decomposition yield of $CH_2{=}CH^{18}F$ from $C_2H_4{}^{18}F$ (and originally ethylene) could be separated from $CH_2=CH^{18}F$ from C_2H_2. The addition reaction of ^{18}F with C_2H_2 is slightly favored (1.20) over reaction with C_2H_4.

Studies with ethylene and acetylene were also undertaken in the presence of various other molecules serving as competitive scavengers for thermal ^{18}F atoms. The usual product analysis-techniques for competition reactions were employed and the results are presented in Table IX (18).

Table IX. Relative Scavenger Efficiencies for Removal of Radicals (18).

Radical	Scavenger	Relative Efficiency
$CH_2{}^{18}FCH_2$	HI	(1.0)
	NO	5.0
	O_2	1.05
	SO_2	0.33
	CO	0.005
$CH^{18}F=CH$	HI	(1.0)
	CO	0.022
	N_2	< 0.003
Relative Rates of Removal of ^{18}F by Reaction at 4000 Torr		
	C_2H_2	(1.0)
	C_2H_4	0.83
	SO_2	0.04
	O_2	<0.005
	NO	<0.01
	CO	<0.01
	N_2	<0.002

The addition of ^{18}F to $CH_3C{\equiv}CH$, as with the alkenes discussed earlier, does not show strong preferential reactions. The analysis for the three ^{18}F-fluoropropenes expected from reaction with methylacetylene is shown in Figure 4. Terminal reaction is favored by about a factor of 3 (the sum of 25.6% and 16.6% versus 13.7%), and the yields of the two possible terminal products reflect the greater stability of the cis compound (17). Although no experiments have yet been reported, a highly selective reaction with one multiple bond in a molecule containing two or more olefinic or acetylenic linkages seems quite unlikely in view of the nonspecific nature of fluorine atom addition in the systems already investigated.

Isotope Effects in Fluorine-Alkene Reactions.

The addition of fluorine atoms to olefins carries the possibility of significant isotope effects at several stages of reaction. The most extensive study of these possibilities has been carried out by Frank and Rowland (7, 15, 49), using ^{18}F added to C_2H_4 and to C_2D_4. Since fluorine atoms react quite indiscriminately, the inital addition step is unlikely to exhibit any appreciable isotopic difference between C_2H_4 and C_2D_4. However, once the radicals have formed, the decomposition rate for $C_2H_4{}^{18}F^*$ is expected to be much faster than that of $C_2D_4{}^{18}F^*$ in analogy with numerous other isotopic radical reaction rate comparisons for which D/H substitution reduces the reaction rate. These expectations have been quantitatively confirmed both by experimental measurements of pressures for half stabilization in SF_6 ($C_2H_4{}^{18}F$, 80 torr; $C_2D_4{}^{18}F$, 13 torr) and by Rice-Ramsperger-Kassel-Marcus calculations indicating that $C_2H_4F^*$ should react about six times more rapidly than $C_2D_4F^*$.

The most interesting observations of isotopic differences between these radicals, however, are found in the collisional de-excitation of these radicals with various bath gases. The data of Figure 5 and Table X show that the ratios of rates of decomposition for $C_2H_4F^*$ and $C_2D_4F^*$ are about 6 for collisions with SF_6 and NF_3, and only 4 for collisions with CF_4 and C_2F_6. The excitation energies of these radicals are determined solely by the thermochemistry of $F + C_2H_4$ and $F + C_2D_4$, independent of the bath gas. The existence of different ratios of decomposition rates requires isotopic differences in the collisional energy losses with different bath gases, and that these energy losses be sufficiently small (i.e. "weak" collisions) that decomposition after one or more collisions will still be appreciable.

The magnitude of collisional energy transfer from chemically activated molecular species is usually evaluated from graphs of decomposition to stabilization (D/S) ratios vs $(\text{pressure})^{-1}$ and can be expressed in terms of a characteristic energy loss, e.g.,

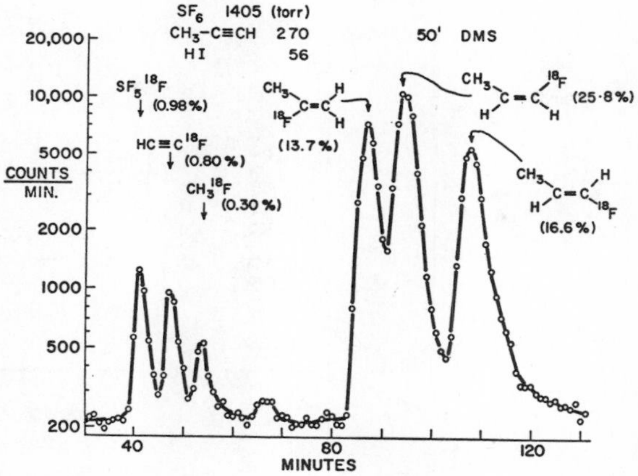

Figure 4. Radio gas chromatographic separation of three ^{18}*F-labeled isomers of* $C_3H_5{}^{18}F$ *from* ^{18}*F reactions in* SF_6*–*$CH_3C{\equiv}CH$*–HI mixtures (27)*

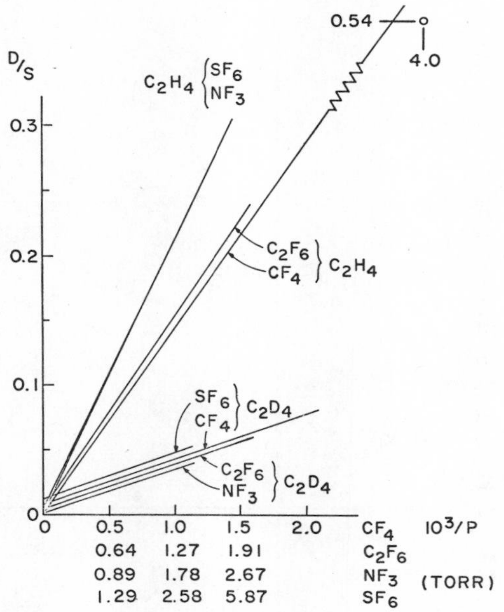

Figure 5. Graph of decomposition/stabilization vs. pressure^{-1} *for* $CH_2{}^{18}FCH_2$ *and* CD_2*-*$^{18}FCD_2$ *in* NF_3*,* CF_4*,* SF_6*, and* C_2F_6*. Different pressure*$^{-1}$ *scales are used for each gas—chosen to give the same visual slope for* $CD_2{}^{18}FCD_2$ *in all four gases (7, 24, 49).*

Table X. Extrapolated Pressures for Half-Stabilization for Fluoroethyl Radicals in NF_3, CF_4, SF_6 and C_2F_6 Bath Gases.

Radical / Bath Gas	Extrapolated Half-Stabilization Pressures (Torr)		Isotope Effect
	$CH_2{}^{18}FCH_2$	$CD_2{}^{18}FCD_2$	Ratio $\frac{H}{D}$
NF_3	117 ±8	19±2	6.2±0.7
CF_4	144 ±10	34±3	4.2±0.4
SF_6	80 ±6	13±1.5	6.2±0.8
C_2F_6	117 ±8	27±2.5	4.3±0.4

"step-size" in a step-ladder de-excitation model (50). With strong collisions and large energy losses, these graphs are straight lines; with small step-sizes, deviations from linearity are observed at very low pressures. Few measurements of step-size exist for isotopic activated species in collision with the same bath gas.

The difference in magnitude of the isotope effect requires that there be an isotope effect in the removal of energy by collision with either CF_4 and C_2F_6, or with SF_6 and NF_3, or with all four. The molecules NF_3, CF_4, SF_6, and C_2F_6 increase in size in that order and should show a corresponding decrease in the lifetimes of excited radical species toward stabilization. However, since the pressure for half-stabilization in CF_4 and C_2F_6 is appreciably higher than for NF_3 and SF_6, CF_4 and C_2F_6 must be "weak colliders", i.e., that $C_2H_4F^*$ and $C_2D_4F^*$ are frequently still able to undergo decomposition despite one collision (or more) with CF_4 or C_2F_6 as bath gases. The change in ratio of the isotopic rates from 6 to 4 requires that the average energy loss per collision with CF_4 or C_2F_6 (corresponding to step size for the step-ladder model) is less for $CD_2{}^{18}FCD_2$ than for its isotopic counterpart $CH_2{}^{18}FCH_2$ (49, 15).

While one weak collision is not sufficient to prevent further decomposition of the excited species, the isotopic ratio of rate constants will be altered only if the rate constants for decomposition after one collision are not negligible and are in a ratio different from the initial rate constants. Thus, a reduction of the isotopic rate constant ratio from 6.2 to 4.2 requires a substantially greater contribution to decomposition after the first collision with CF_4 from $CD_2{}^{18}FCD_2$ than from $CH_2{}^{18}FCH_2$. The data are roughly consistent with average energy losses sufficient to reduce the decomposition rate constant by a factor of two per collision for $CF_2{}^{18}FCD_2$ and a factor of three per collision for $CH_2{}^{18}FCH_2$ (49, 15).

Assuming a stepladder cascade mechanism, the energetics of the system are such that the average energy loss per collision with CF_4 by $CD_2{}^{18}FCD_2$ cannot be larger than about 2 kcal/mole in order to obtain numerical agreement with the data. The basic requirement is simply that collisions with CF_4 or C_2F_6 are "weaker" than with SF_6 or NF_3, and the numerical values are obtained by assuming that the strong collision assumption is satisfactory for SF_6 and NF_3. This assumption is not very stringent for the range of experimental conditions of Figure 5, since an average energy loss of 5 kcal/mole is sufficient to make negligible any decomposition after a single collision. The general question of strong vs. weak collisions with polyatomic bath molecules is an active subject of investigation and some controversy. Energy losses of such small magnitude are not inconsistent with at least some of the current literature estimates for collisions involving molecular species of comparable size (50-55).

Equilibration of Energy During the Reaction of Fluorine Atoms with Ethylene.

Several recent studies by different experimental techniques have provided information about the degree of equilibration of energy during the reactions of fluorine atoms with ethylene. Each of these techniques, however, provides information about a different stage of the process, and fulfill complementary roles in describing the entire reaction sequence. Our own experiments demonstrating lifetimes of 10^{-9} sec for species with 45 kcal/mole excitation energy meet the usual criteria for application of the R.R.K.M. theory, including complete energy equilibration within the $C_2H_4F^*$ molecule. Indeed, our R.R.K.M. calculations describe quite satisfactorily the observed decomposition rates for both $C_2H_4F^*$ and $C_2D_4F^*$. On the other hand, evidence for non-equilibration of energy in the fluorine plus ethylene system has been gained from the molecular beam experiments of Lee and co-workers (34, 35, 56) and from the infrared chemiluminescence experiments of McDonald et al. (57). Each of these measurements, however, pertains to a different step in the reaction sequence below.

The rate of decomposition of $C_2H_4F^*$ is controlled by the rate of passage into the transition state, $C_2H_4F^{\ddagger}$, and this step is the one calculated by R.R.K.M. theory.

$$F + C_2H_4 \longrightarrow C_2H_4F^* \longrightarrow C_2H_4F^{\ddagger} \longrightarrow H + C_2H_3F(v=n)$$

$$\longrightarrow C_2H_3F(v=o)$$

The beam experiments of Lee et al. (34, 35) have shown a non-equilibrium distribution of translational energy in the CH_2=CHF or CD_2=CDF molecules as they arrive at the detector after a collision-free trajectory. This translational energy does not represent, however, $C_2H_4F^{\ddagger}$ at the moment of reaction, but the energies of C_2H_3F and H after separation. The measured 1-2 kcal/mole of kinetic energy for all C_2H_3F(or C_2D_3F) molecules signifies the presence of a 1-2 kcal/mole barrier in the exit channel forcing all dissociating CH_2-CHF-H complexes to convert this barrier energy into translational energy of the separating products. These same beam experiments have shown through their forward-backward symmetry that the C_2H_4F complex has a lifetime $>10^{-11}$ sec., long enough for several rotational periods. Even when the initial F atom has kinetic energy of 15 kcal/mole, the forward-backward symmetry is maintained (56).

The infrared chemiluminescence experiments of McDonald et al. demonstrate that the infrared quanta emitted (time scale $\sim 10^{-3}$ sec.) by the vibrationally-excited C_2H_3F or C_2D_3F products from F + ethylene are not characteristic of molecules with energy fully equilibrated in all degrees of freedom. This lack of energy equilibration within the molecule, however, applies to C_2H_3F species vibrationally-excited with about 10 kcal/mole of excitation energy, and not to the $C_2H_4F^{*}$ species with about 50 kcal/mole excitation energy prior to H atom loss. Experiments with more highly vibrationally excited C_2H_3F molecules (e.g. from F + C_2H_3Br) do indicate equilibration of energy prior to infrared emission (57).

Summary and Comments.

The technique of using thermalized ^{18}F atoms for the study of fluorine atom reactions has proven very useful with unsaturated hydrocarbons and halocarbons, providing data on mechanisms, relative rate constants and factors controlling such reactions. The characteristic difficulties of macroscopic fluorine chemistry are often avoided at tracer levels, and analysis by radio gas chromatography can be quite straightforward. However, experiments at pressures below 0.1 atmosphere are relatively difficult, and most of the usual analytical methods are inapplicable at product mole fractions $<10^{-10}$.

Finally, many other classes of compounds can be readily substituted for alkenes and alkynes with little variation in equipment and technique. The extension to study of ^{18}F atom reactions with organometallic compounds is only one example of the broad applicability of tracer ^{18}F studies. As interest in the reactions of atomic fluorine continues to intensify, the radioactive ^{18}F technique promises to be one of the important complementary approaches through which progress will continue.

Acknowledgment. This work has been supported by U.S.E.R.D.A. Contract No. AT(04-3)-34, Project Agreement No. 126.

Literature Cited

1. Fettis, J. C., Knox, J. H., Progr. React. Kinet. (1960), 2, 1.
2. Foon, R., Reid, G. P., Trans. Faraday Soc. (1971), 67, 3513.
3. Homann, K. H., Solomon, W. C., Warnatz, J., Wagner, H. G., Zetzsch, C., Ber. Bunsenges. Phys. Chem. (1970), 74, 585.
4. Kapralova, G. A., Margolin, A. L., Chaikin, H. M. Kinet. Catal. (1970), 11, 669.
5. Smail, T., Miller, G., Rowland, F. S., J. Phys. Chem. (1970), 74, 3464.
6. Smail, T., Iyer, R. S., Rowland, F. S., J. Amer. Chem. Soc. (1972), 94, 1041.
7. Williams, R. L., Rowland, F. S., ibid. (1972), 94, 1047.
8. Iyer, R. S., Ph. D. Thesis, University of California, Irvine, 1974. This thesis, entitled "Hot and Thermal Reactions of Atomic ^{18}F in the Gas Phase", contains the most complete description of ^{18}F recoil chemistry as practiced by Rowland and coworkers.
9. Colebourne, N., Todd, J. F. J., Wolfgang, R., Chemical Effects of Nuclear Transformations, Vol.1, International Atomic Energy Agency, Vienna, 1965, p 149.
10. Todd, J. F. J., Colebourne, N., Wolfgang, R., J. Phys. Chem. (1967), 71, 2875.
11. Lee, J. K., Lee, E. K. C., Musgrave, B., Tang , Y. -N., Root, J. W., Rowland, F. S., Anal. Chem. (1962), 34, 741.
12. Williams, R. L., Iyer, R. S., Rowland, F. S., J. Amer. Chem. Soc., (1972), 94, 7192.
13. Williams, R. L., Rowland, F. S., J. Phys. Chem. (1972), 76, 3509.
14. Frank, J. P., Rowland, F. S., J. Phys. Chem. (1974), 78, 850.
15. Frank, J. P., Rust, F., Rowland, F. S., Physical Chemistry Paper 87, 170th Meeting of the American Chemical Society, Chicago, August 1975.

16. Flores, A. L., Darwent, Deb. B., J. Phys. Chem. (1969), 73, 2203.
17. Rowland, F. S., Cramer, J. A., Iyer, R. S., Milstein, R., Williams, R. L., "Synthesis of ^{18}F-Labelled Compounds by Direct Reactions of Atomic ^{18}F in Radiopharmaceuticals and Labelled Compounds", Vol. 1, International Atomic Energy Agency, Vienna, 1973.
18. Milstein, R., Williams, R. L., Rowland, F. S., J. Phys. Chem. (1974), 78, 857.
19. Flores, A. L., Darwent, D. deB., J. Phys. Chem. (1968), 72, 3407.
20. Anbar, M., Neta, M., J. Chem. Phys. (1962), 37, 2757.
21. Spicer, L., Todd, J. F. J., Wolfgang, R., J. Amer. Chem. Soc. (1968), 90, 2425.
22. Tang, Y.-N., Rowland, F. S., J. Phys. Chem. (1967), 71, 4576.
23. Tang, Y.-N., Smail, T., Rowland, F. S., J. Amer. Chem. Soc. (1969), 91, 2130.
24. Frank, J. P., Rust, F., Rowland, F. S., unpublished experiments.
25. Grant, E. R., Ph. D. Thesis, University of California, Davis, 1975, Grant, E. R., Root, J. W., Chem. Phys. Lett. (1974), 27, 484; Manning, R. G., Grant, E. R., Merrill, J. C., Parks, N. J., Root, J. W., J. Chem. Kinetics (1975), 8, 39.
26. Williams, R. L., Rowland, F. S., J. Phys. Chem. (1971), 75, 2709.
27. Williams, R. L., Rowland, F. S., J. Phys. Chem. (1973), 77, 301.
28. Wagner, H. Gg., Warnatz, J., Zetzsch, C., An. Asoc. Quim. Argent., (1971), 59, 169.
29. The chemistry of ^{19}F is similar to that of ^{18}F.
30. Milstein, R., Rowland, F. S., Physical Chemistry Paper 134, 167th Meeting of the American Chemical Society, Los Angeles, April 1974.
31. Grant, E. R., Root, J. W., J. Chem. Phys. (1975), 63, 2970.
32. Foon, R., Kaufman, M., Progr. React. Kinet. (1975), 8, 81.
33. Parks, N. J., Krohn, K. J., Root, J. W., J. Chem. Phys. (1971), 55, 2690.
34. Parson, J. M., Shobatake, K., Lee, Y. T., Rice, S. A., Disc. Faraday Soc.(1973), 55, 344.

35. Parson, J. M., Lee, Y. T., J. Chem. Phys. (1972), 56, 4658.
36. Bumgardner, C. L., Lawton, E. L., McDaniel, K. G., Carmichael, H., J. Amer. Chem. Soc. (1970), 92, 1311.
37. Frank, J. P., later improved unpublished results; see references cited in (12).
38. Cvetanovic, R. J., Advan. Photochem. (1963), 1, 157.
39. Tyler, S., Frank, J. P., Rowland, F. S., unpublished results.
40. Harrington, R. E., Rabinovitch, B. S., Frey, H. M., J. Chem. Phys. (1960), 33, 1271.
41. Kerr, J. A., Parsonage, M. J., "Evaluated Kinetic Data on Gas Phase Addition Reactions", Butterworths, London, 1972, p 12.
42. Haszeldine, N., Steele, B. R., J. Chem. Soc., (1957), 2800.
43. Libit, L., Hoffmann, R., J. Amer. Chem. Soc. (1974), 96, 1370.
44. Lee, F. S. C., University of California, Irvine, private communication.
45. Bumgardner, C. L., McDaniel, K. C., J. Amer. Chem. Soc. (1969), 91, 1032.
46. Bumgardner, C. L., Lawton, E. L., Acc. Chem. Res. (1974), 7, 14.
47. Williams, R. L., Rowland, F. S., unpublished results.
48. Lee, F. S. C., Rowland, F. S., unpublished results.
49. Frank, J. P., Rowland, F. S., J. Phys. Chem. (1974), 78, 850.
50. Setser, D. W. in "MTP International Review of Science, Physical Chemistry", Vol. 9, Butterworths, London, 1972, p 1.
51. Rabinovitch, B. S., Thrush, B. A., J. Phys. Chem. (1971), 75, 3376.
52. See, for example, Georgakakos, J. H., Rabinovitch, B. S., McAlduff, E. J., J. Chem. Phys. (1970), 52, 2143.
53. Atkinson, R., Thrush, B. A., Proc. Roy. Soc. A 316, (1970), 123, 131, 143.
54. Topor, M. G., Carr, Jr., R. W., J. Chem. Phys. (1973), 58, 757.
55. Mutch, W., Root, J. W., private communication; Mutch, W., Ph. D. Thesis, University of California, Davis, 1973.
56. Farrar, J. M., Lee, Y. T., J. Chem. Phys. (1976), 65, 1414.
57. Moehlman, J. G., Gleaves, J. T., Hudgens, J. W., McDonald, J. D., J. Chem. Phys. (1974), 60, 4790.

3

Radiotracer Studies of Thermal Hydrogen Abstraction Reactions by Atomic Fluorine

SIU-HONG MO, EDWARD R. GRANT,* FRANK E. LITTLE,† RONALD G. MANNING,‡ CHESTER A. MATHIS,† GERALD S. WERRE, and JOHN W. ROOT

Department of Chemistry and Crocker Nuclear Laboratory, University of California, Davis, CA 95616

Crossed beam studies of near-thermal F + D_2 and F + CD_4 reactions yielded direct mechanisms with predominantly backward scattering of DF (1-3).

$$F + D_2 \longrightarrow DF + D \qquad (1)$$

Process 1 proceeds through close (*ca.* 1 Å) repulsive encounters that favor the collinear FDD critical configuration. Available *ab initio* (4) and semi-empirical (5) potential energy surfaces (PES) also favor direct collinear encounters for isotopic variants of this thermal reaction.

The microscopic product energy disposal has been characterized for thermal F atom hydrogen abstraction reactions through chemiluminescence and chemical laser techniques.

$$F + RH \longrightarrow HF + R \qquad (2)$$

Information theory analysis showed that structurally related reactants exhibit essentially the same HF product vibrational disequilibrium, and that all known examples of reaction 2 in non-pi-bonded systems involve direct microscopic dynamics on repulsive early barrier PES with mixed energy release (6).

Although classical kinetics studies cannot yield energy disposal details, accurate Arrhenius parameters provide a useful basis for semi-empirical PES energy scaling (4,5) and for other theoretical characterizations of bimolecular

---

*Present Address: Department of Chemistry, Cornell University, Ithaca, NY 14850.

†Association of Western Universities Graduate Research Fellows 1975-76. (F.E.L.) and 1977-78 (C.A.M).

‡Present Address: Stanford Research Institute, Menlo Park, CA 94025.

reactions (7-11). Kinetic Isotope Effect (KIE) measurements in equilibrium systems at normal temperatures have been extensively utilized for modeling purposes (12-15). Low temperature KIE experiments have provided a limited amount of microscopic energy consumption information (16) together with most of the available definitive evidence pertaining to the occurrence of tunneling in chemical reactions (17). For all of the above reasons it is important to obtain accurate experimental Arrhenius parameters and KIE results for thermal F atom hydrogen abstraction reactions in organic and inorganic substances.

Recent comprehensive literature reviews include those by Foon and Kaufman (18) and Jones and Skolnik (19). Hydrogen abstraction reactions by atomic fluorine exhibit unusually large exothermicities with activation energies in the range 0-2500 cal mole^{-1}. Since the values for many substances are less than 500 cal mole^{-1}, exceptional experimental sensitivity is required. Severe kinetic complications further increase the difficulty of obtaining precise and accurate thermochemical kinetics results for these reactions (18-21).

To date the most successful competitive reaction techniques that have permitted the elucidation of Arrhenius parameters utilized (i) an SF_6 discharge flow reactor followed by simultaneous mass spectrometric detection of HF and DF (7, 16, 22) or (ii) a moderated nuclear recoil ^{18}F procedure introduced from this laboratory that involves direct $H^{18}F$ product analysis (23-26). Experimental KIE results obtained for the F + $H_2(D_2)$ reactions using these procedures exhibited semi-quantitative agreement at an enhanced level of sensitivity (22, 24, 26). Because the flow reactor technique has been based upon well established equilibrium kinetic methodology, this agreement suggested that the moderated recoil ^{18}F atom distribution had achieved complete temperature equilibrium. However, Foon and Kaufman have noted an apparent discrepancy between preferred 300°K absolute rate constants (18) vis-a-vis ^{18}F results (25) for H_2, CH_4 and CHF_3. Relevant kinetic data for these substances have been summarized in Table I.

In the present work we have sought to resolve this discrepancy based upon new experimental results and to provide a critique of the nuclear recoil technique for the investigation of thermal hydrogen abstraction reactions by atomic fluorine.

Experimental.

The basic methods of sample preparation, irradiation and analysis have been described elsewhere (23, 26, 44, 45). Recent experiments have incorporated several refinements as detailed below.

Table I. Absolute Rate Constants for 300°K Thermal Hydrogen Abstraction Reactions by Atomic Fluorine.

Reactant	k_1 (cm^3 $mole^{-1}$ sec^{-1})	A x 10^{-13}	E_a (cal $mole^{-1}$)	Reference
H_2	3.8	-	-	(27)
	1.8 ± 0.6	-	-	(28)
	1.52	9.3	1080 ± 170	(29)
	1.5 ± 0.8[a]	-	-	(30)
	1.5 ± 0.4	-	-	(18, 31)
	1.1	16	1600	(32)
	1.0	-	2150 ± 230	(33)
	1.00 ± 0.08	-	-	(34)
	0.4 ± 0.1	-	-	(35)
	0.14	-	-	(36)
	0.077	4.8 ± 0.3	2470 ± 30	(37)
CH_4	>6.0	-	-	(38)
	4.8[a]	33	1150	(39)
	4.3	-	-	(27)
	3.6 ± 3.6	-	-	(30)
	1.3	10	1210	(40)
	0.32	2.5	1210	(41)
	0.18 ± 0.06	4.0	1850 ± 230	(36)
CHF_3	0.019	-	-	(38)
	0.019	0.75	2200 ± 400	(42)
	0.012 ± 0.003[b]	0.50[b]	2200 ± 170[b]	(43)
	0.012 ± 0.018[a]	0.63 ± 0.01[a]	230 ± 750[a]	(30)
	0.0090 ± 0.0008	-	-	(34)

a. Preferred Values Cited in (<u>18</u>).
b. Preferred Values provided by H. G. Wagner. Ref. (<u>43</u>) supplants ref. (<u>42</u>).

Sample Preparation. Gas phase samples have routinely been prepared in cylindrical Kimax KG-33 ampoules (23) using a pressure-drop procedure to determine mixture compositions (24-26, 46). Pressure measurements for H_2/C_3F_6 and CH_4/C_3F_6 samples were carried out to an absolute accuracy of ± 0.05 Torr by means of a Wallace and Tiernan 0-200 Torr model FA-145 mechanical gauge. More recent experiments have utilized a Barocell model 1174 capacitance manometer provided with a model 570-D-1000T-1B2-H5 temperature stabilized 0-1000 Torr sensor. Individual component pressures measured with the calibrated Barocell had an absolute accuracy of ± 0.003 Torr. Unless noted otherwise the total sample pressure corresponding to the irradiation temperature was always 1000 ± 10 Torr. Gravimetric checks (23, 47) of the total pressures were routinely carried out.

All organic materials had minimum purities of 98 mole %. Apart from routine vacuum degassing, these substances were used without further purification. Prepurified H_2 (≥ 99.9%), Extra Dry O_2 (≥ 99.6%), High Purity Cl_2 (≥ 99.5%), C.P. CH_4 (≥ 99.0%), C.P. C_2H_6 (≥ 99.0%), Freon 116 C_2F_6 (≥ 99.6%), Genetron 23 CHF_3 (≥ 98.0%), and I.G. SF_6 (≥ 99.97%) were obtained from Matheson Gas Products. The C_3F_6 (≥ 99.0%) was provided by Pierce Chemical Co.

Irradiations. Our standard irradiation procedure has also been modified. Dedicated bombardments carried out at the Crocker Laboratory chemical target facility have previously employed 27.5 MeV (d,n) reactions on thick beryllium targets (23, 48). During the past year parasitic irradiations have been utilized in conjunction with the Crocker ^{123}Xe isotope production procedure, which has been based upon the 65 MeV $^{127}I(p,5n)^{123}Xe$ nuclear reaction (49). Protons emerging from the NaI primary target were stopped in a graphite Faraday cup located ca. 2 cm from the base of our sample oven. Fast neutrons suitable for ^{18}F production via the $^{19}F(n,2n)^{18}F$ nuclear reaction were emitted from both the NaI and graphite targets. These parasitic neutrons exhibited both a reduced intensity and a different laboratory kinetic energy distribution relative to the standard dedicated irradiation conditions. Since the parasitic distribution extended to 65 MeV, the relative production rate of ^{11}C isotopic impurity was significantly increased. The ^{11}C activity was routinely reduced to the noise level of the radioassay through the use of extended irradiations and post-irradiation delays (50).

An updated controlled temperature irradiator (51) incorporated an asbestos housing; multiple cartridge heaters installed within an aluminum sample oven; individual ports for a maximum of 31 samples; internally mounted thermistor sensor and thermocouple monitor probes; an adjustable resistance bridge Triac controller (52) sensitive to ± 0.005°K; and prior calibration (± 0.05°K) as well as continuous monitoring (± 0.15°K) during the irradiation of the control point temperature by means of thermocouple potentiometry. Because of susceptibility to radiation damage, each thermistor sensor was used only once. The thermal contact between oven and glass ampoules was relatively inefficient, so that 12 hour warmup periods were required in order to insure satisfactory temperature equilibration. Measurements with a Hewlett-Packard quartz thermometer demonstrated that the temperature uniformity of the oven was within ± 0.01°K. The estimated absolute uniformity and accuracy of sample irradiation temperatures was within ± 0.25°K of the control point value.

Processing of Samples. Following a 0.5 hr radiation cooling period, samples were partitioned into volatile organic, volatile inorganic and non-volatile inorganic aliquots (23). Under normal conditions the trace level $H^{18}F$ product was quantitatively removed from the gas phase through physical adsorption on the KG-33 sample vessel walls. After recovery of the gaseous sample contents, the adsorbed $H^{18}F$ was recovered for radioassay by means of a 0.3 hr continuous extraction with 0.2 molar K_2CO_3 solution. This separation technique has been employed in all of our previous ^{18}F experiments as well as in analogous hydrogen abstraction studies with nuclear recoil ^{38}Cl (53-55), ^{39}Cl and ^{34m}Cl (56). Kimble KG-33 glass was found to be more effective than fused quartz for $H^{18}F$ adsorption. Vycor type glasses are porous and should probably be avoided (57). Provided that the KG-33 vessels were chemically cleaned and preconditioned in vacuo through repeated slow (ca. 10 min) heatings up to the annealing temperature, the $H^{18}F$ adsorption separation was both rapid and quantitative.

Sources of possible interference with this product analysis procedure include glass passivation, heterogeneous isotopic exchange or catalytic reaction, or the presence of other adsorbable labeled substances such as $Cl^{18}F$, $NO^{18}F$ or $COF^{18}F$. Substantial ($\geq$ 90%) prevention of $H^{18}F$ (26, 56) or $NO^{18}F$ (58) adsorption has been achieved through prior exposure of the glass vessel to the products from slow in situ radiolysis of fluorinated hydrocarbons. Adsorbed $H^{38}Cl$ on fused quartz was rapidly and quantitatively displaced following the addition of dry gaseous HCl carrier. Similarly, the addition of dry Cl_2 was accompanied by rapid heterogeneous isotopic exchange (54, 55). Although comparable gas phase experiments have not been carried out with adsorbed $H^{18}F$, polar diatomics and reactive or exchangeable substances have been excluded from the ^{18}F samples under normal conditions. As discussed below, interference from adsorption of another labeled substance was observed during the present moderated CHF_3/C_3F_6 competitive experiments. Difficulty in eluting the adsorbed $H^{18}F$ was previously encountered with unmoderated H_2/C_3F_6 samples at irradiation temperatures in excess of 375°K (59). Rapid recovery was achieved with 5% aqueous HF rinsing solution under these conditions, and the problem did not arise when C_2F_6 or SF_6 moderator was present. Because the organic yield fractions were not affected in unmoderated experiments, we suspect that the anomalous behavior may have been caused by radiolytically initiated thermal C_3F_6 polymerization leading to partial physical entrapment of the adsorbed $H^{18}F$.

Unopened samples were connected to the gas transfer apparatus by means of 40 cm lengths of 1.11 x 0.80 cm o.d. x i.d. Tygon tubing, which were tightly packed with 10.0 g of 30/40 mesh granular $K_2CO_3 \cdot 1.5H_2O$ held in place by means of Pyrex wool plugs. Unadsorbed $H^{18}F$,† products from scavenging reactions of $C_nF_{2n}{}^{18}F$ radicals with O_2 (23), inorganic substances such as $SiX_3{}^{18}F$[X = F, Cl, or alkyl radical] (56), and possibly also $F^{18}F$ (23) reacted rapidly and efficiently with the K_2CO_3 stripper. Less reactive molecular halogens such as $Cl^{38}Cl$ did not react under these conditions (54, 55). Distilled water was percolated through the recovered stripper column in order to insure the extraction of all ^{18}F activity for the radioassay.

Over the years the stripper activities have demonstrated systematic variations depending upon the exact procedures utilized for reagent purification, glass ampoule preconditioning and sample degassing. Since ^{18}F atoms are essentially unreactive toward O_2 (18, 19) and since large oxygen containing perfluoro-organic molecules are not adsorbed on glass, these variations have not interfered with the $H^{18}F$ kinetic analysis (26). It is difficult to insure the rigorous exclusion of trace O_2 from the samples, particularly in the presence of 77°K noncondensible reagents. Cesium gettering has been employed for H_2, D_2, CH_4 and CD_4 reagent purification (26, 46).‡ Trace concentrations of O_2 may also be introduced as a result of radiolytic degradation of the KG-33 sample vessels, and we suspect that the observed stripper yield variations have resulted from scavenging of labeled aliphatic organic radicals by impurity O_2.

Until the behavior of carrier free $F^{18}F$ under these conditions has been better established, the previous speculative assignment (23) of stripper yields to fluorine abstraction channels should be regarded as tentative. In recoil ^{38}Cl experiments with $H_2/C_2H_4/CF_2Cl_2$ mixtures, neither $Cl^{38}Cl$ nor $F^{38}Cl$ products were detected. The analogous $Cl^{18}F$ forming chlorine abstraction reactions with chlorocarbons are thermally accessible (18, 19). Molecular halogens (61) including F_2 (18) add to pi-bonded substances under homogeneous conditions. Such additions to olefins have been observed to be both rapid and quantitative for radiolabeled Br_2 (62) and I_2 (63, 64) species in the presence of molecular halogen carriers. Few experiments

†Incomplete $H^{18}F$ adsorption was observed in condensed phase experiments carried out in small KG-33 capillaries (45, 47, 60).

‡Alkali metal gettering cannot be used with halogenated substances.

have been conducted under carrier-free conditions, but radiobromine addition to olefins apparently does not occur at large specific radioactivity (62, 65). Although we suspect that carrier-free $F^{18}F$ would not undergo efficient homogeneous addition to olefins, further research is needed to settle this question.

Radioassay. The coincident gamma-ray spectroscopic radioassay utilized a pair of identical Harshaw 16MB16/5A Matched Line NaI(Tℓ) scintillation detectors mounted coaxially within a lead vault provided with four inch thick walls. The sample aliquot preparation and handling methods have been described above and elsewhere (23). Two different counting procedures were employed in the recent experiments. A modernized version of the original differential pulse height analysis system has been depicted schematically in Figure 1. This highly automated device was based upon a Nuclear Data Model 4410 multichannel analyzer provided with segmented memory, real time clock, cassette magnetic tape program storage and data output, and fast paper tape punch output. Raw signal data required external processing in order for the ND-4410 to be utilized for coincidence measurements. Its count rate response characteristics were markedly superior to those of the original (23) ND-120 machine. The inactive period associated with input-output transactions was also drastically shortened, thus minimizing ^{18}F radioactivity losses during the analysis procedure. Simultaneous coincidence/noncoincidence measurements of the full pulse height spectra provided continuous monitoring for isotopic impurities and for counting efficiency variations. Post-irradiation decay corrections were based upon the real time clock, which was automatically read and recorded along with the radioactivity and sample identification data for each sample. Although it was quite expensive to acquire and maintain, the ND-4410 system was both convenient and versatile.

In the confirmed absence of radioisotopic impurity complications, the routine measurement of detailed pulse height spectra is not really required in experiments of the present type. The ND-4410 proved to be subject to frequent failure - causing the loss of costly cyclotron irradiations - and to excessive maintenance costs. For these reasons a greatly simplified version of this apparatus was developed, which has been employed in all of the recent experiments. As shown in Figure 2 the optimizable coincidence timing system was retained. However, the ND-4410 was replaced using parallel single channel analyzers together with a multiple channel scaler/timer readout system that either consisted of the listed buffered memory Ortec components or of an

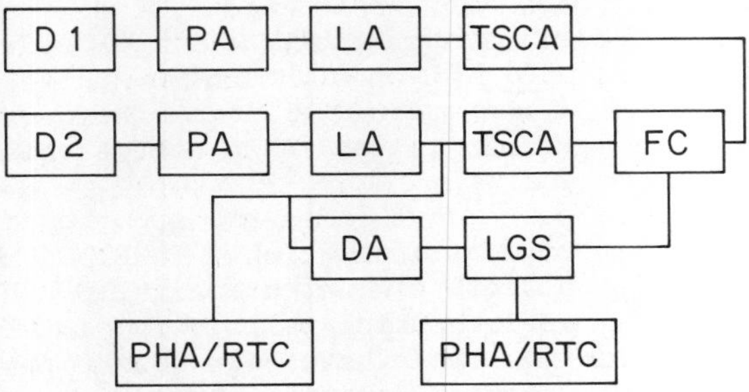

Figure 1. Schematic of coincidence/noncoincidence γ-ray pulse height analyzer. Key to symbols: D1, D2, Harshaw 16MB16/5A detector; PA, Ortec 113 preamplifier; LA, Ortec 485 linear amplifier; TSCA, Ortec 551 timing single channel analyzer; FC, Ortec 414A fast coincidence; DA, Ortec 489 delay amplifier; LGS, Ortec 442 linear gate and stretcher; PHA/RTC, nuclear data pulse height analyzer with segmented memory and real time clock.

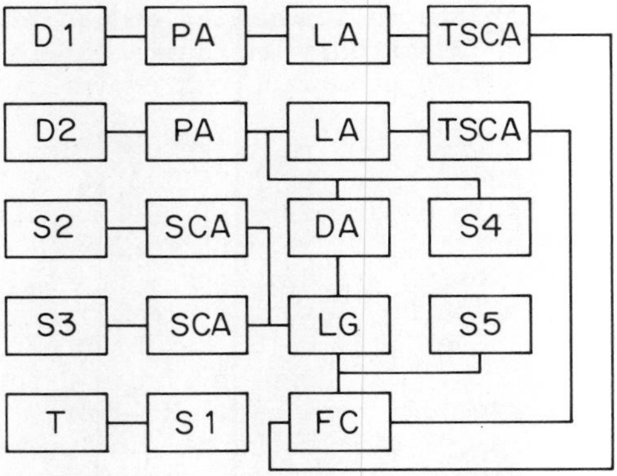

Figure 2. Schematic of coincidence-limited pulse height analyzer with real time clock. Key to symbols: SCA, Ortec 406A single channel analyzer; T, Ortec 708 Buffered Memory Timer; S1–S5, Ortec 707 buffered memory scaler; others as in Figure 1.

equivalent solid state device developed at this laboratory.† Either data acquisition system was outputted through a standard Teletype combination page lister and paper tape punch. Although the buffered memory feature was not strictly required for the present measurements, this same apparatus was also utilized for radiochromatographic analysis (output channels S1 and S5, cf. Figure 2).

The absolute efficiency of the optimized coincidence detector was only ca. 1% at 0.511 MeV under the present conditions of inefficient counting geometry (50). The maximum response rate for either scaler was ca. 15 MHz with <50 microsecond losses between readout cycles. The linear response range of the single channel analyzer system extended to ca. 1 MHz, which greatly exceeded present requirements. The real time (S1), the integrated coincident 0.511 MeV photopeak activity (S2), the integrated coincident spectrum including the photopeak (S3), and the non-coincident total activity were monitored, usually for sufficient 30 sec counting cycles to yield 10,000 or more counts per aliquot in channel S2.‡ These data provided the necessary bases for post-irradiation decay corrections, for monitoring the photopeak counting efficiency and for detecting the presence of gamma ray emitting radioisotopic impurities. The paper tape output was processed off line by means of a Wang Laboratories model 720C programmable electronic calculator provided with model 703 accessory tape editor.

Results.

Thermal F-to-HF Reactions. In all of our thermal competitive experiments ^{18}F addition to C_3F_6 has been utilized as the standard reference reaction.

†The latter apparatus consisted of a four-channel six-decade buffered memory multiple scaler provided with optional coincidence gating; with a readout interval timer preselectable from 1 to 9.9×10^5 sec, which also served as a digital real time clock; with a Teletype output format generator; with coded oscillographic integrating rate meters, which served as backup systems for the digital channels; and with internal power supplies. Originally developed under the auspices of the U.S. Air Force Office of Scientific Research (Contract AF-AFOSR-68-1493), the scaler can still be produced using existing printed circuit diagrams. A standard teletype unit and strip chart recorders are required for monitoring the digital and integrated ratemeter outputs. Several such units have proven essentially maintenance free during a ten year period of heavy use.

‡Typical background count rates in channels S2 and S3 were 15 cpm and 60 cpm, respectively.

$$^{18}F + RH(R_1) \xrightarrow{k_1} H^{18}F + R \qquad (3)$$

$$^{18}F + C_3F_6(R_2) \xrightarrow{k_2} C_3F_6{}^{18}F \qquad (4)$$

At 1000 Torr in the presence of 95 mole % of C_2F_6 or SF_6 moderator, unimolecular loss of ^{18}F following reaction 4 is unimportant (26). All of the $C_3F_6{}^{18}F$ radicals are deposited in the volatile activity fractions along with products from energetic ^{18}F reactions with the moderator. Plots of composition dependent $H^{18}F$ absolute yield data for C_2F_6 moderated H_2, D_2, CH_4, C_2H_6 and CHF_3 have been shown in Figure 3. Equation 5 follows from the kinetic analysis for these systems.

$$\left(\frac{T}{Y_1}\right) = \left(\frac{k_2}{k_1}\right)\left(\frac{1}{X_1}\right) + \left(\frac{k_1 - k_2}{k_1}\right) \qquad (5)$$

Here T denotes the 0.86 ± 0.01 fraction of the initially produced energetic ^{18}F atoms failing to undergo hot reaction (24-26), Y_1 the absolute yield of $H^{18}F$, X_i the mole fraction of reactant R_i ignoring the presence of the thermally inert moderator, and k_i the corresponding thermal bimolecular rate constant.

Plots of (T/Y_1) vs. $(1/X_1)$, which are predicted by Eq. 5 to be linear, have been shown in Figure 4. According to this mechanism consistency test the H_2, D_2, CH_4 and C_2H_6 competitive systems were well behaved over the full range of investigated mixture compositions and reaction temperatures. However, the Eq. 5 plot for the CHF_3/C_3F_6 system exhibited pronounced nonlinearity at C_3F_6 mole fractions in excess of ca. 0.40. This behavior, which accounts for the discrepancy in the published CHF_3 relative rate constant (25) noted by Foon and Kaufman (18), has received further consideration in the discussion. The present 308°K CHF_3/C_3F_6 absolute yield results, which cannot be analyzed in the above fashion, have been listed in Table II.

Comparison of Parasitic Vs. Dedicated Irradiation Techniques. Results obtained via the parasitic procedure were evaluated through a series of unmoderated CH_3CF_3/C_3F_6 experiments conducted at 300°K. The systemization of results for this nonthermal system was rather complicated to carry out because of the number of available reaction channels.

$$^{18}F^*(^{18}F) + CH_3CF_3(R_1) \xrightarrow{\kappa_{11}} H^{18}F(P_{11}) + CH_2CF_3 \qquad (6)$$

$$\xrightarrow{\kappa_{1T}} \text{Organic Products} + F^{18}F + H^{18}F \qquad (7)$$

$$^{18}F^*(^{18}F) + C_3F_6(R_2) \xrightarrow{\kappa_2} \text{Organic Products} + F^{18}F \qquad (8)$$

Here the asterisk (*) denotes excess translational excitation.

The phenomenological kinetic analysis, which has been considered elsewhere (44), required the inclusion of three composite reaction channels. The present experimental method cannot differentiate between corresponding energetic and thermal channels, which have thus been included in composite nonthermal rate coefficients κ_i (59). Very small pressure dependent $H^{18}F$ yield contributions were formed via secondary unimolecular decomposition following energetic ^{18}F organic product forming reactions with CH_3CF_3.

Table II. Averaged Absolute Yield Results For C_2F_6 Moderated CHF_3/C_3F_6 Competitive Experiments At 308°K and 1000 Torr Pressure.

Component Mole Fractions[a]		Absolute Yields[b] (%)	
$[CHF_3]$	$[C_3F_6]$	CHF_3/C_3F_6[c]	$CHF_3/C_3F_6/O_2$[d]
0.9034	0.0966		5.14, 30.27
0.9001	0.0999	64.86, 31.07	
0.8005	0.1995		5.30, 20.80
0.8001	0.1999	73.49, 21.13	
0.7003	0.2997		5.40, 16.91
0.6996	0.3004	78.27, 16.23	
0.5981	0.4019		4.84, 12.63
0.5964	0.4036	83.27, 10.02	
0.5007	0.4993	81.85, 11.08	
0.5000	0.5000		4.62, 11.61
0.3997	0.6003	81.92, 11.19	
0.3994	0.6006		5.04, 10.40
0.3007	0.6993		4.42, 9.43
0.2999	0.7001	84.23, 5.31	
0.2001	0.7999		4.77, 8.92
0.1976	0.8024	85.41, 8.30	
0.0999	0.9001	84.67, 8.63	

a. Calculated including thermal competitors only.
b. Listed in the order Y_1, Y_2 with Y_1 = volatile organic, and Y_2 = nonvolatile inorganic yields.
c. $[C_2F_6]$ = 0.950 ± 0.001 absolute mole fraction.
d. $[O_2]$ = 0.0100 ± 0.0001; $[C_2F_6]$ = 0.940 ± 0.001 absolute mole fractions. Total pressure 1010 ± 10 Torr.

The complete mechanism has been established for the energetic ^{18}F vs. CH_3CF_3 system throughout the range $0.1 \leq P/Z < 600$ atm by means of experiments that incorporated detailed organic product analysis (44, 45, 47, 60, 66).† The integrated kinetic rate expression for the unmoderated CH_3CF_3/C_3F_6 system was as follows:

$$\left(\frac{1}{Y_1}\right) = \left(\frac{\kappa_{2T}}{\kappa_{11}}\right)\left(\frac{1}{X_1}\right) + \left(\frac{\kappa_{1T}-\kappa_{2T}}{\kappa_{11}}\right) \qquad (9)$$

In Eq. 9 κ_{11} and κ_{iT} denote composite phenomenological rate coefficients associated with reactions (6) and with total R_i processes. The quantity $(1/Y_{11})$ exhibited direct proportionality to $\log_e(1/X_1)$ for CH_3CF_3/C_3F_6 mixtures containing up to 70 mole % C_3F_6.

$$\left(\frac{1}{Y_{11}}\right) = A \log_e(1/X_1) + B \qquad (10)$$

The significance of Eq. 10 has been considered elsewhere (44).

The direct comparison of results from parasitic vs. dedicated irradiations for the unmoderated CH_3CF_3/C_3F_6 system has been shown in Figure 5 and Table III. The quantitative agreement between these data sets shows that no detectable kinetic differences resulted from the use of widely different fast neutron distributions. A less extensive comparison for the 273°K C_2F_6 moderated H_2/C_3F_6 mixture system also yielded identical results for the two modes of irradiation.

Table III. Equation 10 Regression Analysis Results For Unmoderated CH_3CF_3/C_3F_6 Experiments.[a]

Irradiation Type	T (°K)	A	B
Dedicated	297.5	2.24 ± 0.11	1.26 ± 0.07
Parasitic	300.0	2.18 ± 0.32	1.17 ± 0.02

a. $0.3 \leq X_1 \leq 1.0$

†It is necessary to take account of gas imperfection over the cited large experimental pressure range. The effective pressure (P/Z) was determined experimentally for each sample. Here Z denotes the gas compressibility factor from the empirical equation of state PV = ZnRT.

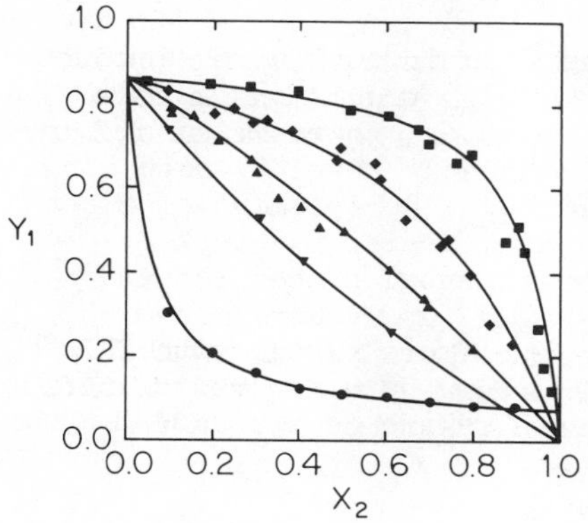

Figure 3. Plots of composition-dependent $H^{18}F$ absolute yield data. Key to symbols: ■ = C_2H_6; ◆ = CH_4; ▲ = H_2; ▼ = D_2; ● = CHF_3.

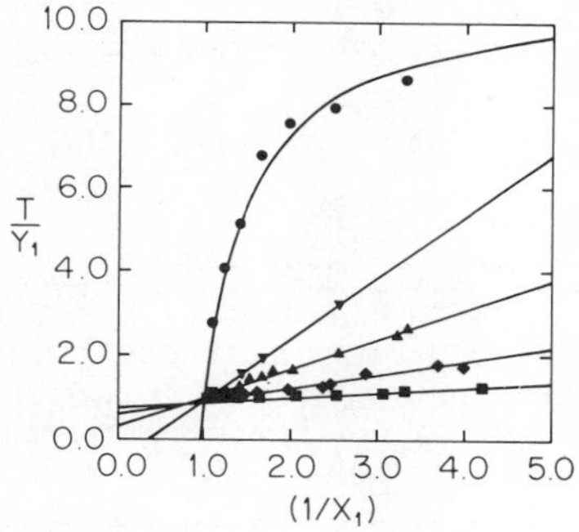

Figure 4. Plots of T/Y_1 vs. $1/X_1$. Key to symbols: same as in Figure 3.

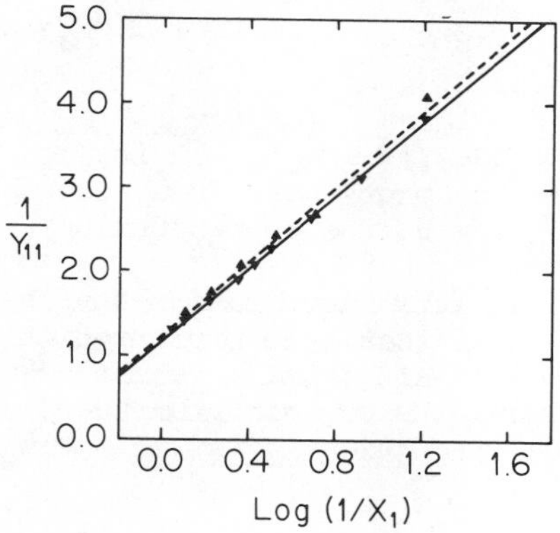

Figure 5. Parasitic vs. dedicated irradiations for the unmoderated CH_3CF_3/C_3F_6 system. Key to symbols: ▲, - - - = dedicated; ▼, —— = parasitic.

Discussion.

Background. It was not our original intention to seek the development of a universal method for high precision (k_2/k_1) measurements. As shown in Table I, however, there are a number of serious discrepancies among the available absolute rate constants for thermal F-to-HF reactions. The achievement of a set of accurate (k_2/k_1) values would greatly assist the task of critical evaluation of available experimental procedures and results.

Because atomic fluorine reacts much more slowly with CHF_3 than with hydrocarbons, reaction 11 (shown here for radiofluorine) has been recommended for adoption as a reference standard (18, 30, 42, 43).

$$^{18}F + CHF_3 \longrightarrow H^{18}F + CF_3 \qquad (11)$$

The competitive technique is poorly suited for situations that involve widely different reaction rates. Because of this factor along with the other experimental difficulties that have been enumerated below, the achievement of accurate (k_2/k_1) results for CHF_3/C_3F_6 competitors using the radiotracer method has proven to be unusually difficult.

Inadequacy of the Two-Channel Model For the Moderated CHF_3/C_3F_6 System. The application of a two-channel kinetic model for the interpretation of thermal RH/C_3F_6 competitive experiments led to the Eq. 5 regression analysis procedure. As shown on Figure 4 the data obtained for the C_2F_6 moderated CHF_3/C_3F_6 system were inconsistent with this method of analysis. The nature of the discrepancy can be illustrated using a rearranged integrated kinetic rate equation that permits the calculation of (k_2/k_1) values from data measured for individual samples.

$$\left(\frac{k_2}{k_1}\right) = \frac{(T-Y_1)X_1}{Y_1(1-X_1)} \qquad (12)$$

Intrasample results have been shown on Figure 6 for several C_2F_6 moderated RH/C_3F_6 competitive systems.† The Eq. 12 (k_2/k_1) values for H_2, D_2, CH_4 and C_2H_6 hydrogen donors were independent of RH/C_3F_6 relative concentration. However, the corresponding

†Because potentially severe error propagation accompanies the use of Eq. 12, the Eq. 5 regression procedure has been preferred in the past for obtaining (k_2/k_1) values in kinetically well behaved systems (24-26). Based upon more extensive experimental results, this recommendation has now been modified (*vide infra*).

two-channel results for CHF_3/C_3F_6 decreased monotonically with increasing C_3F_6 concentration, actually vanishing in the moderated pure C_3F_6 limit! Similar behavior has also been noted for CH_3CF_3, CHF_2CHF_2 and C_2HF_5 hydrogen donors, but RH/C_3F_6 systems containing alkane and cycloalkane competitors are well behaved.

The Eq. 5 regression procedure has been utilized for competitive systems that involve nonpolar hydrogen donors. Its inadequacy in experiments with fluorinated alkanes suggests the possible importance of translational non-equilibrium effects (18), of impurity reactions, or of other sources of mechanism failure.

Non-Equilibrium Complications. Quantitatively identical KIE Arrhenius parameters for F-to-HF reactions with methane and molecular hydrogen have recently been reported based upon nuclear recoil (24, 26) and conventional flow reactor (16, 22) methods. This agreement between widely different techniques suggested that energetic ^{18}F reaction contamination was probably not important in the nuclear recoil experiments. The unprecedented sensitivity levels achieved in these investigations and the pronounced reaction rate increases associated with translational excitation (44, 59, 67) constituted compelling aspects of this argument.

The present anomalous (k_2/k_1) behavior observed for the C_2F_6 moderated CHF_3/C_3F_6 system resulted from the failure of the nonvolatile inorganic yield to vanish in the moderated pure C_3F_6 limit (cf. Figure 3). If the significance of the agreement between experimental KIE results is discounted, then the assumed importance of energetic reaction contamination could account for qualitative decreases in (k_2/k_1) with decreasing CHF_3/C_3F_6 relative concentration. For systems having true (k_2/k_1) values larger than ca. 5, the $H^{18}F$ yield from energetic contamination would fall off more slowly with increasing C_3F_6 concentration than that from the corresponding thermal process. Composition dependent apparent (k_2/k_1) ratios would then result in qualitative accord with the behavior shown on Figure 6.

The above analysis, however, has been misleadingly oversimplified. The $H^{18}F$ yields from both thermal and energetic F-to-HF processes must vanish at the moderated pure C_3F_6 limit (44, 45, 59)! In the absence of pronounced second-order collision density effects (45) non-equilibrium phenomena cannot account for the large residual apparent $H^{18}F$ yield observed in the absence of CHF_3 reactant. A detailed analysis of the probable magnitudes of non-equilibrium effects in experiments of this type has been provided elsewhere (26).

Simple Chemical Impurity Models. Three-channel kinetic models were derived in which RH and olefinic impurities were included in the various sample components. Residual $H^{18}F$ yields were predicted at the C_3F_6 limit when impurity RH was included in the C_3F_6 or C_2F_6. Respective upper limiting H_2 impurity concentrations of ca. 2.0 and 0.2 mole % were assigned to these reagents based upon comparisons between the experimental C_2F_6 moderated H_2/C_3F_6 results (24, 26) vs. computer simulations generated for three-channel impurity models.† When combined with literature k_1 data, the cited maximum impurity levels could not account for the magnitude of the residual apparent $H^{18}F$ yield observed at the moderated C_3F_6 limit. Even without this additional constraint, neither three-channel model was capable of reproducing the experimental results shown in Table II and Figure 3.

Spurious Apparent $H^{18}F$ Product Formation in the Moderated CHF_3/C_3F_6 System. An important insight followed from an investigation of moderated pure CHF_3 samples. In the absence of C_3F_6 competitor, the measured 0.910 ± 0.007 nonvolatile inorganic yield exceeded the reported 0.86 ± 0.01 thermal reaction fraction at 1000 Torr (24-26). A spurious contribution to the apparent $H^{18}F$ yield was thus indicated that must have originated from the energetic ^{18}F reaction fraction.

Further experiments were performed in order to corroborate the spurious yield hypothesis and to elucidate its kinetic consequences. The inclusion of a constant 1.00 mole % concentration of O_2 scavenger in moderated CHF_3/C_3F_6 experiments at 1000 Torr insured the availability of an efficient homogeneous reaction pathway for labeled aliphatic free radical products. As shown in Table II and Figures 7 and 8, the measured nonvolatile inorganic yields were not affected by the inclusion of O_2 in the samples, and the oxygen containing scavenged reaction products were quantitatively recovered with the volatile inorganic yields (23, 47). These results demonstrate that the spurious nonvolatile inorganic yield component could not have been formed from a labeled aliphatic free radical precursor.

The $H^{18}F$ yields shown in Figures 7 and 8 were not measurably depleted by the presence of O_2 additive. No evidence has been obtained to indicate that atomic fluorine undergoes bimolecular reactions with O_2 (18, 19). Systematic losses of ^{18}F atoms from the thermal reaction fraction in the presence of O_2 would readily have been detected here with a sensitivity at the 99% confidence level corresponding to an absolute yield of less than 1%. The absence of such an effect in 5 mole % CHF_3 scavenged C_2F_6 samples serves to establish the conservatively estimated upper limiting rate constant value of ca. 6×10^9 cm^3 $mole^{-1}$ sec^{-1} for the 300°K thermal $^{18}F + O_2$ reaction. This value, which is about one hundred fold smaller than previous limiting estimates (18, 19), has been based upon the preferred

†The cited concentration limits were based upon an assumed 1.5×10^{13} cm^3 $mole^{-1}$ sec^{-1} impurity k_1, corresponding to the recommended 300°K value for H_2 (18).

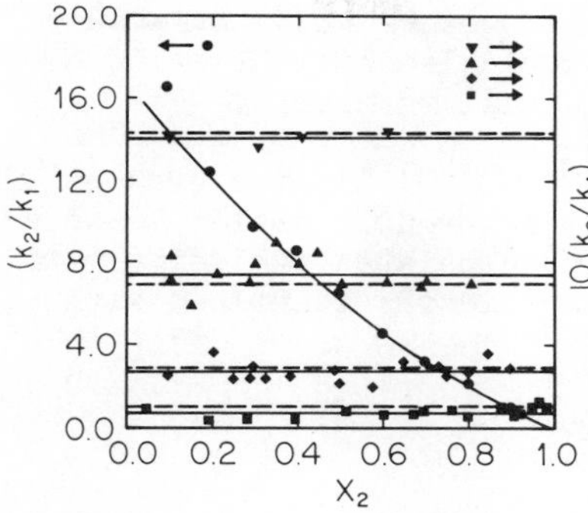

Figure 6. Several C_2F_6-moderated RH/C_3F_6 competitive systems. Key to symbols: --- = previous Equation 5 analysis; —— = Equation 12 result; others same as in Figure 3.

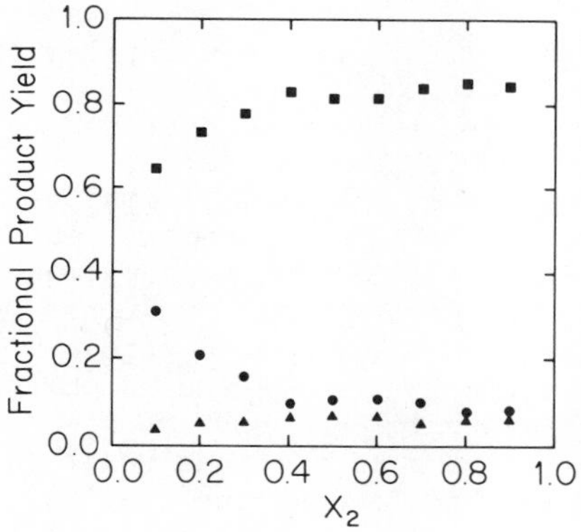

Figure 7. $CHF_3/C_3F_6/C_2F_6$ at 1000 Torr and 308°K. Key to symbols: ■ = organic; ● = nonvolatile inorganic activities; ▲ = volatile inorganic activities.

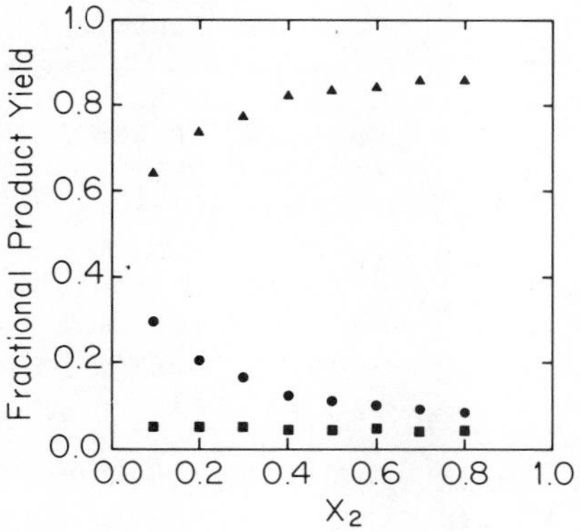

Figure 8. $CHF_3/C_3F_6/C_2F_6/O_2$ at 1010 Torr and 308°K. Key to symbols: same as in Figure 7.

k_1 value for CHF_3 listed in Table I together with the assumed absence of a catalytic cycle that would lead to the eventual redeposition of ^{18}F atoms in the thermal reservoir.

Energetic ^{18}F Reactions With C_2F_6. Within the present experimental sensitivity, the volatile organic yields measured in the presence of O_2 (cf. Figure 8) consisted exclusively of stable molecular products from energetic ^{18}F reactions with C_2F_6.

$$^{18}F^{*} + C_2F_6 \longrightarrow (C_2F_5{}^{18}F)^{\dagger} + F \quad (13)$$

$$\longrightarrow (CF_3{}^{18}F)^{\dagger} + CF_3 \quad (14)$$

In Eqs. 13 and 14 the dagger (†) denotes that 0.66 ± 0.05 and 0.30 ± 0.05 fractions (44, 68) of the F-for-F and F-for-CF_3 primary products contained sufficient internal excitation to induce secondary decomposition at low pressure.

$$(C_2F_5{}^{18}F)^{\dagger} + M \longrightarrow C_2F_5{}^{18}F + M \quad (15)$$

$$(C_2F_5{}^{18}F)^{\dagger} \longrightarrow (CF_2{}^{18}F)^{\dagger} + CF_3 \quad (16)$$

$$\longrightarrow (C_2F_4{}^{18}F)^{\dagger} + F \quad (17)$$

$$\longrightarrow C_2F_5 + {}^{18}F \quad (18)$$

$$(CF_2{}^{18}F)^{\dagger} + M \longrightarrow CF_2{}^{18}F + M \quad (19)$$

$$(CF_2{}^{18}F)^{\dagger} \longrightarrow :CF^{18}F + F \quad (20)$$

$$(CF_2{}^{18}F)^{\dagger} \longrightarrow :CF_2 + {}^{18}F \quad (21)$$

$$(C_2F_4{}^{18}F)^{\dagger} + M \longrightarrow C_2F_4{}^{18}F + M \quad (22)$$

$$(C_2F_4{}^{18}F)^{\dagger} \longrightarrow C_2F_3{}^{18}F + F \quad (23)$$

$$(C_2F_4{}^{18}F)^{\dagger} \longrightarrow C_2F_4 + {}^{18}F \quad (24)$$

$$(CF_3{}^{18}F)^{\dagger} + M \longrightarrow CF_3{}^{18}F + M \quad (25)$$

$$(CF_3{}^{18}F)^{\dagger} \longrightarrow CF_2{}^{18}F + F \quad (26)$$

$$(CF_3{}^{18}F)^{\dagger} \longrightarrow CF_3 + {}^{18}F \quad (27)$$

Here the M notation symbolizes cascading deactivation sequences that inhibit product decomposition through collisional energy transfer to host gas molecules.

As determined from radio gas chromatographic product analysis, the constant 0.049 ± 0.003 organic yield measured at

1000 Torr in the presence of O_2 was comprised of stabilized $C_2F_5{}^{18}F$, $C_2F_3{}^{18}F$ and $CF_3{}^{18}F$ from reactions 15, 23 and 25. Based upon the reported 0.14 ± 0.01 total hot reaction yield with C_2F_6 at this pressure, a constant 0.09 ± 0.01 hot yield must have been partitioned between the inorganic yield fractions.† This latter quantity thus provides the source for the spurious apparent $H^{18}F$ yield measured for C_2F_6 moderated pure CHF_3 samples.

The nature of this spurious process receives further consideration in the following two sections. The reader who is mainly interested in the final kinetics results may wish to omit this material.

Possible Mechanistic Origin of the Spurious Nonvolatile Inorganic Yield. From the constant organic yield behavior shown on Figure 8, all of the $(C_2F_5{}^{18}F)^†$ and $(CF_3{}^{18}F)^†$ secondary decomposition processes were unaffected by the CHF_3/C_3F_6 relative concentration. Only $:CF^{18}F$ forming reaction 20 could have contributed to the spurious nonvolatile inorganic yield under these conditions, presumably through stoichiometric process 28 (69, 70), which does not represent an elementary reaction.

$$:CF^{18}F + \tfrac{1}{2}O_2 \longrightarrow COF^{18}F \qquad (28)$$

Decomposition channels 19, 22 and 26 produce aliphatic radicals, which were recovered with the volatile inorganic activity in the presence of O_2. The ^{18}F atom loss channels 18, 21, 24 and 27 simply contributed to the apparent thermal reaction fraction (26).

The formation of $:CH^{18}F$ and $:CF^{18}F$ following energetic ^{18}F atomic substitution reactions in fluorinated methanes (71-73) and ethanes (44, 60, 66) has been well documented. Lindner and Pauwels obtained direct evidence for $COF^{18}F$ production in unpublished recoil ^{18}F experiments with fluoromethanes (74, 75). Although $:CF^{18}F$ was identified as the $COF^{18}F$ precursor through competitive experiments with C_3F_6 scavenger, the detailed mechanism for process 28 was not established.

$$:CF^{18}F + C_3F_6 \longrightarrow CF_3\text{-}\underset{\diagdown\ \diagup \atop CF_2}{CF\text{-}CF^{18}F} \qquad (29)$$

Product analysis evidence has been obtained for reaction 29 (56, 74, 75) as well as for analogous olefinic additions with C_2F_4 and with C_4 hydrocarbon olefins (72, 73, 76). With fluorinated olefins such $:CF^{18}F$ trapping reactions are much less efficient than insertions into HI or HBr (72).

$$:CF^{18}F + HX \longrightarrow CHXF^{18}F \qquad (30)$$

†The total hot yield exhibits an apparent pressure dependence because of ^{18}F loss decomposition channels 18, 21, 24 and 27.

Extensive COF_2 formation has been reported, again by an unspecified mechanism probably involving $:CF_2$, from the 400°K pyrolysis of trifluoromethylfluorophosphoranes in the presence of O_2 (69, 70).

There is general agreement (77-81) that the electronic ground state for $:CF_2$ is of the singlet $^1A_1$ type. The best available theoretical values for the lowest lying singlet to triplet energy separation in this species occur in the range 42 ± 5 kcal mole^{-1} (78-80).† The rather meager existing quantitative kinetic information for $^1A_1$ $:CF_2$ has been summarized elsewhere (76, 81).

Carbene chemistry tends to be rather complex, and the chemical behavior of corresponding vibrationally cold singlet and triplet species is difficult to establish with certainty (77). Significant discrepancies (*vide supra*) have been noted with respect to the reactivity of $:CF_2$ toward O_2 under different experimental circumstances. Oxygen should not interfere with $^1A_1$ $:CF_2$ reactions. Although quantitative kinetic data are not available, $^3B_1$ $:CF_2$ is likely to be more reactive toward O_2 than the vibrationally cold $^1A_1$ species (81).

Based upon the cited large $^1A_1$-$^3B_1$ energy separation, collision induced singlet to triplet intersystem crossing would be highly improbable for vibrationally cold $:CF^{18}F$. Although the relevant physical details have not been fully elucidated (77), photodissociation precursor processes often produce vibrationally excited $:CH_2$ which is readily capable of undergoing intersystem crossing. Energetic ^{18}F atomic substitution reactions lead to the deposition of many eV of internal excitation in the primary species whose secondary decompositions produce $:CF^{18}F$. For example, roughly 50% of F-for-F produced $CF_3{}^{18}F$ is detected at 1000 Torr as $:CF^{18}F$ (56, 71, 72). This sequential unimolecular process has a *minimum* endothermicity of 9.3 eV (68, 82). It seems evident that nuclear recoil production methods would lead to highly excited $:CF^{18}F$ for which the accessibility of intersystem crossing offers a plausible explanation for $COF^{18}F$ production *via* process 28.‡ Although the need for further research is clearly indicated, we tentatively conclude that $:CF^{18}F$ from reaction 20 was mainly scavenged through C_3F_6 addition reaction 29 in C_2F_6 moderated RH/C_3F_6 experiments with nonpolar hydrogen donors, and that process 28

†The cited $^1A_1$-$^3B_1$ energy separation for $:CF_2$ may be subject to further theoretical refinement. These computational methods yielded 10-25 kcal mole^{-1} for the corresponding $:CH_2$ energy separation. Although this quantity has not been well established, it is generally thought to occur within the range 1-10 kcal mole^{-1} (77).

‡A similar proposal has been offered by Pauwels in order to account for the extreme sensitivity of measured $COF^{18}F$ yields to the presence of I_2 additive (74).

leading to adsorbed $COF^{18}F$ took place to a significant extent in the moderated CHF_3/C_3F_6 reaction system.

Further Tests - Detection of C_2F_6 Impurity Reactions. Additional experiments were conducted at 1000 Torr in which Cl_2 additive was included in the samples (cf. Table IV). Thermal reaction 31 is believed to occur with a rate that approaches the gas kinetic collision frequency (18, 19).

$$^{18}F + Cl_2 \longrightarrow Cl^{18}F + Cl \qquad (31)$$

At large specific radioactivity the $Cl^{18}F$ product was exclusively recovered in the nonvolatile inorganic yield fraction. The measured 0.862 ± 0.005 yield from reaction 31 in 5-10 mole % Cl_2 scavenged C_2F_6 provided an improved determination of the magnitude of the thermal reaction fraction.

Insertion of $:CF_2$ into Cl_2 has been observed under conventional reaction conditions (69, 70). The efficacy of this $:CF^{18}F$ trapping process has been confirmed in the present work through radio gas chromatographic product analysis measurements in which the disappearance of the spurious nonvolatile inorganic yield in 5.0 mole % CHF_3 scavenged C_2F_6 was accompanied by the appearance of a quantitatively identical yield of $CCl_2F^{18}F$.

$$:CF^{18}F + Cl_2 \longrightarrow CCl_2F^{18}F \qquad (32)$$

The mixture composition dependence of the spurious apparent $H^{18}F$ yield must be precisely known in order to permit the development of a modified kinetic analysis scheme. From Figure 7 the 0.076 ± 0.004 extrapolated nonvolatile inorganic yield corresponding to the moderated pure C_3F_6 limit exceeded the spurious apparent $H^{18}F$ yield by 0.028 ± 0.010. An additional set of measurements was carried out in which the partial pressures of CHF_3, C_3F_6 and O_2 remained unchanged, but the total pressure was increased to 10,000 Torr through the addition of C_2F_6. The purposes for this experiment included the following: The C_2F_6 concentration was increased by roughly tenfold in order (i) to enhance the possible importance of $H^{18}F$ producing moderator impurity reactions, (ii) to diminish the likelihood for energetic ^{18}F reaction contamination, and (iii) to suppress the $:CF^{18}F$ yield from reaction 20 through increased collisional deactivation.

Table IV. Absolute Yield Results for Scavenged C_2F_6 Samples At 273°K And 1000 Torr Pressure.

Component Mole Fractions[a]	Absolute Yields[b] (%) Organic	Nonvolatile Inorganic
$[C_2F_6] = 0.950$	12.47 ± 0.17	85.66 ± 0.16
$[Cl_2] = 0.050$	11.53 ± 0.15	87.01 ± 0.14
	11.84 ± 0.14	86.74 ± 0.13
	11.67 ± 0.14	85.61 ± 0.13
$[C_2F_6] = 0.900$	12.92 ± 0.25	85.51 ± 0.30
$[Cl_2] = 0.100$	12.79 ± 0.28	86.49 ± 0.32
Average Value[c]	12.20 ± 0.60	86.17 ± 0.66
$[C_2F_6] = 0.940$	4.73 ± 0.11	91.46 ± 0.16
$[CHF_3] = 0.050$	5.06 ± 0.06	91.03 ± 0.08
$[O_2] = 0.010$	5.07 ± 0.10	91.32 ± 0.12
	4.95 ± 0.06	90.01 ± 0.08
Average Value[c]	4.95 ± 0.16	90.96 ± 0.66

a. Calculated Including Thermal Competitors Only.
b. Uncertainties Represent Radioactive Statistics Standard Errors.
c. Uncertainties Represent Standard Deviations From The Mean.

The 10,000 Torr results have been shown in Table V and Figure 9. The nonvolatile inorganic yield at the moderated CHF_3 limit decreased to 0.88 ± 0.01. In the presence of O_2 this change was accompanied by an equivalent increase in the organic yield, which again exhibited the same value at all CHF_3/C_3F_6 relative concentrations. The spurious nonvolatile inorganic yield residue thus was significantly reduced at 10,000 Torr.

Detailed mechanistic investigations that incorporated organic product analysis have been completed for the energetic ^{18}F vs. CH_3CF_3 and CH_3CHF_2 reaction systems (44, 45, 47, 60, 66). The only secondary decomposition processes that were significantly inhibited at 10,000 Torr were those analogous to reactions 20 and 26. The observation that four different primary reaction channels of each of these types exhibited closely similar phenomenological pressure dependences strongly supports the above analysis based upon reaction 20.†

†Secondary decomposition analogous to reaction 20 was not observed following energetic F-for-αH substitution in CH_3CHF_2.

Perhaps the most striking feature of the 10,000 Torr results is that the extrapolated nonvolatile inorganic yield corresponding to the moderated C_3F_6 limit increased to a value of 0.18 ± 0.01. Based upon the observed 0.068 ± 0.020 constant organic yield and the cited thermal reaction fraction, only 0.071 ± 0.007 of the yield from energetic reaction sequence 13-27 was distributed between the inorganic yield fractions. In the C_3F_6 limit, therefore, most of the nonvolatile inorganic yield residue must have consisted of $H^{18}F$ formed from a trace impurity contained in the C_2F_6 moderator.

Table V. Absolute Yield Results For C_2F_6 Moderated CHF_3/C_3F_6 Competitive Experiments at 308°K and 10,010 Torr Pressure.

Component Mole Fractions[a]		Absolute Yields (%)	
$[CHF_3]$	$[C_3F_6]$	Organic	Nonvolatile Inorganic
0.9000[b]	0.1000	37.57	60.61
0.9000	0.1000	7.17	58.50
0.7999	0.2001	6.72	46.40
0.6995	0.3005	6.70	39.25
0.5999	0.4001	6.92	33.04
0.5011	0.4989	6.65	27.64
0.3996	0.6004	7.00	25.88
0.2998[b]	0.7002	76.41	20.62
0.2982	0.7018	6.98	23.71
0.2002	0.7998	6.71	22.22

a. Calculated Including Thermal Competitors Only. Unless Noted All Samples Include $[O_2] = (1.00 \pm 0.01) \times 10^{-3}$, $[C_2F_6] = 0.994 \pm 0.001$ Absolute Mole Fractions.
b. No O_2 Additive.

Modified Kinetic Analysis For The Moderated CHF_3/C_3F_6 System. Provisions must be included in the revised kinetic analysis for the spurious apparent $H^{18}F$ yields that arose from energetic ^{18}F reactions with C_2F_6 and from impurity reactions. The organic yields measured for O_2 containing samples at constant (P/Z) did not vary with CHF_3/C_3F_6 relative concentration. The yield of the precursor species that led *via* energetic reactions to the production of spurious apparent $H^{18}F$ was thus constant. Nonvolatile inorganic yields measured at constant (P/Z) and mixture composition were insensitive to the presence of O_2 additive, suggesting that the energetic reaction contributions to the apparent $H^{18}F$ yields were probably independent of CHF_3/C_3F_6 relative concentration. The 0.048 ± 0.009 magnitude of this constant spurious yield at 1000 Torr was established as the difference between the nonvolatile inorganic yield measured for 5 mole % CHF_3 scavenged C_2F_6 and the revised thermal reaction fraction (cf. Table IV). The first step in the revised kinetic analysis thus involved correction of the measured nonvolatile inorganic yields through subtraction of this experimentally determined quantity.

The corrected total $H^{18}F$ yields were then interpreted using a three-channel kinetic model that included reactions 3 and 4 together with the C_2F_6 impurity process. The integrated kinetic rate equation was as follows:

$$\frac{Y_1 + Y_3}{T} = \frac{X_1 + X_3(k_3/k_1)}{X_1 + X_2(k_2/k_1) + X_3(k_3/k_1)} \tag{33}$$

In Eq. 33 subscripts 1-3 designate reactions 3, 4 and the impurity process, respectively. Since C_2F_6 is thermally inert (26), its concentration has been omitted in these X_i calculations.

From Figures 7-9 the apparent $H^{18}F$ yields were inversely proportional to X_2 throughout the range $0.50 \leq X_2 \leq 1.00$. An 0.028 ± 0.010 experimental Y_3^0 value followed from the usual spurious yield correction applied to the 0.076 ± 0.004 linear regression analysis intercept corresponding to the C_3F_6 limit at 1000 Torr. Neither the identity of the impurity nor its concentration were known a priori. A 0.0012 ± 0.0004 estimate of its mole fraction in the C_2F_6 reagent followed from the above Y_3^0 value together with the assumption that the impurity was H_2.† This estimate, which corresponds to an X_3 value of 0.022 ± 0.008, is consistent with the previously described impurity sensitivity tests carried out for the moderated H_2/C_3F_6 system.

A nonlinear regression procedure permitted the specification of (k_2/k_1) from the corrected total $H^{18}F$ yield results. The impurity terms in Eq. 33 contain the quantity k_3X_3, which was accurately maintained at a constant value throughout the present experiments and which was evaluated from Y_3^0 and k_2 (cf. Table VI).‡

$$k_3X_3 \doteq \frac{k_2Y_3^0}{T-Y_3^0} \tag{34}$$

†The k_2 and k_3 values (cf. Tables I and VII) employed in this calculation were 1.0×10^{13} and 1.5×10^{13} cm^3 $mole^{-1}$ sec^{-1}, respectively. Corresponding concentration limits for other impurity substances follow from Eq. 34.

‡The k_3X_3 values calculated via approximate Eq. 34 were better than 95% accurate.

It would thus appear that the experimental determination of k_3X_3 could be utilized in order to reduce Eq. 33 to a pseudo two-channel problem. Unfortunately, accuracy limitations precluded the adoption of this procedure. The nonlinear regression technique proved to be extremely sensitive to the assumed magnitude of k_3X_3 within the range corresponding to the statistically allowed limits on Y_3^o. Based upon the available results, Y_3^o (and hence k_3X_3) could not be established with sufficient precision to permit an unambiguous assignment of (k_2/k_1).

Table VI. Nonlinear Regression Analysis Of Results For C_2F_6 Moderated CHF_3/C_3F_6 Competitive Experiments.[a]

Trial (k_2/k_1)	ΔY_3^o	r.m.s. variance
40	-0.300	0.88
50	-0.138	0.72
60	0.023	0.62
70	0.184	0.52
80	0.296	0.49
82	0.312	0.49
84	0.312	0.48
86	0.328	0.48
88	0.344	0.48
90	0.360	0.48
92	0.376	0.48
100	0.440	0.48
120	0.504	0.48
160	0.600	0.50
200	0.663	0.52

a. Temperature 308°K. Pressure 1010 ± 10 Torr.
$\Delta Y_3^o = Y_3^o(\text{trial}) - Y_3^o(\text{exp})$.

The nonlinear regression procedure was based upon a modification of Eq. 12.

$$G(X_1) = \frac{[T-(Y_1 + Y_3)]X_1}{(Y_1 + Y_3)\ (1-X_1)} \qquad (35)$$

Here the quantity $(Y_1 + Y_3)$ represents the corrected total $H^{18}F$ yields obtained as described above. In the absence of the impurity process, $G(X_1)$ reduces to the invariant (k_2/k_1) relative rate constant. For finite (k_3/k_1) $G(X_1)$ decreases monotonically with decreasing X_1, vanishing in the moderated C_3F_6 limit. In the present fitting procedure experimental $G(X_1)$ values were compared against trial and error calculated results obtained from

Eq. 33. In order to characterize the sensitivity and possible uniqueness of this procedure, the calculations were optimized with respect to both (k_2/k_1) and k_3X_3. The full allowed range of Y_3^o was included with an incremental Y_3^o value of 0.0002. For each trial k_3X_3 value from Eq. 34, many trial (k_2/k_1) values were utilized in order to predict the corresponding total $H^{18}F$ yields _via_ Eq. 33.†

Finally, the experimental _vs._ calculated $G(X_1)$ results obtained from Eq. 35 were compared with the aid of three fitting criteria, thus serving to establish the preferred (k_2/k_1) value.

(i) The best possible agreement with the experimental Y_3^o value was sought;

(ii) The experimental $G(X_1)$ were required to be randomly distributed with respect to the calculated results; and

(iii) The r.m.s. variance between experimental and calculated results was minimized. This procedure has been illustrated in Table VI and Figure 10 in which each experimental data point represents the average from four or more replicate determinations. Based upon the minimum variance criterion, a single preferred (k_2/k_1) value was obtained corresponding to each k_3X_3. These optimized (k_2/k_1) results, which have been given in the table, showed that on variance grounds alone the data fits could be eliminated for which Y_3^o differed by more than ± 15% from the experimental value and also for which (k_2/k_1) was smaller than _ca._ 80. However, the fits corresponding to the range $80 < (k_2/k_1) < 140$ were roughly equivalent, serving to establish the 99.5% confidence level range of possible values. The preferred 84 ± 10 (k_2/k_1) result was chosen in order to yield both the minimum variance and the closest agreement with the measured Y_3^o.‡

Although this multiparameter procedure is rather complex, it does not constitute an empirically optimized parametric "fit" to an arbitrarily constructed mathematical model. The necessities for the utilization of nonvolatile inorganic yield corrections and of the present three-channel impurity kinetic model have been experimentally established. Although the variance minimization modeling criterion clearly is important, it has not been employed to produce artificial agreement with the 300°K literature k_1 value for CHF_3. Only Y_3^o has been subjected to parametric optimization, and the final value of 0.031 differs negligibly from the measured 0.028 ± 0.010 result. One of the most important aspects of the regression procedure is that it provided a quantitative basis for determining the standard error of estimate for (k_2/k_1).

---

†The regression calculations were based upon X_1 and X_2 values obtained directly from the sample filling data (_i.e._, the _ca._ 10^{-2} magnitude of X_3 was ignored).

‡The cited standard error of estimate for (k_2/k_1) corresponds to the 68% confidence level.

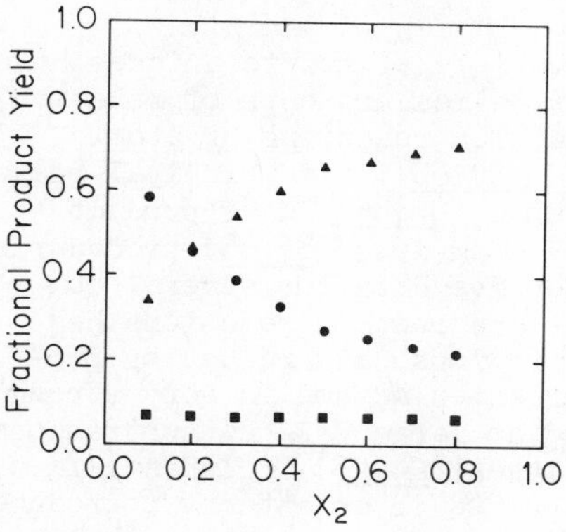

Figure 9. $CHF_3/C_3F_6/C_2F_6/O_2$ *at 10,010 Torr and 308°K. Key to symbols: same as in Figure 7.*

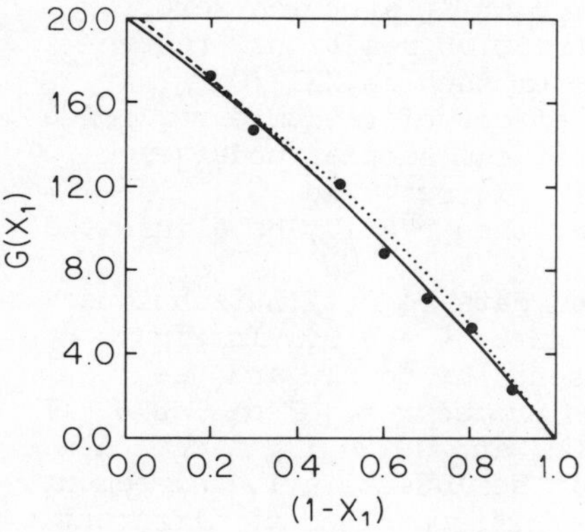

Figure 10. Minimized r.m.s. variance between experimental and calculated results. Key to symbols: ——, $(k_2/k_1) = 84$, $Y_3° = 3.138\%$; · · ·, $(k_2/k_1) = 60$, $Y_3° = 2.810\%$; – – –, $(k_2/k_1) = 120$, $Y_3° = 3.314\%$.

A final consistency test for the present impurity kinetic model requires an intercomparison of Y_3^0 results at 1000 and 10,000 Torr. Based upon the above described procedure, the 0.16 ± 0.02 10,000 Torr result for Y_3^0 corresponded to an estimated H_2 impurity mole fraction in the C_2F_6 of 0.0009 ± 0.0001, in semiquantitative agreement with the corresponding 0.0012 ± 0.0004 1000 Torr result.

Relative Rate Constants. A variety of new techniques have been developed for obtaining accurate measurements of relative rate constants for hydrogen abstraction reactions by atomic fluorine (7,16,18,19,22,24-27,83-85,87,88). Of principal present interest are the HF chemiluminescence (84,85), $H^{18}F$ product analysis (24-26), and indirect ^{18}F atom loss (87,88) procedures.

The availability of more extensive data for several competitive systems that involve fast abstraction reactions has necessitated modification of our previous data reduction procedure. The unweighted Eq. 5 regression method is very strongly influenced by data points measured at large C_3F_6 concentrations. Our previous Eq. 5 regression analyses (24-26) included $(1/X_2)$ weighting of the individual data points in an attempt to reduce the significance of this type of bias. The alternative Eq. 12 sample-by-sample (k_2/k_1) calculation technique suffers from extreme sensitivity to random error noise under conditions corresponding to the extrema in either sample compositions or measured $H^{18}F$ yields. For competitive systems that involve rate constant differences larger than ca. fivefold, pronounced $H^{18}F$ yield changes occur in the vicinity of a composition extremum (cf. Figure 3). Throughout the remainder of the composition range the quantities $(T-Y_1)$ or Y_1 approach zero for cases involving fast and slow abstraction reactions, respectively. In a really difficult system such as CHF_3/C_3F_6, it becomes mandatory to minimize the effect of random error noise through the processing of 5-10 identical samples under each set of investigated conditions (cf. Figures 7 and 8). The occasional spurious measurement can then readily be eliminated upon statistical grounds.

A direct comparison of several data reduction techniques has been given in Table VII and Figures 4 and 6. In Figure 6 the solid and dashed curves represent the Eq. 12 and the previous Eq. 5 weighted regression procedures. From Table VII the results obtained for H_2, D_2, CH_4 and CD_4 using these data reduction methods generally exhibit semiquantitative agreement. However, the intercomparison of results obtained at different temperatures reveals that the Eq. 12 rate constants are somewhat less reliable unless many data points have been utilized for their determination. Although we now prefer the uniform sample weighting provided by Eq. 12, the Eq. 5 weighted regression procedure is the method of choice unless 10 or more data points have been obtained at each temperature. Equation

Table VII. Comparison Between Data Reduction Methods.

Substance (No. of Samples)	Temperature (°K)	Calculated (k_2/k_1) Values non-weighted Eq. 5	weighted Eq. 5	non-weighted Eq. 12
D_2(3)	273	1.912 ± 0.020	1.899 ± 0.026	1.897 ± 0.019
D_2(4)	303	1.454 ± 0.020	1.437 ± 0.020	1.408 ± 0.016
D_2(4)	345	0.839 ± 0.033	0.861 ± 0.029	0.950 ± 0.030
D_2(3)	405	0.649 ± 0.012	0.652 ± 0.008	0.610 ± 0.013
D_2(7)	457	0.441 ± 0.021	0.432 ± 0.022	0.457 ± 0.028
H_2(4)	273	0.832 ± 0.011	0.823 ± 0.012	0.805 ± 0.012
H_2(14)	303	0.693 ± 0.011	0.707 ± 0.013	0.748 ± 0.039
H_2(3)	353.5	0.407 ± 0.012	0.405 ± 0.013	0.355 ± 0.015
H_2(4)	405	0.348 ± 0.020	0.341 ± 0.022	0.321 ± 0.017
H_2(7)	475	0.297 ± 0.012	0.342 ± 0.017	0.350 ± 0.017
CD_4(14)	273	0.482 ± 0.026	0.504 ± 0.024	0.562 ± 0.043
CD_4(5)	303	0.449 ± 0.011	0.438 ± 0.013	0.401 ± 0.012
CD_4(5)	357.5	0.347 ± 0.011	0.348 ± 0.010	0.387 ± 0.018
CD_4(8)	429	0.161 ± 0.018	0.180 ± 0.018	0.251 ± 0.040
CH_4(3)	273	0.254 ± 0.002	0.254 ± 0.002	0.276 ± 0.007
CH_4(16)	303	0.317 ± 0.012	0.294 ± 0.033	0.282 ± 0.024
CH_4(6)	357.5	0.189 ± 0.002	0.189 ± 0.002	0.204 ± 0.007
CH_4(10)	429	0.215 ± 0.010	0.219 ± 0.009	0.214 ± 0.023
CH_4(8)	475	0.154 ± 0.026	0.162 ± 0.026	0.207 ± 0.029

12 should always be used in order to facilitate the rejection of spurious data. The number of data points required in order to ensure the precise specification of (k_2/k_1) increases rapidly as the competing reaction rates become more different.

$$N \approx 15 + 2(k_2/k_1);\ k_2 > k_1 \tag{36}$$

$$N \approx 15 + 2(k_1/k_2);\ k_2 < k_1 \tag{37}$$

Provided that the above guidelines are followed and that the intrinsic reproducibility of individual measurements is comparable to that shown in Table IV, then the standard relative error (68% confidence level) in (k_2/k_1) should be given approximately as $(N)^{1/2}$.

With C_2F_6 moderator $:CF_2$ corrections analogous to those described above for CHF_3/C_3F_6 but much smaller in magnitude have been found to be important for isotopic variants of C_2H_6 and neo-C_5H_{12} reactants (83). Within the range $0.2 < (k_2/k_1) < 5.0$ the neglect of these corrections could introduce 5-10% positive systematic errors in the calculated (k_2/k_1) results. This effect becomes even more serious outside of this range because of the increased numerical importance of the extrema in Y_1 (vide supra). Sufficient data have not been obtained in the present H_2, D_2, CH_4 and CD_4 experiments to permit the use of Eq. 12 or the precise quantitative specification of $:CF_2$ corrections. Accordingly, the preferred rate constants and Arrhenius parameters for these substances have been based upon weighted regression analysis via Eq. 5.

A comparison of 300°K relative rate constants obtained using chemiluminescence and $H^{18}F$ product analysis methods is presented in Table VIII. The (k_2/k_1) values listed in column three have been based upon the revised thermal reaction fraction (cf. Table V), upon new sample data in many instances, upon the modified data reduction procedures described above, and upon the Arrhenius parameters listed in Table XII. These updated results are intended to supplant the F-to-HF rate constants reported previously from this laboratory (25).

The available chemiluminescence measurements (84,85) of F-to-HF relative rate constants have utilized CH_4 as internal standard.

$$F + CH_4 \xrightarrow{k_{38}} HF + CH_3 \tag{38}$$

A useful consistency test for the data listed in Table VIII involves the intercomparison of calculated (k_2/k_{38}) ratios. With the exception of the single discrepant case of CH_2F_2 reactant, the results obtained using these markedly different techniques exhibit good agreement. More complete $H^{18}F$ product analysis measurements obtained for the moderated CH_3CF_3/C_3F_6 system revealed the occurrence of complications similar but less severe in magnitude to those described above for CHF_3/C_3F_6. The available results are not sufficiently complete to permit application of the Eq. 35 nonlinear regression procedure, so that an updated (k_2/k_1) value has not been assigned to CH_3CF_3.†

An analogous comparison has been given in Table IX between 283°K results obtained via the $H^{18}F$ product analysis and ^{18}F atom loss (87,88) methods.‡ In this instance the appropriate consistency test involves intercomparions between calculated (k_2/k_{39}) ratios.

$$^{18}F + C_2H_2 \xrightarrow{k_{39}} C_2H_2{}^{18}F \qquad (39)$$

Table IX. Intercomparison of $H^{18}F$ Product Analysis Vs. ^{18}F Loss F-to-HF Relative Rate Constants at 283°K.

Substance	(k_2/k_1) (This Work)[a]	(k_{39}/k_1) Ref. (88)	(k_2/k_{39}) (Calculated)
CH_3CF_3		18.9 ± 4.3	
D_2	1.79 ± 0.03	12.5 ± 3.1	0.143 ± 0.036
H_2	0.85 ± 0.01	7.1 ± 1.0	0.120 ± 0.017
C_2H_2		7.1 ± 3.6	
CD_4	0.49 ± 0.01	4.1 ± 0.7	0.119 ± 0.021
HI		2.7 ± 0.4	
CH_4	0.30 ± 0.03	2.4 ± 0.2	0.126 ± 0.014
H_2S		0.77 ± 0.05	
C_2H_6	0.066 ± 0.009	0.71 ± 0.06	0.092 ± 0.015
Nonweighted Average (k_2/k_{39})			0.120 ± 0.018
Weighted Average[b] (k_2/k_{39})			0.115 ± 0.008
Calculated[c] k_{39} (cm^3 $mole^{-1}$ sec^{-1})			$(9.2 \pm 0.7) \times 10^{13}$
Literature[d] k_{39} (cm^3 $mole^{-1}$ sec^{-1})			1.0×10^{14}

a. Temperature Corrections Based Upon Table XII.

b. Ref. (86).

c. Based Upon Weighted Average (k_2/k_{39}).

d. Ref. (21).

†Based upon a comparison with results reported in Ref. (88) the previous (25) (k_2/k_1) value for CH_3CF_3 is probably in error by ca. 300%.

‡Because the preliminary (k_1/k_{39}) results reported in ref. (87) were subsequently revised, only the updated values have been quoted in Table IX.

Table VIII. Intercomparison of $H^{18}F$ Product Analysis Vs. Chemiluminescence F-to-HF Relative Rate Constants At 300°K.

Substance	(k_2/k_1) (This Work)[a]	(k_{38}/k_1) Ref. (85)	(k_2/k_{38}) (Calculated)
CHF_3	88 ± 10	330 ± 50	0.264 ± 0.050
CH_2F_2	2.05 ± 0.22	> 33	< 0.062
CH_3NO_2		11 ± 2	
H_2O_2		10 ± 2	
HCl		7.7 ± 1.2	
		5.3 ± 0.6[b]	
D_2	1.48 ± 0.02		
CH_3CN		5.3 ± 1.7	
H_2	0.73 ± 0.01	2.7 ± 0.4	0.265 ± 0.040
CH_3F	0.55 ± 0.02[c]	2.1 ± 0.3	0.263 ± 0.041
	0.54 ± 0.02[d]		
CD_4	0.45 ± 0.01		
C_2H_4		1.8 ± 0.3	
CH_3Cl		1.7 ± 0.3	
HBr		1.5 ± 0.2	
		1.2 ± 0.2[b]	
HI		1.4 ± 0.2	
		0.83 ± 0.07[b]	
CH_4	0.29 ± 0.02	(1.00 ± 0.15)	0.285 ± 0.049
C_6H_6		1.0 ± 0.2	
CH_3COCH_3		0.63 ± 0.09	
CH_3CHO		0.63 ± 0.09	
H_2S		0.50 ± 0.08	
C_2D_6	0.117 ± 0.007		
cyclo-C_3H_6	0.102 ± 0.009	0.42 ± 0.06	0.245 ± 0.043
neo-C_5D_{12}	0.102 ± 0.002[d]		
C_2H_6	0.074 ± 0.010	0.39 ± 0.06	0.188 ± 0.039
CH_3OCH_3		0.37 ± 0.06	
neo-C_5H_{12}	0.063 ± 0.005	0.22 ± 0.03	0.283 ± 0.047
cyclo-C_5H_{10}	0.047 ± 0.005		
cyclo-C_6H_{12}	0.045 ± 0.007	0.23 ± 0.04	0.193 ± 0.040
GeH_4		0.17 ± 0.03	
SiH_4		0.15 ± 0.02	
Nonweighted Average (k_2/k_{38})			0.248 ± 0.038
Weighted Average[e] (k_2/k_{38})			0.244 ± 0.015
Calculated[f] k_{38} (cm^3 $mole^{-1}$ sec^{-1})			$(4.3 \pm 0.3) \times 10^{13}$
Literature[g] k_{38} (cm^3 $mole^{-1}$ sec^{-1})			4.8×10^{13}

Table VIII. Continued.

a. Temperature Corrections Based Upon Table XII.

b. Ref. (84).

c. Temperature Correction Based Upon The Assumption That $E_{CH_3F} = E_{CH_4}$.

d. 303 ± 8°K.

e. Ref. (86).

f. Based Upon Weighted Average (k_2/k_{38}).

g. Ref. (18).

Once again, the apparent agreement between these techniques is essentially quantitative within the framework of the available standard errors.†

Absolute Rate Constants. The substances for which reliable 300°K absolute k_1 values are available include CHF_3, HCl, H_2 and CH_4 (18). A principal objective of the present study has involved the development of accurate relative-to-absolute rate constant scaling procedures based upon these standards.‡ The relevant data have been summarized in Table X. The present 300°K (k_2/k_1) values for CHF_3, H_2 and CH_4 have been combined with 14 independent k_1 measurements in order to obtain a $(1.06 \pm 0.03) \times 10^{13}$ cm^3 $mole^{-1}$ sec^{-1} absolute k_2. This result, which can be compared to a 1.2×10^{13} cm^3 $mole^{-1}$ sec^{-1} literature estimate (21), provides the necessary basis for (k_2/k_1) scaling.

From Table X the present temperature corrected results exhibit semi-quantitative consistency with absolute k_1 measurements reported by Clyne and coworkers (30), by Wagner and coworkers (32,39,43), and by various others (27-29,31,33,34,38). Respective standard deviations from the mean calculated k_2 obtained for these experiment groupings were 3%, 26%, and 30%. The enhanced quality of the agreement with the results of Clyne and coworkers suggests that their standard error assignments may have been somewhat conservative and that the 300°K k_2 value can now be regarded as established to an absolute accuracy (99.5% confidence level) of *ca.* ± 10%. Based upon the number and variety of the experimental techniques represented in Table X together with the overall consistency of the results, we further conclude that 300°K absolute k_1 values for CHF_3, H_2 and CH_4 have been established to absolute accuracies of *ca.* ± 30% or better.

Many additional scaling relations and calculated k_1 results follow from the available relative rate constants. The 300°K k_2 from Table X was utilized in Table VIII for the estimation of k_{38} from the calculated average (k_2/k_{38}) ratio. An analogous determination of k_{39} at 283°K followed in Table IX from the Arrhenius parameters for reaction 4 (*cf.* Table XII).∓

†Based upon comparison with a literature absolute k_1 result for hydrogen abstraction from C_2H_2 (*cf.* Table XI), the ^{18}F atom loss procedure is apparently subject to enhanced uncertainty when the competing reactions have rate constants that differ by more than tenfold.

‡Owing to experimental complications (*vide supra*), HCl has not been investigated by means of the $H^{18}F$ product analysis technique.

∓Because the HF chemiluminescence (84,85) and ^{18}F atom loss (87,88) techniques have been less extensively calibrated against the CHF_3, HCl, H_2 and CH_4 standards, calibrations analogous to that shown in Table X have not been attempted using data obtained *via* these methods.

Table X. Intercomparison of $H^{18}F$ Product Analysis (k_2/k_1) Results *Vs*. Absolute k_1 Values For CHF_3, H_2 and CH_4 at 300°K.

Substance	(k_2/k_1) (This Work)[a]	$k_2 = k_1 \times (k_2/k_1)$ ($cm^3\ mole^{-1}\ sec^{-1}$) $\times 10^{-13}$ Ref. (30)	Refs. (32,39,43)	Refs. (27-29,31,33,34,38)
CHF_3	88 ± 10	1.056	1.056	1.672, 0.792
H_2	0.726 ± 0.011	1.089	0.799	2.759[b], 1.307, 1.104, 1.089, 0.726, 0.726
CH_4	0.285 ± 0.024	1.026	1.368	1.225
Nonweighted Group Averages[c]		1.057 ± 0.032	1.074 ± 0.285	1.080 ± 0.329
Nonweighted Average k_2 ($cm^3\ mole^{-1}\ sec^{-1}$)				$(1.07 \pm 0.27) \times 10^{13}$
Weighted Average[d] k_2 ($cm^3\ mole^{-1}\ sec^{-1}$)				$(1.06 \pm 0.03) \times 10^{13}$

a. Temperature Corrections Based Upon Table XII.

b. Ref. (27) Value For H_2 Excluded From Averages.

c. Cited Uncertainties Represent Standard Deviations From The Mean.

d. Ref. (86).

Table XI. Summary of Absolute Rate Constants For Thermal Hydrogen Abstraction and Olefin Addition Reactions By Atomic Fluorine.[a]

Substance	Competitor	Reference	k_1 $(cm^3\ mole^{-1}\ sec^{-1}) \times 10^{-13}$
CHF_3	-	(18)	0.012[b]
	-	(30)	0.012 ± 0.018
	-	(43)	0.012 ± 0.003
	CH_4	(85)	0.013 ± 0.002
	C_3F_6	(This Work)	[0.0120 ± 0.0014]
$CHCl_2F$	-	(18,89)	0.09
$CHClF_2$	-	(38)	0.15
	-	(18,89)	0.05
$CHCl_3$	-	(30)	[0.32 ± 0.08]
C_2H_2	-	(19,90)	[0.35 ± 0.10]
	C_2H_2	(88, This Work)	1.3 ± 0.7[c]
CH_3NO_2	CH_4	(85)	[0.39 ± 0.06]
H_2O_2	CH_4	(85)	0.43 ± 0.08
CH_3CF_3	C_2H_2	(88)	[0.49 ± 0.12]
CH_2F_2	-	(38)	1.1
	CH_4	(85)	0.13 ± 0.02
	C_3F_6	(This Work)	[0.52 ± 0.06]
CH_2Cl_2	-	(30)	[0.58 ± 0.14]
H_2O	-	(42)	0.66
HCl	-	(18)	0.65[b]
	-	(38)	0.73
	-	(18,89)	0.65
	-	(19,91)	0.55
	CH_4	(84)	0.83 ± 0.10
	CH_4	(85)	0.56 ± 0.09
	Average		[0.66 ± 0.10]
D_2	-	(29)	1.3
	C_2H_2	(88)	0.73 ± 0.19[c]
	C_3F_6	(This Work)	0.52 ± 0.04[c]
	C_3F_6	(This Work)	[0.71 ± 0.04]
CH_3CN	CH_4	(85)	[0.82 ± 0.14]
C_3F_6[d]	H_2, CH_4, C_2H_2	(21)	1.2[c]
	H_2, CH_4, CHF_3	(This Work)	0.94 ± 0.05[c]
	H_2, CH_4, CHF_3	(This Work)	[1.06 ± 0.03]
CH_3Cl	-	(30)	[1.4 ± 0.4]
	CH_4	(85)	2.6 ± 0.4
H_2	-	(18)	1.5[b]
	-	(30)	1.5 ± 0.8
	C_2H_2	(88)	1.3 ± 0.2[c]
	C_3F_6	(This Work)	1.10 ± 0.06[c]
	CH_4	(85)	1.6 ± 0.3
	C_3F_6	(This Work)	[1.46 ± 0.07]

Table XI. Continued.

Substance	Competitor	Reference	k_1 $(cm^3\ mole^{-1}\ sec^{-1}) \times 10^{-13}$
CFCl=CFCl[d]	C_2H_2	(21)	1.7[c]
	C_2H_2	(21, This Work)	1.5[c]
CH_3F	-	(38)	5.3
	CH_4	(85)	2.1 ± 0.3
	C_3F_6	(This Work)	1.9 ± 0.1
	Weighted Average	(85, This Work)	[1.91 ± 0.10]
CD_4	C_2H_2	(88)	2.2 ± 0.4[c]
	C_3F_6	(This Work)	1.93 ± 0.08[c]
	C_3F_6	(This Work)	[2.4 ± 0.1]
C_2H_4	CH_4	(85)	[2.6 ± 0.4]
HBr	-	(91)	4.0
	CH_4	(84)	3.8 ± 0.6
	CH_4	(85)	3.0 ± 0.5
	Average	(84,85,91)	[3.6 ± 0.5]
HI	C_2H_2	(88)	3.4 ± 0.5
	CH_4	(84)	5.2 ± 0.6
	CH_4	(85)	3.2 ± 0.5
	Weighted Average	(84,85,88)	[3.8 ± 0.3]
C_6H_6	CH_4	(85)	[4.2 ± 0.7]
CH_4	-	(18)	4.8[b]
	-	(30)	3.6 ± 3.6
	C_2H_2	(88)	3.8 ± 0.5[c]
	C_3F_6	(This Work)	3.1 ± 0.3[c]
	C_3F_6	(This Work)	3.7 ± 0.3
	Table VIII	(85, This Work)	[4.3 ± 0.3]
CH_3COCH_3	CH_4	(85)	[6.9 ± 1.1]
CH_3CHO	CH_4	(85)	[6.9 ± 1.1]
C_2H_4[d]	C_2H_2	(21)	8.4[c]
	C_2H_2	(21, This Work)	[7.6 ± 0.6][c]
H_2S	C_2H_2	(88)	11.9 ± 1.3[c]
	CH_4	(85)	[8.7 ± 1.4]
C_2D_6	C_3F_6	(This Work)	[9.0 ± 0.6]
C_2H_2[d]	CH_4	(21)	10.2[c]
	Table IX	(21, This Work)	[9.2 ± 0.7][c]
CHCl=CHCl[d]	CH_4	(21)	10.2[c]
	RH, RD, C_3F_6	(21, This Work)	[9.2 ± 0.7][c]
cyclo-C_3H_6	CH_4	(85)	10.4 ± 1.6
	C_3F_6	(This Work)	10.4 ± 1.0
	Weighted Average	(85, This Work)	[10.4 ± 0.8]
neo-C_5D_{12}	C_3F_6	(This Work)	[10.4 ± 0.4][e]
CH_3OCH_3	CH_4	(85)	[11.7 ± 2.0]
C_2H_6	C_2H_2	(88)	12.8 ± 1.5[c]
	C_3F_6	(This Work)	14.3 ± 2.0[c]
	CH_4	(85)	13.0 ± 2.2

Table XI. Continued.

Substance	Competitor	Reference	k_1 (cm^3 $mole^{-1}$ sec^{-1}) x 10^{-13}
C_2H_6 (Cont'd)	C_3F_6	(This Work)	14.3 ± 2.0
	Weighted Average	(85, This Work)	[13.7 ± 1.5]
neo-C_5H_{12}	CH_4	(85)	19.5 ± 3.2
	C_3F_6	(This Work)	16.8 ± 1.3
	Weighted Average	(85, This Work)	[17.2 ± 1.2]
cyclo-C_6H_{12}	CH_4	(85)	18.7 ± 3.1
	C_3F_6	(This Work)	23.5 ± 3.5
	Weighted Average	(85, This Work)	[20.8 ± 2.3]
cyclo-C_5H_{10}	C_3F_6	(This Work)	[22.5 ± 2.5]
GeH_4	CH_4	(85)	[26.0 ± 4.3]
SiH_4	CH_4	(85)	[28.2 ± 4.6]

a. 300°K Unless Noted Otherwise. Recommended k_1 Values Enclosed in Brackets. Also cf. Table I.

b. Recommended Literature Survey Value.

c. 283°K.

d. Olefin Addition Rate Constant.

e. 303°K.

Appendix I.

The Arrhenius parameters used in the above described calculations have been listed in Table XII. Further discussion of these results will be provided in a future publication.

Table XII. Arrhenius Parameters For Thermal Hydrogen Abstraction And Olefin Addition Reactions By Atomic Fluorine.[a]

Substance	(A_2/A_1)	A_1	(E_1-E_2)	E_1	Reference
		$(cm^3\ mole^{-1}\ sec^{-1}) \times 10^{-13}$	(cal mole^{-1})		
D_2		4.9		790 ± 180	(29)
				2750	(33)
		3.0		2600	(93)
		6.3		2160	(94,95)
	0.062 ± 0.012	[120 ± 35]	1860 ± 130	[3050 ± 150]	(This Work)
H_2[b]		12.6		2340	(94,95)
	0.043 ± 0.016	[180 ± 80]	1615 ± 225	[2800 ± 235]	(This Work)
CHF_3[b]		[0.50]		[2200 ± 170]	(43)
CD_4		3.7		2080	(93)
	0.101 ± 0.018	[76 ± 21]	890 ± 100	[2075 ± 130]	(This Work)
$CHCl_2F$		[1.09]		[1510 ± 500]	(89)
CH_4[b]		33		1150	(39)
	0.082 ± 0.014	[93 ± 26]	615 ± 100	[1800 ± 130]	(This Work)

Table XII. Continued.

Substance	(A_2/A_1)	A_1 $(cm^3\ mole^{-1}\ sec^{-1}) \times 10^{-13}$	(E_1-E_2) (cal mole^{-1})	E_1	Reference
C_3F_6[c]		[7.6 ± 1.6]		[1190 ± 80]	(This Work)
HCl		[2.8]		[880]	(39)
neo-C_5H_{12}		0.38		0	(41)
		5.0		0	(40)
	0.23 ± 0.02	[33 ± 8]	-755 ± 75	[435 ± 105]	(This Work)
C_2D_6	0.87 ± 0.11	[8.8 ± 2.2]	-1190 ± 80	[0 ± 80]	(This Work)

a. Recommended 273°K-450°K Values Enclosed in Brackets.
b. *Cf.* Table I Summaries for H_2, CHF_3 and CH_4.
c. Olefin Addition Reaction.

A compilation of k_1 results including recommended values has been presented in Table XI. Literature k_1 data have generally not been endorsed in the absence of cited standard errors. Reported rate constants exhibiting either serious internal inconsistencies or gross disagreement with established results have not been listed. Completeness has deliberately been sacrificed here in the interest of providing a critical data evaluation. Certain absolute k_1 techniques such as HF chemiluminescence (38), electron spin resonance (34,35), and laser pulse decay (27,92) have consistently yielded large discrepancies *vis-a-vis* the other methods. More comprehensive k_1 tabulations have been provided elsewhere (18,19).

With the exception of the single case of CH_4, the present procedure has yielded essentially quantitative agreement with the standard k_1 values endorsed by Foon and Kaufman (18). Thirty-six recommended values have been listed including many reactions that are too fast to be accurately characterized using presently available direct techniques. Although the tabulated results include a few discrepancies, the consistent agreement between scaled k_1 values indicates that an encouraging degree of order has begun to emerge within this experimentally difficult area of classical gas kinetics research.

Acknowledgement.

This research has been supported by the U.S. Energy Research and Development Administration under contract number AT-(04-3)-34, agreement 158. The authors thank Professor W. E. Jones for providing a preprint of his review article (19) in advance of its publication.

Literature Cited.

1. Schafer, T. P., Siska, P. E., Parson, J. M., Tully, F. P., Wong, Y. C., and Lee, Y. T., J. Chem. Phys. (1970), 53, 3385.
2. Farrar, J. M., and Lee, Y. T., this volume.
3. Parson, J. M., and Lee, Y. T., J. Chem. Phys. (1972), 56, 4658.
4. Bender, C. F., and Schaefer, H. F., this volume.
5. Muckerman, J. T., J. Chem. Phys. (1972), 57, 3388.
6. Bogen, D., and Setser, D. W., this volume.
7. Persky, A., J. Chem. Phys. (1974), 60, 49.
8. Klein, F. S., Ann. Rev. Phys. Chem. (1975), 26, 191.
9. Truhlar, D. G., and Wyatt, R. E., Ann. Rev. Phys. Chem. (1976), 27, 1.
10. Weston, R. E., Science (1967), 158, 332.
11. Menzinger, M., and Wolfgang, R., Angew. Chem. internat. Edit. (1969), 8, 438.
12. LeRoy, R. L., J. Phys. Chem. (1969), 73, 4338.
13. Greene, E. F., and Kuppermann, A., J. Chem. Ed. (1968), 45, 361.
14. Melton, L. A., and Gordon, R. G., J. Chem. Phys. (1969), 51, 5449.
15. Malerich, C. J., and Davis, D. R., J. Chem. Phys. (1971), 55, 4141.
16. Klein, F. S., and Persky, A., J. Chem. Phys. (1974), 61, 2472.
17. Goldanskii, V. I., Ann. Rev. Phys. Chem. (1976), 27, 85.
18. Foon, R., and Kaufman, M., Progress In Reaction Kinetics (1975), 8(2), 81.
19. Jones, W. E., and Skolnik, E. G., Chem. Revs. (1976), 76, 563.
20. Appelman, E. H., and Clyne, M.A.A., this volume.
21. Rowland, F. S., Rust, F., and Frank, J. P., this volume.
22. Persky, A., J. Chem. Phys. (1973), 59, 3612, 5578.
23. Parks, N. J., Krohn, K. A., and Root, J. W., J. Chem. Phys. (1971), 55, 2690.
24. Grant, E. R., and Root, J. W., Chem. Phys. Letters (1974), 27, 484.
25. Manning, R. G., Grant, E. R., Merrill, J. C., Parks, N. J., and Root, J. W., Int. J. Chem. Kinetics (1975), 7, 39.
26. Grant, E. R., and Root, J. W., J. Chem. Phys. (1975), 63, 2970.
27. Kompa, K. L., and Wanner, J., Chem. Phys. Letters (1972), 12, 560.
28. Dodonov, A. F., Lavrovskaya, G. K., Morozov, I. I., and Tal'roze, V. L., Dokl. Akad. Nauk SSSR (1971), 198, 622; Dokl. Phys. Chem. Eng. Trans. (1971), 198, 440.

29. Igoshin, V. I., Kulakov, L. V., and Nikitin, A. I., Krakt. Soobshich. Fiz. (1973), 1, 3; Chem. Abstr. (1973), 79, 149944; Kvantovaya Elektron Moscow (1973), 50; Sov. J. Quant. Electron. (1974), 3, 306.
30. Clyne, M.A.A., McKenney, D. J., and Walker, R. F., Can. J. Chem. (1973), 51, 3596.
31. Bozzelli, J., Thesis (Princeton University, Princeton, N.J., 1972).
32. Homann, V. K., Solomon, W. C., Warnatz, J., Wagner, H. G., and Zetzsch, C., Ber. Bunsenges. Phys. Chem. (1970), 74, 585.
33. Kapralova, G. A., Margolina, A. L., and Chaikin, A. M., Kinet. Katal. (1970), 11, 669; Dokl. Phys. Chem. Eng. Trans. (1971), 197, 281; ibid. (1971), 198, 452.
34. Goldberg, I. B., and Schneider, G. R., J. Chem. Phys. (1976), 65, 147.
35. Rabideau, S. W., Hecht, H. G., and Lewis, W. B., J. Magn. Reson. (1972), 6, 384.
36. My, L. T., Peyron, M., and Puget, P., J. Chim. Phys. (1974), 71, 377.
37. Foon, R., and Reid, G. P., Trans. Faraday Soc. (1971), 67, 3513.
38. Pollock, T. L., and Jones, W. E., Can. J. Chem. (1973), 51, 2041.
39. Wagner, H. G., Warnatz, J., and Zetzsch, C., Anales Assoc. Quim. Argentina (1971), 59, 169.
40. Fettis, G. C., Knox, J. H., and Trotman-Dickenson, A. F., Can. J. Chem. (1960), 38, 1643.
41. Fettis, G. C., Knox, J. H., and Trotman-Dickenson, A. F., J. Chem. Soc. (1960), 1064.
42. Zetzsch, C., Ph.D. Dissertation (Goerg-August University, Gottingen, 1971).
43. Wagner, H. G., private communication.
44. Manning, R. G., Mo, S. H., and Root, J. W., J. Chem. Phys. (1977), 67, 636.
45. Manning, R. G., and Root, J. W., J. Phys. Chem. (1977), 81(21), 0000.
46. Grant, E. R., Ph.D. Dissertation (University of California, Davis, 1974, University Microfilms No. 75-15434).
47. Manning, R. G., Ph.D. Dissertation (University of California, Davis, 1975, University Microfilms No. 76-1795).
48. Lucas, L. L., and Root, J. W., J. Appl. Phys. (1972), 43, 3886.
49. Wilkins, S. R., Shimose, S. T., Hines, H. H., Jungerman, J. A., Hegedus, F., and DeNardo, G. L., Int. J. Appl. Rad. Isotopes (1975), 26, 279.
50. Bennett, C. W., M.S. Thesis (University of California, Davis, 1977).

51. Werre, G. S., M.S. Thesis (University of California, Davis, 1977).
52. Pettijohn, R. R., Mutch, G. W., and Root, J. W., J. Phys. Chem. (1975), 79, 2077.
53. Stevens, D. J., and Spicer, L. D., J. Chem. Phys. (1976), 64, 4798.
54. Stevens, D. J., Ph.D. Dissertation (University of Utah, 1976).
55. Stevens, D. J., and Spicer, L. D., private communication.
56. Grant, E. R., Knierim, K. D., Krohn, K. A., Little, F. E., Manning, R. G., Mathis, C. A., Mo, S. H., Pettijohn, R. R., and Root, J. W., unpublished results.
57. Root, J. W., Ph.D. Dissertation (University of Kansas, Lawrence, 1964, University Microfilms No. 65-7004).
58. Welch, M. J., and Krohn, K. A., private communication.
59. Grant, E. R., and Root, J. W., J. Chem. Phys. (1976), 64, 417.
60. Manning, R. G., and Root, J. W., J. Chem. Phys. (1976), 64, 4926; J. Phys. Chem. (1975), 79, 1478.
61. Hudlicky, M., Chemistry of Organic Fluorine Compounds (Pergamon Press, New York, 1961).
62. Gadeken, O. C., Ayres, R. L., and Rack, E. P., Anal. Chem. (1970), 42, 1105.
63. Ayres, R. L., Michejda, C. J., and Rack, E. P., J. Amer. Chem. Soc. (1971), 93, 1389.
64. Ayres, R. L., Ph.D. Dissertation (University of Nebraska, Lincoln, 1970).
65. Ayres, R. L., private communication.
66. Krohn, K. A., Parks, N. J., and Root, J. W., J. Chem. Phys. (1971), 55, 5771, 5785.
67. Feng, D. F., Grant, E. R., and Root, J. W., J. Chem. Phys. (1976), 64, 3450.
68. Manning, R. G., Krohn, K. A., and Root, J. W., Chem. Phys. Letters (1975), 35, 544.
69. Seyferth, D., in Carbenes, vol. II, Moss, R. A., and Jones, M., Eds. (Wiley-Interscience, New York, 1975), pp. 101ff.
70. Mahler, W., Inorg. Chem. (1963), 2, 230.
71. Tang, Y. N., Smail, T., and Rowland, F. S., J. Amer. Chem. Soc. (1969), 91, 2130.
72. Smail, T., and Rowland, F. S., J. Phys. Chem. (1970), 74, 1866.
73. Smail, T., Miller, G. E., and Rowland, F. S., J. Phys. Chem. (1970), 74, 3464.
74. Pauwels, E.K.J., Ph.D. Dissertation (University of Amsterdam, 1971).
75. Lindner, L., private communication.
76. Moss, R. A., in Carbenes, vol. I, Moss, R. A., and Jones, M., Eds. (Wiley-Interscience New York, 1973), pp. 153ff.

77. Gaspar, P. P., and Hammond, G. S., in Ref. 69, pp. 207ff.
78. Takabe, T., Takahashi, M., and Fukutome, H., Prog. Theor. Phys. (1976), 56, 349.
79. Staemmler, V., Theoret. Chim. Acta (Berl.) (1974), 35, 309.
80. Harrison, J. F., J. Amer. Chem. Soc. (1971), 93, 4112.
81. Hsu, D. Y., Umstead, M. E., and Lin, M. C., this volume.
82. McKnight, C. F., and Root, J. W., J. Phys. Chem. (1969), 73, 4430.
83. Mo, S. H., Manning, R. G., and Root, J. W., manuscript in preparation.
84. Jonathan, N., Melliar-Smith, C. M., Okuda, S., Slater, D. H., and Timlin, D., Mol. Phys. (1971), 22, 561.
85. Smith, D. J., Setser, D. W., Kim, K. C., and Bogan, D. J., J. Phys. Chem. (1977), 81, 898.
86. Meyer, S. L., Data Analysis For Scientists and Engineers (Wiley, New York, 1975).
87. Williams, R. L., and Rowland, F. S., J. Phys. Chem. (1971), 75, 2709.
88. Williams, R. L., and Rowland, F. S., J. Phys. Chem. (1973), 77, 301.
89. Warnatz, J., Wagner, H. G., and Zetzsch, C., Report T-0240/92410/01017 to the Fraunhofer Gesellschaft (1972).
90. Wolfrum, J., Ph.D. Dissertation (Georg-August University, Gottingen, 1968).
91. Warnatz, J., Ph.D. Dissertation (Georg-August University, Gottingen, 1968).
92. Pearson, R. K., Cowles, J. O., Hermann, G. L., Gregg, D. W., and Creighton, J. R., I.E.E.E. J. Quant. Elec. (1973), 9, 879.
93. Foon, R., Reid, G. P., and Tait, K. B., J. Chem. Soc., Faraday Trans. I (1972), 68, 1131.
94. Jaffe, R. L., and Anderson, J. B., J. Chem. Phys. (1971), 54, 2224.
95. Jaffe, R. L., and Anderson, J. B., J. Chem. Phys. (1972), 56, 682.

Classical Kinetics: Organic Radicals and Ions

4

Kinetics and Mechanism of the Addition of Fluorine-Containing Radicals to Olefins

J. M. TEDDER and J. C. WALTON

Department of Chemistry, Purdie Building, University of St. Andrews, St. Andrews KY16 9ST, Scotland

In spite of sophisticated theories to explain directive effects in organic chemistry the factors which govern the preferred site of attack to an unsymmetrical olefin by a free radical remain obscure. A solution to this problem represents in itself a considerable challenge to mechanistic organic chemistry. There is however an important experimental feature which makes any conclusions drawn from such studies have a much wider interest. Free radical addition reactions can be studied over a wide range of temperatures entirely in the gas phase. Furthermore both the forward and the reverse reactions are capable of investigation. The complete absence of a solvent means that activation parameters determined experimentally should be capable of direct theoretical interpretation. Directive effects determined from such studies represent effects either transmitted through the molecule or through space and cannot be due, as so often in solution, to change wrought by the substituent on the solvation of the transition state. Extensive gas phase kinetic data for a series of radicals and a series of olefins will provide a real opportunity to test theoretical approaches without the unknown, and as yet uninterpretable effect of the solvent.

Sources of Fluoroalkyl Radicals

The first extensive studies of trifluoromethyl radicals depended on the photolysis of trifluoromethyl iodide as the radical source (1). Trifluoromethyl iodide is cleaved into trifluoromethyl radicals and iodine atoms by the 253.7 nm mercury line and trifluoromethyl bromide undergoes similar fission when irradiated by the 185 nm mercury line

$$CF_3I \xrightarrow{h\nu = 253.7} CF_3\cdot + I\cdot \quad ^2P_{\frac{1}{2}}$$

$$CF_3Br \xrightarrow{h\nu = 185} CF_3\cdot + Br\cdot \quad ^2P_{\frac{1}{2}}$$

These reactions are more complicated than many authors give credit, since even if the iodine atom is electronically excited ($^2P_{\frac{1}{2}}$), the trifluoromethyl radical must be thermally excited. The excess energy in the photolysis of trifluoromethyl iodide at 250 nm is about 37 kcal mol^{-1} while the excess energy in the photolysis of trifluoromethyl bromide at 185 nm is 85 kcal mol^{-1}. Fass and Willard have shown that in the former reaction most of the excess energy appears as translation, whereas in the latter process most appears as vibration, so that "hot" trifluoromethyl radicals formed in the photolysis of trifluoromethyl iodide are less selective than the "hot" radicals formed in the photolysis of the bromide (2).

If the subsequent reactions involve the regeneration of trifluoromethyl radicals by iodine abstraction, e. g. in the reactions with ethylene;

$$CF_3\cdot + CH_2=CH_2 \longrightarrow CF_3CH_2CH_2\cdot$$

$$CF_3CH_2CH_2\cdot + CF_3I \longrightarrow CF_3CH_2CH_2I + CF_3\cdot$$

the nature of the primary step is unimportant if the chains are long. However, if there is no chain or the chains are short the presence of thermally excited radicals can make kinetic studies extremely difficult to interpret.

The most extensively used source of trifluoromethyl radicals is the photolysis of hexafluoroacetone. The primary process has been investigated in some detail by Kutschke and co-workers. Two mechanisms of photolysis were observed, one from thermally excited singlet and the other from the lowest triplet state (3). The quantum yield is temperature dependent, the activation energy for the decomposition of the triplet being estimated at 16 kcal mol^{-1}. There is evidence that the excited singlet hexafluoroacetone can undergo cyclic addition with perfluoroolefins to form the corresponding oxetane (4). This reaction involves addition of the triplet state of the ketone to the electron deficient double bond of the fluoroolefin(5).

$$[CF_3COCF_3]^3 + CF_3CF=CF_2 \longrightarrow \begin{array}{c} O \!-\!\!-\! C(CF_3)_2 \\ |\qquad\quad | \\ CF_2\!-\!CFCF_3 \end{array}$$

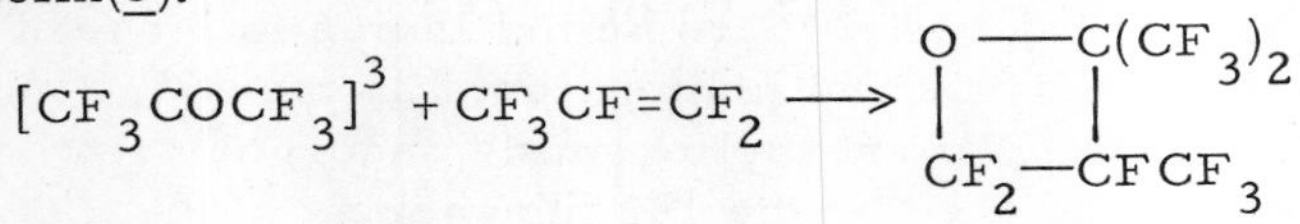

Another complication is that trifluoromethyl radicals are capable of adding to hexafluoroacetone (6). Gordon found that in addition to hexafluoroethane and carbon monoxide, the

$$(CF_3)_2CO \xrightarrow{h\nu} (CF_3)_2CO^* \longrightarrow 2CF_3\cdot + CO$$

$$2CF_3\cdot \longrightarrow C_2F_6$$

$$CF_3\cdot + (CF_3)_2CO \longleftrightarrow (CF_3)_3CO\cdot$$

$$(CF_3)_3CO\cdot + \cdot CF_3 \longrightarrow (CF_3)_3COCF_3$$

photolysis of hexafluoroacetone by itself led to the formation perfluoro(methyl-t-butyl)ether. He estimated the activation energy for the addition step to be 9.7 ± 0.26 kcal mol^{-1} (for the reverse step E = 30.6 ± 1.3 kcal mol^{-1}), this is within the range for the addition of alkyl radicals to olefins so that this reaction could be a serious side reaction in radical addition studies (6).

Hexafluoroazomethane has been less studied. Photolysis at 365 nm has a quantum yield of nearly unity at low pressures but this falls with increasing pressure indicating quenching of an excited state. To account for the experimental observations two different modes of decomposition of the photoactivated molecule have been postulated (7). Like hexafluoroacetone, hexafluoroazomethane will add trifluoromethyl radicals:

$$CF_3N_2CF_3 \xrightarrow{h\nu} CF_3N_2CF_3^* \longrightarrow 2CF_3\cdot + N_2$$

$$2CF_3\cdot \longrightarrow C_2F_6$$

$$CF_3\cdot + CF_3N_2CF_3 \longrightarrow (CF_3)_2N\dot{N}CF_3$$

$$CF_3\cdot + (CF_3)_2N\dot{N}CF_3 \longrightarrow (CF_3)_2N_2(CF_3)_2$$

Recently Whittle and co-workers have developed hexafluoroacetic anhydride as a source of trifluoromethyl radicals (8)

$$(CF_3CO)_2O \xrightarrow[255\text{-}165\ nm]{h\nu} 2CF_3\cdot + CO + CO_2$$

There is little evidence of quenching of electronically excited anhydride molecules by added scavengers and unlike hexafluoroacetone and hexafluoroazomethane trifluoromethyl radicals do not appear to add to the anhydride.

The higher perfluoroalkyl radicals C_nF_{2n+i} have mostly been obtained by the photolysis of the corresponding iodides (9), although the perfluorocarboxylic acid anhydrides, which are fairly readily available, look a good alternative source (10).

The fluoro-chloroalkyl radicals and the fluorobromoalkyl radicals have both been produced by the photolysis of appropriate iodides or bromides (11). The photolysis of dibromodifluoromethane has been investigated in some detail (12). An alternative method of production of fluorochloromethyl and fluorobromomethyl radicals has been the use of benzoyl peroxides as an initiator (13). More recently di-t-butylperoxide has been used in orientation studies (14).

Difluoromethyl radical has been produced using excited mercury atoms as photosensitisers (15), but so far this

$$Hg^{*} + CF_2HCl \longrightarrow CF_2H\cdot + Cl\cdot + Hg\ (+HgCl)$$

method has not been used in kinetic studies.

The partially fluorinated radicals have been prepared by photolysis of the fluoroalkyl iodides (16, 17) and by the photolysis of the appropriate fluoroalkyl ketone (e.g. CH_2FCOCH_2F) (18). Fluoromethyl and difluoromethyl radicals have also been formed from the fluoroalkyl iodides using di-t-butyl peroxide as initiator (16, 17). Problems can arise in kinetic studies with these radicals because the chain carrying step may involve hydrogen as well as iodine abstraction. In practice this only seems to be of major importance when electronically excited species are present.

Experimental Methods

In the addition of a radical source R_fX to an unsymmetrical alkene $\underline{E}$ there are two possible products formed from addition to the least substituted end (k_2) and the most substituted end (k_2'):

$$R_f\cdot + \underline{E} \longrightarrow R_f\underline{E}\cdot \qquad (k_2)$$
$$R_f\cdot + \underline{E} \longrightarrow \underline{E}R_f\cdot \qquad (k_2')$$
$$R_f\underline{E}\cdot + R_fX \longrightarrow R_f\underline{E}X + R_f\cdot \qquad (k_3)$$
$$\underline{E}R_f\cdot + R_fX \longrightarrow X\underline{E}R_f + R_f\cdot \qquad (k_3')$$

It can readily be shown that the Orientation ratio OR defined as k_2'/k_2 is given by:

$$OR = k_2'/k_2 = [X\underline{E}R_f]/[R_f\underline{E}X]$$

provided the chains are long and the addition steps are not reversible.

Relative rates of addition of a given radical to series of alkenes can be determined by means of the rates of consumption of the alkenes, but in practice most studies have utilised product analysis. In a system such as the one described above, where the main termination involves combination of $R_f\cdot$, the rate constant for addition to a given site in the alkene can be measured relative to the rate constant for combination (k_c) by analysis of the products for the adduct and the dimer:

$$k_2/k_c^{\frac{1}{2}} = V(R_f\underline{E}X)/V(R_fR_f)^{\frac{1}{2}}[\underline{E}]$$

Alternatively, the rates of addition to a series of alkenes can be measured relative to the rate of addition to one specific alkene (usually ethylene) as a reference reaction. The relative rate constants are then given by:

$$k_2/k_{2e} = \frac{V(R_f\underline{E}X)\ [CH_2{=}CH_2]}{V(R_fCH_2CH_2X)[\underline{E}]}$$

where k_{2e} represents the rate constant for addition to ethylene, and the equation applies for small percentage conversion of the alkenes. If the addition steps are reversible the expression becomes more complex:

$$\frac{V(R_f\underline{E}X)}{V(R_fCH_2CH_2X)} = \frac{k_2[\underline{E}]\ (1 + k_{-2e}/k_{3e}[R_fX]\ )}{k_{2e}[CH_2{=}CH_2]\ (1 + k_{-2}/k_3[R_fX])}$$

and the ratio of the rates of formation of the two adducts depends on the concentration of the radical source, the ratio of the rate constants for decomposition to transfer (k_{-2}/k_3) for each adduct radical as well as on the ratio of the rate constants for addition to the two alkenes.

Szwarc and co-workers used another method for the measurement of relative rates of addition of trifluoromethyl radicals to olefins, in which the olefin was mixed with an excess of a hydrocarbon (usually 2, 3-dimethylbutane) and hexafluoroazomethane was utilised as the radical source (19, 20, 21)

$$CF_3N_2CF_3 + h\nu \longrightarrow 2CF_3\cdot + N_2$$

$$CF_3\cdot + RH \xrightarrow{k_d} CF_3H + R\cdot$$

$$CF_3\cdot + \underline{E} \xrightarrow{k_2} CF_3\underline{E}\cdot$$

By comparing the ratio of fluoroform to nitrogen produced in the absence of alkene with that in the presence of alkene the relative rate constant ratio k_2/k_d could be determined. This technique gives the overall rate of addition to an olefin, and cannot distinguish between the two ends of an unsymmetrical alkene, and it also relies on the assumption that radicals are lost only in the addition step, although appropriate corrections can be made where side reactions are identified.

Non-stationary state techniques are required for the determination of absolute rate constants and, almost invariably in the field of free-radical chemistry, absolute rate determinations have been made directly on combination reactions. Ultimately, therefore, absolute rate constants for radical addition reactions depend on a knowledge of the rate of addition relative to the rate of combination of the radical involved.

Absolute Rates

The rate constants of radical combination reactions, which form the basis from which absolute rates of addition reactions are determined, have been the subject of a vigorous controversy in recent years. That of methyl radicals is well established at 2×10^{10} l. mol^{-1} s^{-1}, but for other alkyl radicals evidence has been presented for both "low" values and "high" values, and the final outcome remains in doubt (22).

Ayscough's original value for the rate constant of combination of trifluoromethyl radicals, (23) which was measured by the intermittent photolysis of hexafluoroacetone, has recently been challenged by several new determinations. Ogawa, Carlson and Pimentel (24) using flash photolysis of CF_3I coupled with rapid scan infra-red detection found the combination rate constant to be both temperature and pressure dependant with values in the range 5.9 to 9.2×10^9 l mol^{-1} s^{-1}. Basco and Hathorn (25) detected the UV absorption spectrum of $CF_3\cdot$ radicals and measured the rate constant as 3×10^9 l mol^{-1} s^{-1} at 25°C from its second order decay. Hiatt and Benson, (26) using the "buffer" technique with CH_3I and CF_3I, coupled with thermochemical estimates of the equilibrium constant, found the rate constant to be in the range 1.8 to 15×10^9 l mol^{-1} s^{-1}, depending on the thermochemical parameters. Okafo and Whittle (27) have recently given an improved value for $D(CF_3{-}I)$ and when this is applied to Hiatt and Bensons' data combination rate constants at the lower end of the range are obtained. An average value of about 3×10^9 l mol^{-1} s^{-1} would seem to be appropriate.

Combination rate constants for difluorochloromethyl radicals (1.2×10^{10} l mol^{-1} s^{-1}) (28) and chlorodecafluorocyclohexyl radicals (3×10^8 l mol^{-1} s^{-1}) (29) are the only others which have been measured for fluoroalkyl radicals. Since these were determined by the intermittent illumination method they obviously need checking by independent techniques.

Sangster and Thynnes' absolute rate constant for the addition of $CF_3\cdot$ radicals to ethylene was obtained from experiments in which the radicals reacted competitively with ethylene and H_2S (30). The relative rate constant was then converted to an absolute rate constant by use of Arthur and Gray's determination of the rate of abstraction of hydrogen from H_2S by $CF_3\cdot$ radicals, (31) which was in turn based on Ayscough's rate constant for $CF_3\cdot$ radical combination. The absolute value for $CF_3\cdot$ addition to ethylene derived from Pearson and Szwarcs' work (20) was also ultimately based on Ayscough's combination rate constant (23). Values of the absolute rate constant of $CF_3\cdot$ addition to ethylene based on the data of these two groups of workers, but corrected with the new value of the combination rate constant of CF_3 radicals are given in table 1.

Table I
Absolute Arrhenius Parameters and Rate Constants at 164°C for Radical Additions to Ethylene

Radical	log k (164°C) $l\ mol^{-1}s^{-1}$	log A $l\ mol^{-1}s^{-1}$	E kcal mol^{-1}	ΔS^{o*} cal mol^{-1} K^{-1}	ΔH^{o*} kcal mol^{-1}	Ref.
$CH_3\cdot$	4.7	8.52	7.7	-30.0	-25.5	(32)
CH_2F	[5.5]	[7.6]	[4.3]	-31.2	[-27]	(33)
$CF_3\cdot$ (1)	7.3	7.8	1.0	-35.3	-27.5	(20)
(2)	6.7	8.11	2.8	"	"	(30)
(3)	6.6	8.0	2.9	"	"	(34)
$CF_2Br\cdot$	[6.5]	[8.0]	[3.1]			(35)
$CCl_3\cdot$	4.7	7.8	6.3	-34.4	-17.4	(36)

Parentheses indicate estimated values. *Thermochemical data from J. A. Kerr and M. J. Parsonage, "Evaluated Kinetic Data on Gas Phase Addition Reactions" Butterworths, 1972

The rate of $CF_3\cdot$ addition to ethylene has recently been determined directly relative to the rate of $CF_3\cdot$ combination and the absolute rate constant deduced from this study is also given in table I (34). This recent determination is in excellent agreement with that of Sangster and Thynne, which enables the following Arrhenius parameters to be recommended for $CF_3\cdot$ addition to ethylene: $\log A = 8.0 \pm 0.3\ l\,mol^{-1}s^{-1}$ $E = 2.9 \pm 0.5$ kcal mol^{-1}.

Absolute rate constants have been estimated for fluoromethyl radicals (33), and for difluorobromomethyl radicals (35). The comparison of these rate constants in table I shows that the pre-exponential factors are all equal, within the experimental error (with the possible exception of the methyl radical), irrespective of the nature of the attacking radical. Trifluoromethyl radicals are the most reactive species and the most electronegative; the activation energies increase in a uniform manner as hydrogen replaces fluorine in the radical.

Kerr and Parsonage have compared the addition rate constants with the unimolecular decomposition rate constants of the product radicals (32). These two rate constants are related through the equilibrium constant which can be calculated by thermochemical methods. The enthalpy and entropy changes in the addition reactions, calculated by statistical mechanics or estimated where necessary by the group aditivity method, are given in table I. The absolute rate constants and the activation energies appear to follow the trend in the enthalpies of reaction (with the exception of trichloromethyl) i. e. the more exothermic the reaction the lower the activation energy.

The Arrhenius parameters for the decomposition of n-propyl radicals given by Benson and O'Neal: $\log (k_{-2}/s^{-1}) = 13.8 - 33.2/2.3RT$ (37) together with the thermodynamic data yield a calculated rate constant for methyl addition to ethylene given by $\log k_2 = 8.9 - 8.1/2.3RT$, in good agreement with the experimental parameters. More recent experimental data on the decomposition of n-propyl leads to rather poorer agreement (38). Experimental data for the decomposition reactions is lacking for the other radicals, but in each case when the addition rate constants are combined with the thermochemical data "reasonable" values are obtained for the decomposition rate parameters (32).

The A-factors for $CF_3\cdot$ and $CH_3\cdot$ radicals have been calculated directly by Transition State Theory (39, 40). The calculated values depend critically on the chosen vibrational frequencies of the transition states. They tend to show that the transition state is loose, as expected from the exothermicity of the reactions and as predicted by SCF-MO calculations, but that association of the radical with one of the alkene carbons and concomitant loosening of the alkene double bond is by no means negligible. (see p.

Relative Rates

The difficulty in measuring absolute rates has limited the

number of radicals for which activation parameters are available. Often however the orientation of addition to an unsymmetric olefin and the relative rate of two different olefins is of more interest than the absolute rates themselves. Table II shows the ratio of the rate constants for the addition of $CH_3\cdot$, $CH_2F\cdot$, $CF_2H\cdot$, and $CF_3\cdot$ to ethylene and tetrafluoroethylene.

Table II

The Relative Activation Parameters for the Addition of $CH_3\cdot$, $CH_2F\cdot$, CHF_2 and $CF_3\cdot$ to ethylene and tetrafluoroethylene

(δE in kcal mol^{-1})

Radical	$A_{C_2F_4}/A_e$	$E_{C_2F_4}-E_e$	$k_{C_2F_4}/k_e(164^o)$
$CH_3\cdot$	0.34	-2.5	6.0
$CH_2F\cdot$	0.76	-1.3	3.4
$CHF_2\cdot$	0.83	-0.2	1.1
$CF_3\cdot$	0.87	+1.7	0.1

At 164°C the methyl radical adds six times faster to tetrafluoroethylene than to ethylene, while the trifluoromethyl radical adds ten times slower. The table shows that this diverging behaviour in the relative rates of the two radicals is almost entirely due to changes in the difference in the activation energies. This kind of divergent behaviour is often interpreted in terms of relative electronegativity, according to which picture the methyl radical is "nucleophilic" when compared to the "electrophilic" trifluoromethyl radical. We shall find a lot of evidence to support the idea that "polarity" plays an important part in determining the relative reactivity of fluoroalkyl radicals. Table II shows a regular progression in the activation energy differences from the negative value for methyl radicals to the positive value for trifluoromethyl radicals with fluoromethyl and difluoromethyl radicals giving values in between.

Tables III and IV show the relative Arrhenius parameters for the addition of a wide variety of radicals to fluoroethylene, 1,1-difluoroethylene, trifluoroethylene and tetrafluoroethylene taking the rate of addition to ethylene as standard [$CF_3\cdot$ (39); $CHF_2\cdot$ (16); $CH_2F\cdot$ (17); $CH_3\cdot$ (40); $CCl_3\cdot$ (41, 36); $CH_2Cl\cdot$ (14); $CBr_3\cdot$ (42); $CHBr_2\cdot$ (42); $CF_2Br\cdot$ (35); $CFBr_2\cdot$ (43); $C_2F_5\cdot$ (9); $(CF_3)_2CF\cdot$ (44); $CF_2I\cdot$ (16)]. Table III shows that the pre-exponential ("A") factors only vary slightly from one radical to another or from one olefin to the next. The smallest "A" factor ratio is one tenth, the largest is forty.

Table III

Relative Arrhenius Parameters for Radical Addition to Olefins

$\text{Log } A_x - \text{Log } A_{CH_2=CH_2}$

Radical	$CH_2=CHF$	$CH_2=CF_2$	$CHF=CH_2$	$CHF=CF_2$	$CF_2=CH_2$	$CF_2=CHF$	$CF_2=CF_2$	$CH_2=CHCH_3$	$CF_2=CFCF$
x	δ log A	δ log A	δ log A	δ log A	δ log A	δ log A	δ log A	δ log A	δ log A
$CF_3\cdot$	-0.1	-0.2	-0.3	-0.6	-0.6	-0.5	-0.1		
$CF_2Br\cdot$	-0.2	+0.0	-0.6	-1.0	-0.1	-0.3	+1.1		
$CFBr_2\cdot$	-0.4	-0.5	-0.4	+1.2	+1.5	+1.4	-0.5		
$CF_2I\cdot$	-0.1	-0.1	-0.5	-0.7	-0.2	-0.3	+0.4		
$C_3F_7\cdot$	-0.1	+0.7	+0.1	+0.2	+1.1	+0.2	+1.6		
$CCl_3\cdot$	-0.2	+0.1	-0.2	+0.8	+0.1	+0.8	+1.4	+1.3	+1.0
$CH_3\cdot$	0.2		-0.7	-0.1	-0.4	-1.5	-0.1	-0.8	
$CHF_2\cdot$	0.0						-0.1		
$CH_2F\cdot$							-0.1		
$(CF_3)_2CF\cdot$	+0.1	0.0	-0.3	-0.4	-0.6	-0.7	-0.2		

log A in $l\ mol^{-1}\ s^{-1}$

Table IV

Relative Arrhenius Parameters for Radical Addition to Olefins

$E_x - E_{CH_2=CH_2}$

Radical	$CH_2=CHF$	$CH_2=CF_2$	$CHF=CH_2$	$CHF=CF_2$	$CF_2=CH_2$	$CF_2=CHF$	$CF_2=CF_2$	$CH_2=CHCH_3$	$CF_2=CFCF_3$
x	δE	δE	δE	δE	δE	δE	δE	δE	δE
$CF_3\cdot$	+0.5	+1.2	+1.9	+1.9	+3.2	+2.7	+1.7		
$CF_2Br\cdot$	+0.4	+1.8	+1.6	+0.2	+4.4	+2.0	+3.9		
$CFBr_2$	+0.3	+.08	+2.3	+4.3	+8.1	+5.6	+0.2		
$CF_2I\cdot$	+0.6	+1.7	+1.2	+0.6	+3.0	+1.8	+1.0		
$C_3F_7\cdot$	+0.8	+2.7	+3.2	+3.1	+8.2	+4.1	+4.7		
$CCl_3\cdot$	+0.1	+1.4	+2.1	+2.9	+5.1	+3.9	+2.9	+0.2	+3.0
$CH_3\cdot$	+0.4		+0.3	-0.7	+2.2	-4.1	-1.8	-0.8	
$CHF_2\cdot$	+0.3						-0.2		
$CH_2F\cdot$							-1.3		
$(CF_3)_2CF\cdot$	+0.8	+1.2	+3.6	+2.8	+6.1	+4.2	+3.4		

E in kcal mol^{-1}

We shall see later that relative rates of addition vary by five orders of magnitude (from 10x to 10^{-4}x the rate of addition to ethylene) so that it is clear that rate and orientation of addition is principally governed by variations in activation energy. Table IV, besides showing the wide variation in activation energy differences (δE varies from -4.1 to +8.2 kcal), shows that only for methyl radicals are the activation energies less for addition to the highly fluorinated olefins than they are for addition to ethylene. This is another manifestation of the methyl radical's "nucleophilic" character.

For every radical the activation energy is greater for addition to CHF-end of vinyl fluoride than for addition to the CH_2- end, and in every case the activation energy for addition to the CH_2- end of vinyl fluoride is greater than that for addition to ethylene. In other words radicals add preferentially to the least substituted end of vinyl fluoride because attack at this site is less deactivated than attack at the substituted end. Not because attack at the CH_2- end is activated. A single fluorine substituent is deactivating to all radicals, including methyl. This conclusion shows that simple resonance pictures in which a substituent fluorine atom is supposed to stabilise an adduct radical is not an adequate picture.

$$R\cdot + CH_2{=}CHF \longrightarrow RCH_2\dot{C}H{-}\ddot{\underset{\cdot\cdot}{F}}:$$
$$\updownarrow$$
$$RCH_2\overset{\ominus}{\ddot{C}}H{-}\overset{\oplus}{\dot{\underset{\cdot\cdot}{F}}}:$$

Addition to the CHF- end of vinyl fluoride and the CF_2- end of 1,1-difluoroethylene both yield adduct radicals with the unpaired electron on a $CH_2\cdot$ group as in the addition to ethylene ($RCH_2\dot{C}H_2$; $RCHF\dot{C}H_2$ and $RCF_2\dot{C}H_2$). Table IV shows that the activation energies increase substantially as the substitution at the site of attack increases, and clearly resonance stabilisation involving the substituent fluorines makes no contribution. An alternative resonance type theory involves hyperconjugation and could be used to explain the increasing activation energy of the above series in terms of reduced opportunities for hyperconjugation. Examination of the activation energies for the addition to the CH_2- end of 1,1-difluoroethylene, the CHF- end of trifluoroethylene and to tetrafluoroethylene provides no support for such a hypothesis. Indeed although the overall results show a very consistent pattern there is no evidence to suggest the results can be correlated in a simple resonance pictures involving either conjugation or hyperconjugation.

Table V shows the relative rates for addition at 150° and most of the data comes from Tables III and IV. However two radicals ($CH_2Cl\cdot$ and $CBr_3\cdot$) are included in this Table for which no accurate temperature variation is available. The general pattern of results confirms the picture built up from the tables of relative Arrhenius Parameters. The relative rates are the results of direct measurement and therefore probably represent a more accurate summary. A very noticeable feature of Table V is the low rates of addition of heptafluoro-iso-propyl radicals to CHF- and CF_2- sites. This strongly suggests that classical "steric hindrance" plays a significant role in free radical addition.

The most comprehensive table of relative results is Table VI in which the orientation ratios for the addition of an extensive range of fluoroalkyl radicals to vinyl fluoride, 1,1-difluoro-ethylene and trifluoroethylene are listed. Although all radicals add preferentially to the CH_2- end of vinyl fluoride and 1,1-di-fluoroethylene, the orientation of addition to trifluoroethylene depends on the radical. The "electrophilic" radicals adding preferentially to the CHF- site and the nucleophilic radicals adding preferentially to the CF_2- site. In the radical series $CF_3\cdot$; $CHF_2\cdot$; $CH_2F\cdot$; $CH_3\cdot$ the orientation ratios show a progressive increase in attack at the most fluorine substituted end of the olefin. This is in accord with a "polar" picture in which the more "nucleophilic" the radical the greater the attack at the fluorine containing site. However in the radical series $CF_3\cdot$, $CF_2Br\cdot$, $CFBr_2\cdot$, $CBr_3\cdot$ the orientation ratios show a decrease in attack at the most fluorine substituted end of the olefin. This series of radicals shows a smaller drop in electro-negativity but $CBr_3\cdot$ is appreciably less electrophilic than $CF_3\cdot$, so that the reverse trend in orientation ratios must be due to steric hindrance. Even greater evidence for steric effects come from the series $CF_3\cdot$, $CF_3CF_2\cdot$, and $(CF_3)_2CF\cdot$. The perfluoro-iso-propyl radical is the most selective radical in table VI. It is important to note (see Table IV) that this selectivity is primarily due to activation energy differences. The A-factor ratios for the reactions involving perfluoro-iso-propyl radicals are no greater than those for other radicals. This is consistent with an "early" transition state for these reactions.

Table V

RELATIVE RATES OF ADDITION OF RADICALS TO SPECIFIC SITES IN FLUOROETHYLENES

Measurements from Gas Phase reactions at 164°C unless otherwise stated.

ALKENE	$CF_3\cdot$	$CHF_2\cdot$	$CH_2F\cdot$	$CH_3\cdot$	$CF_2Br\cdot$	$CF_2I\cdot$	$CFBr_2\cdot$	$CCl_3\cdot$	$C_3F_7\cdot$	$(CF_3)_2CF\cdot$	$CH_2Cl\cdot$ +	$CBr_3\cdot$ +
*$CH_2=CH_2$	1.00	1.00	1.00	1.00	1.00	1.00	1.00	1.00	1.00	1.00	1.00	1.00
*$CH_2=CHF$	0.48	0.32	0.28	0.57	0.39	0.38	0.29	0.56	0.32	0.48	0.40	0.27
*$CH_2=CF_2$	0.15	0.10	-	-	0.14	0.11	0.14	0.25	0.07	0.22	0.10	0.19
*$CHF=CH_2$	0.056	0.06	-	0.11	0.037	0.08	0.027	0.06	0.03	0.007	0.06	0.01
*$CHF=CF_2$	0.031	0.15	-	1.00	0.079	0.10	0.10	0.22	0.04	0.015	0.18	0.05
*$CF_2=CH_2$	0.006	0.016	-	0.02	0.004	0.02	0.003	0.004	0.001	0.0002	0.014	
*$CF_2=CHF$	0.016	0.14	-	1.90	0.046	0.06	0.040	0.07	0.01	0.0015	0.18	0.012
*$CF_2=CF_2$	0.12	1.10	3.4	9.51	0.13	0.79	0.23	0.89	0.18	0.0098	0.60	0.15

* indicates site of addition. + Relative rates from sealed tube experiments at 150°C

Table VI
Orientation Ratios (α:β) for the Addition of Alkyl Radicals to Vinyl Fluoride, 1,1-Difluoroethylene and Trifluoroethylene at 150°

Radical	α β CH_2=CHF	α β CH_2=CF_2	α β CHF=CF_2
CF_3	1 : 0.09	1 : 0.03	1 : 0.50
CHF_2	1 : 0.19	1 : 0.15	1 : 0.95
CH_2F	1 : 0.30	1 : 0.44	1 : 2.04
CH_3	1 : 0.20	-	1 : 2.10
CCl_3	1 : 0.07	1 : 0.01	1 : 0.29
CH_2Cl	1 : 0.18	1 : 0.14	1 : 1.03
CBr_3	1 : 0.04	-	1 : 0.24
$CHBr_2$	1 : 0.06	-	1 : 0.31
CF_3	1 : 0.09	1 : 0.03	1 : 0.50
CF_2Br	1 : 0.09	1 : 0.03	1 : 0.47
$CFBr_2$	1 : 0.08	1 : 0.02	1 : 0.37
CBr_3	1 : 0.06	-	1 : 0.31
CF_3	1 : 0.09	1 : 0.03	1 : 0.50
CF_3CF_2	1 : 0.05	1 : 0.01	1 : 0.29
$(CF_3)_2CF$	1 : 0.02	1 : 0.001	1 : 0.06
CF_3CF_2	1 : 0.054	1 : 0.011	1 : 0.29
$CF_3(CF_2)_2$	1 : 0.050	1 : 0.009	1 : 0.25
$CF_3(CF_2)_3$	1 : 0.050	1 : 0.007	1 : 0.24
$CF_3(CF_2)_6$	1 : 0.049	1 : 0.007	1 : 0.23
$CF_3(CF_2)_7$	1 : 0.043	1 : 0.006	1 : 0.22

Theoretical Treatment

Linear Free Energy Relationships The observation that familiar polar effects influence the orientation ratios for radical addition to olefins prompts the application of the well established free-energy relationships. The orientation data can be accommodated in a Hammett plot by assigning a polar constant, σ, to each radical (14). This is defined as the sum of the Hammett sigma constants for the groups attached to the radical centre. In practice, Taft σ° constants have been used (45).

Good correlations were observed between log OR and the values of the radicals for the data of table VI. This encouraged us to apply to our results the "patterns of reactivity" approach utilised by Bamford and Jenkins for interpreting polymer radical reactivities (46). A simple extension of the "patterns" approach leads to the equation (14):

$$\log OR = \log k_2'/k_2 = \sigma\ (\alpha' - \alpha) + (\beta' - \beta)$$

for the orientation ratio, where σ characterises the radical, and α, β are constants characteristic of one end of the alkene and α', β' are constants characteristic of the other end. Values of $\alpha' - \alpha$ determined from the gas phase orientation data compare very well with these from polymer reactivities in solution. Correlation of log OR with $\sigma(\alpha' - \alpha) + (\beta' - \beta)$ gives an excellent straight line with a gradient close to unity and a correlation coefficient of 0.98 (14).

The success of this modified "patterns" treatment shows that orientation of free-radical addition to olefins is governed to a major extent by familiar polar forces which are given quantitative expression by the Hammett equation. The correlations indicate that both the polarity of the olefin and the polar character of the radical are important.

Unfortunately, Taft steric substituent constants E_s are available for only a few radicals (47), but for these few log OR correlates well with E_s. The diameters of the radicals were calculated from the covalent atomic radii, and significant correlations were found between radical diameters so calculated and log OR for each set of results (42). So, besides polar effects, radical orientation is governed to some extent by classical steric hindrance.

Secondary Deuterium Kinetic Isotope Effects

Addition of trifluoromethyl and methyl radicals (48), and several other radical species (49) to deuteriated olefins gives rise to an inverse isotope effect k_D/k_H of between 1.05 and 1.20 depending on the radical and substrate. Earlier attempts to interpret the significance of this isotope effect were based on Streitwieser's suggestion (50) that its main cause was the frequency increase in one of the C-H vibrations of the olefin when the carbon atom was changed from sp^2 to sp^3 hybridization on addition of the radical.

The transition state theory calculations of Safarik and Strausz (49) (51) show that this simple interpretation is not valid. They find that the most important contribution to the isotope effect is the net gain in the isotope-sensitive normal vibrational modes during passage from the reactants to the transition state. For the addition of a polyatomic radical to ethylene the gain in normal modes is six, of which one will coincide with the reaction coordinate. From the remaining five, at least one, the CH_2 twist, will always be isotope-sensitive and will generate an appreciable isotope effect. In the ethylene molecule there is a substantial barrier to the twist motion, which is considerably lowered in the transition state. The results obtained with the model calculations showed that reproduction of the experimental isotope effect implied a significant link between the radical and the olefin, and consequent deviation from the planar ethylene structure in the transition state (51).

Model Calculations of A-factors by Transition State Theory

The A-factors for addition of $CF_3\cdot$ (39) and $CH_3\cdot$ (40) radicals to fluoro-alkenes have been calculated by transition state theory using the expression derived by Herschbach, Johnston, Pitzer and Powell (52):

$$A_2 = 1.26\times10^{13}T^2\exp(\Delta S^{\ddagger}/R)\ \mathrm{cm^3mol^{-1}s^{-1}}$$

The fundamental vibration frequencies of ethylene and its fluorinated derivatives are well established, and those of the two radicals are known from matrix isolation studies. Two models of the transition state were considered. In the first, the vibration frequencies of the transition state were based on those of the radical and alkene with a minimum interaction thus giving a loose model. In the second, the vibration frequencies of the transition state were based on those of the product radical which were in turn estimated from fluoro-substituted propanes; thus giving a tight model.

The results of these calculations have already been referred to. The calculations predict only small variations in the A-factors, as is observed experimentally. Both models predict the highest A-factor for addition to ethylene and lower A-factors for addition to carbons carrying one or two fluorine substituents, in agreement with experiment. Generally the tight transition state gives a better agreement with the experimental order of the A-factors. The implication being that significant bond formation has occurred in the transition state, accompanied by appreciable loosening of the alkene double bond. This is essentially in agreement with the conclusions of Safarik and Strausz from consideration of the secondary deuterium isotope effect. Both models also predict higher A-factors for addition to carbon in tetrafluoroethylene than to other doubly fluorinated carbon atoms in unsymmetrical olefins. This is a consequence of the relatively low entropy of tetrafluoroethylene, due to its high symmetry number, and again agrees with the experimental observations for both radicals.

Molecular Orbital Treatment

The experimental activation energies may be compared with charge densities, q_μ, free valence indicies F_μ, and localization energies L_μ calculated for each atom μ of the olefin by HMO theory. For all the radicals of table IV there is little correlation with either free valence or charge density, but rather scattered correlations can be observed for most of the fluoroalkyl radical activation energies with localization energy. The activation energies determined by Szwarc and co-workers for trifluoromethyl radical addition to alkyl-substituted olefins (20, 21) also show a reasonable correlation with L_μ (53). Better correlation with L_μ is to be expected since it represents an attempt to model the transition state, unlike q_μ or F_μ which simply refer to the ground state of the olefin. The chief drawback of these reactivity indicies is that they are calculated from the properties of the alkenes alone, whereas the experimental data clearly shows that the nature of the attacking radical is also highly significant. In order to allow for the manifest polar character of the reaction we tried including a term δQ_μ which represents the net atom charge (54):

$$E(\text{obs}) = A(L_\mu + B\delta Q_\mu)$$

When B is set equal to 0.5 the correlations for $CF_3\cdot$, $CCl_3\cdot$, i-$C_3F_7\cdot$ and other fluoroalkyl radicals are a marked improvement over plots involving simple localization energies.

However this type of expression can never account for the reversal in orientation observed for the less electrophilic radicals. Even if the constant B changes sign, the proposed expression will still not give a good correlation with experimental methyl radical activation energies.

More sophisticated calculations of the semi-empirical SCF and ab initio types have been used to plot partial potential energy surfaces for methyl radicals with alkenes (55) (56) (57). They suggest that the lowest energy approach of the radical is from above the plane of the alkene directly above one or other of the carbons. The geometry of the transition state was found to be little changed from that of the reactants. So far this type of computation has not been attempted with fluoroalkyl radicals.

Literature Cited

1. Banus, J., Emeleus, H. J., and Haszeldine, R. N., J. Chem Soc., (1950), 3041.
2. Fass, R. A., and Willard, J. E., J. Chem. Phys., (1970), 52, 1874.
3. Whytock, D. A., and Kutschke, K. O., Proc. Roy. Soc. A, (1963), 306, 503; Gandini, A., and Kutschke, K. O., Proc. Roy. Soc. A, (1963), 306 511; Gandini, A., Whytock, D. A., and Kutschke, K. O., Proc. Roy. Soc., A, (1963), 306, 529, 537, 541.
4. Harris, J. F., and Coffman, D. D., J. Amer. Chem. Soc., (1962), 84, 1553.
5. Knipe, R. H., Gordon, A. S., and Ware, W. R., J. Chem. Phys. (1969), 51, 840.
6. Gordon, A. S., J. Chem. Phys., (1962), 36, 30.
7. E-Chung Wu and Rice, O. K., J. Phys. Chem., (1968), 72, 542.
8. Chamberlain, G. A., and Whittle, E., J. C. S. Faraday I, (1972), 68, 88,. 96.
9. Ashton, D. S., Mackay, A. F., Tedder, J. M., Tipney, D. C., and Walton, J. C., J. C. S. Chem. Comm., (1973), 496.
10. Chamberlain, G. A. and Whittle, E., J. C. S. Faraday, I, (1972), 68, 96; (1975), 71, 1978.
11. Walling, C. and Huyser, E. S., Organic Reactions, (1963), 13, 91.
12. Walton, J. C., J. C. S. Faraday I, (1972), 68, 1559.
13. Tarrant, P. and Lovelace, A. M., J. Amer. Chem. Soc., (1955), 77, 768, Tarrant, P., Lovelace, A. M., and Lilyquist, M. R., J. Amer. Chem. Soc., (1955), 77, 2783; Coscia, A. T., J. Org. Chem., (1951), 26, 2995.

14. McMurray, N., Tedder, J. M., Vertommen, L. L. T., and Walton, J. C., J. C. S. Perkin II, (1976), 63.
15. Bellas, M. G., Strausz, O. P. and Gunning, H. E., Canad. J. Chem., (1965), 43, 1022.
16. Sloan, J. C., Tedder, J. M., and Walton, J. C., J. C. S. Perkin II, (1975), 1841.
17. Sloan, J. C., Tedder, J. M., and Walton, J. C., J. C. S. Perkin II, (1975), 1846.
18. Pritchard, G. O., Venugopalan, M., and Graham, T. F., J. Phys. Chem., (1964), 68, 1786.
19. Dixon, P. S., and Szwarc, M., Trans. Faraday Soc., (1963), 59, 112.
20. Pearson, J. M., and Szwarc, M., Trans. Faraday Soc., (1964), 60, 553, 564.
21. Owen, G. E., Pearson, J. M., and Szwarc, M., Trans. Faraday Soc., (1965), 61, 1722.
22. Nonhebel, D. C., and Walton, J. C., Organic Reaction Mechanisms, (1974), 69.
23. Ayscough, P. B., J. Chem. Phys., (1956), 24, 944.
24. Ogawa, T., Carlson, G. A., and Pimental, G. C., J. Phys. Chem., (1970), 74, 2090.
25. Basco, N., and Hathorn, F. G. M., Chem. Phys. Letters, (1971), 8, 291.
26. Hiatt, R., and Benson, S. W., Int. J. Chem. Kinetics, (1972), 4, 479.
27. Okafo, E. N., and Whittle, E., Int. J. Chem. Kinetics, (1975), 7, 273.
28. Majer, J. R., Olavesen, C., and Robb, J. C., Trans. Faraday Soc., (1969), 65, 2988.
29. Bertrand, L., De Maré, G. R., Huybrechts, G., Olbrechts, J., and Toth, M., Chem. Phys. Letts., (1970), 5, 183.
30. Sangster, J. M., and Thynne, J. C. J., J. Phys. Chem., (1969), 73, 2746.
31. Arthur, N. L., and Gray, P., Trans. Faraday Soc., (1969), 65, 434.
32. Kerr, J. A., and Parsonage, M. J., "Evaluated Kinetic Data a Gas Phase Addition Reactions", Butterworths, London, 1972.
33. Sangster, J. M., and Thynne, J. C. J., Trans. Faraday Soc., (1969), 65, 2110.
34. Low, H. C., Tedder, J. M., and Walton, J. C., J. C. S. Faraday I, (1976), 72, 1300.
35. Tedder, J. M., and Walton, J. C., J. C. S. Faraday I, (1974), 70, 308.

36. Sidebottom, H. W., Tedder, J. M., and Walton, J. C., Int. J. Chem. Kinetics, (1972), 4, 249.
37. Benson, S. W., and O'Neal, H. E., NSRDS-NBS, (1967), 21.
38. Papic, M, M., and Laidler, K. J., Canad. J. Chem., (1971), 49, 549.
39. Cape, J. N., Greig, A. C., Tedder, J. M., and Walton, J. C., J. C. S. Faraday I, (1975), 71, 592.
40. Low, H. C., Tedder, J. M., and Walton, J. C., J. C. S. Faraday I, (1976), 72, 1707.
41. Tedder, J. M. and Walton, J. C., Trans. Faraday Soc., (1964), 60, 1969; (1966), 62, 1859.
42. Ashton, D. S., Shand, D. J., Tedder, J. M., and Walton, J. C., J. C. S. Perkin II, (1975), 320.
43. Sloan, J. P., Tedder, J. M. and Walton, J. C., J. C. S. Faraday I, (1973), 69, 1143.
44. Tedder, J. M., Vertommen, L. L. T. and Walton, J. C., unpublished work.
45. Tedder, J. M. and Walton, J. C., Acc. Chem. Res., (1976), 9, 183.
46. Bamford, C. H., and Jenkins, A. D., Trans. Faraday Soc., (1963), 59, 530.
47. Taft, R. W., "Steric Effects in Organic Chemistry", Ed. M. S. Newman, p. 633 Wiley, N. Y., 1956.
48. Feld, M., Stefani, A. P., and Szwarc, M., J. Amer. Chem. Soc., 84, 4451, (1962).
49. Strausz, O. P., Safarik, I., O'Callaghan, W. B., and Gunning, H. E., J. Amer. Chem. Soc., (1972), 94, 1828.
50. Streitweiser, A., Jagow, R. H., Fahey, R. C., and Suzuki, S., J. Amer. Chem. Soc., (1958), 80, 2326.
51. Safarik, I., and Strausz, O. P., J. Phys. Chem., (1972), 76, 3613.
52. Herschbach, D. R., Johnston, H. S., Pitzer, K. F., and Powell, R. E., J. Chem. Phys., (1956), 25, 736.
53. Nonhebel, D. C., and Walton, J. C., "Free Radical Chemistry" p. 236, Cambridge, 1974.
54. Tedder, J. M., Walton, J. C., and Winton, K. D. R., J. C. S. Faraday I, (1972), 68, 160.
55. Basilevsky, M. V., and Chlenov, I. E., Theor. Chim. Acta, (1969), 15, 174.
56. Hoyland, J. R., Theor. Chim. Acta, (1971), 22, 229.
57. Paterson, I., Tedder, J. M., and Walton, J. C., unpublished work.

5

Kinetics and Mechanisms of Reactions of CF, CHF, and CF_2 Radicals

DAVID S. Y. HSU,* M. E. UMSTEAD and M. C. LIN

Chemistry Division, Naval Research Laboratory, Washington, DC 20375

This chapter reviews briefly methods for the production of CF, CHF and CF_2, and in more detail, the reactions of these interesting and important radicals. Although a considerable, but not extensive, amount of work has been done on the reactions of CF_2, little of the chemistry of CF and CHF is known. This chapter also includes the preliminary results of some experiments carried out in this Laboratory on the dynamics of some of the reactions involving these radicals. These results were largely arrived at through investigations of the degree of vibrational excitation of the HF and CO reaction products, determined by HF and CO laser emission and CO laser resonance absorption measurements. The coverage of this review is restricted to the gas phase chemistry of these radicals, and does not include their addition reactions to olefins.

I. CF Radical Reactions

The presence of the CF radical often has been observed spectroscopically in the dissociation of fluorocarbons in electric discharges and in the reactions of F atoms with organic compounds (1-6). Jacox and Milligan (7) produced the radical by the vacuum-ultraviolet (VUV) photolysis of CH_3F, and were able to stabilize it in a nitrogen or argon matrix at 14°K in sufficient concentration to obtain its infrared spectrum.

Simons and Yarwood (8,9) reported that the flash photolysis of $CHFBr_2$ and $CFBr_3$ in thin-walled quartz tubes ($\lambda \geq 160$ nm) produced the CF radical. The presence of CF was detected by the transient appearances in absorption of the $X^2\Pi \rightarrow A^2\Sigma$ transition. They reported that the production of CF was greater from $CHFBr_2$ than from $CFBr_3$. In a similar experiment with the former com-

*NRC/NRL Postdoctoral Research Associate

© 0-8412-0339-7/78/47-066-128$10.00/0

pound carried out in a quartz tube by Merer and Travis (10), however, CF was not detected. Additionally, we found no significant difference between the vibrational population of the CO formed in the flash photolysis of a $CHFBr_2$-SO_2 mixture in the UV and VUV above 165 nm. The CO formed in the $CHFBr_2$ - SO_2 system was found to be considerably colder than that formed in the $CFBr_3$ - SO_2 system either in a Suprasil or in a quartz tube (see below).

Very little has been reported of the chemistry of the CF radical. In a shock tube study, Modica and Sillers (11) measured the rate of the reaction

$$CF + F + M \rightarrow CF_2 + M \qquad (1)$$

and reported that $k_1 = 6.57 \times 10^{26}\ T^{-2.85} ml^2\ mole^{-2}\ sec^{-1}$ when M = Ar. Schatz and Kaufman (12), in a study of chemiluminescence excited by atomic fluorine, observed the emission of bands of the Cameron system, $a^3\Pi \rightarrow X^1\Sigma^+$, of CO up to $v' = 5$. They speculated that these might arise from the following reaction:

$$O + CF \rightarrow CO + F \qquad (2)$$

On the basis of the known heat of formation for the CF radical, $\Delta Hf^\circ_{298} = 59 \pm 2$ kcal/mole (13-16), the exothermicity of reaction (2) is 126 ± 3 kcal/mole, which is 36 kcal/mole below CO ($a^3\Pi$, $v' = 5$).

Reaction (2) is believed to be responsible for the CO laser emission observed in the flash photolysis of a $CFBr_3$ - SO_2 mixture above 165 nm (17). Over 40 vibration-rotation lines between 2 → 1 and 14 → 13 have been identified. The laser output of this system increases very rapidly with the $SO_2/CFBr_3$ ratio and reaches a peak value at $SO_2/CFBr_3 = 2$, similar to that observed in the analogous SO_2 - $CHBr_3$ CO laser system (18). However, unlike the latter system, the power output of the former does not decrease significantly with $SO_2/CFBr_3$ ratios as high as 22, indicating that the CF + SO_2 reaction is not as rapid. Additionally, the laser output of the $CFBr_3$ system is considerably weaker than that of the $CHBr_3$ system, although the energetics of the two differ only slightly.

In order to understand the dynamics of the O + CF reaction occurring in the SO_2 - $CFBr_3$ chemical CO laser system, we carried out CO laser resonance absorption experiments to measure the vibrational energy distribution of the CO formed in the reaction. A detailed description of the flash-photolytic CO laser-probing system can be found in reference (19). Experiments were carried out using both Suprasil ($\lambda \geq 165$ nm) and quartz ($\lambda \geq 200$ nm) flash tubes for mixtures of $CFBr_3$ and SO_2 with He as a diluent. With the Suprasil tube, 10-torr samples of a 1:1:98/$CFBr_3$:SO_2:He mixture were used, whereas with the quartz tube 5-torr samples of both the 1:1:98 and 1:1:48 mixtures were used.

The initial relative vibrational populations were obtained from the absorption traces using an extrapolation method described in reference (20). The results are shown in Fig. 1. In these experiments, the appearance time of the strongest absorption occurred at about 6 μsec. Flash profiles taken with a fast photodiode show that the flash is over within 5 μsec. The secondary vibrational excitation of the CO product that might result from the $X^1\Sigma^+ \rightarrow a^3\Pi$ absorption would therefore not occur to any significant extent. This is supported by the fact that the results obtained from both the quartz and Suprasil tubes agree closely.

The vibrational population of the CO decreases monotonically with a slight but distinct drop in population near v = 12, as the vibrational energy of the CO increases. This is in sharp contrast to the bell-shaped distribution observed in the O + CS reaction (23-25), which could be accounted for by an impulsive model (26) indicating the possibility of the absence of a significant well in the triplet OCS intermediate. In the O + CF reaction, however, the observed CO population was found to lie close to that predicted by a simple statistical model (20), taking the total available reaction energy, $E_{tot} \cong 126 + 2.5RT \cong 128$ kcal/mole. On the basis of this model, the relative vibrational population of CO can be estimated from the following simple expression (20):

$$\frac{N_v}{N_o} = \frac{\Sigma P(E_{tot} - E_v)}{\Sigma P(E_{tot})} \qquad \text{(I)}$$

where E_v and N_v are the vibrational energy and population of CO at the vth level, and $\Sigma P(E)$, the total energy level sum of the FCO intermediate with both CO and CF stretches excluded. The CO stretch corresponds to the mode of vibration leading to the product CO vibrational excitation, and the CF stretch corresponds to the reaction coordinate. Equation (I) can be evaluated by either the direct-count method or the approximation of Whitten and Rabinovitch (27):

$$\frac{N_v}{N_o} = \left(\frac{E_{tot} - E_v + aE_z}{E_{tot} + aE_z}\right)^{s\ddagger} = \left[1 - E_v/(E_{tot} + aE_z)\right]^{s\ddagger} \qquad \text{(II)}$$

where E_z is the zero point energy and $s\ddagger$ is the effective number of vibrations ($s\ddagger$ = 3N-7 for a linear complex, and 3N-8 otherwise), and "a" is a correction factor which has a value between 0 and 1, depending on the amount of energy, $E_{tot} - E_v$. For simple, very exothermic reactions, such as O + CF, $E_{tot} \gg aE_z$, and the classical expression (20):

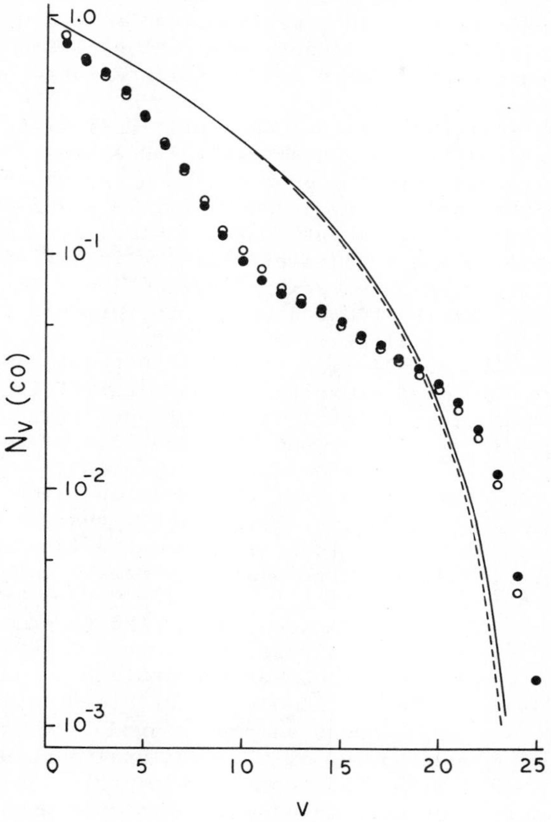

Figure 1. Vibrational distribution of CO formed in the flashed-initiated $CFBr_3$–SO_2 system. Open circles = 1:1:98/$CFBr_3$:SO_2:He mixture in a quartz tube, filled circles = the same mixture in a Suprasil tube; solid and broken curves = statistical distributions based on Equations (II) and (III), respectively.

$$N_v/N_o = (1 - E_v/E_{tot})^{s\ddagger} \equiv (1 - f_v)^{s\ddagger} \qquad \text{(III)}$$

is adequate. This is shown by the results given in Fig. 1 for the O + CF reaction. In this calculation, $s^{\ddagger} = 2$ was used for Eqs. (I)-(III), assuming a linear F---CO dissociation complex. The direct count method based on Eq. (I) predicted exactly the same distribution.

The fact that the observed CO vibrational distribution is nearly statistical, with partial inversion between all levels, is not inconsistent with the presence of a rather deep well. The depth of this well is about 30 kcal/mole based on ΔHf°(FCO) $\cong$ -34 kcal/mole (21) and about 60 kcal/mole according to the value ΔHf° (FCO) $\cong$ -66 kcal/mole(22). The latter seems to be more in line with the observed CO distribution. The effect of the stability of the FCO radical on the product CO vibrational energy distribution will be examined in the future.

The cause of the slight decrease in population near v = 12 is not clear. The exothermicity of reaction (2) is insufficient to excite either CO or F to higher electronic states, and the slight energy difference between $^2P_{1/2}$ and $^2P_{3/2}$ of the ground electronic state of the F atom is not expected to affect dynamically the production of CO from the reaction which proceeds through a complex. It is thus quite likely that the appearance of colder CO populations at lower levels may arise from a side reaction that generates vibrationally colder CO molecules. The most likely candidate for this is O + CFBr → CO + BrF(or F + Br), due to the incomplete photodetachment of the Br atoms from $CFBr_3$. The energetics of this reaction are not known, but are probably much higher than those of the analogous reaction, O + CF_2, to be discussed later. Another possible reaction, O + $CFBr_2$, probably does not have enough energy to produce vibrationally excited CO. The overlapping of a vibrationally colder population from O + CFBr with that from reaction (2), a hotter, near statistical distribution, could account for the deviation occurring at the lower levels of the observed distribution.

The absence of a complete population inversion in the flash-initiated $CFBr_3$ - SO_2 system accounts for the weak CO laser intensity, compared with that detected in the $CHBr_3$ - SO_2 flash (18). In the $CHBr_3$ - SO_2 system, the primary pumping reaction was assumed to be O + CH, which probably proceeds through the HCO complex with a shallower depth of about 20 kcal/mole. The dynamics of this highly exothermic reaction (ΔH° = -176 kcal/mole) are now under investigation.

II. CHF Radical Reactions

Little is known of the chemistry of the CHF radical, largely because of the lack of clean methods for its production. Merer and Travis (10) obtained the absorption spectrum of CHF, produced by the flash photolysis of $CHFBr_2$ in a quartz vessel, and made a rotational analysis of the bands. The spectrum consisted of a single progression of complex bands in the region 430 - 600 nm. They concluded that the ground state of CHF was a singlet, similar to that of CHCl. Although Simons and Yarwood (8,9) reported the production of CF from the flash photolysis of $CHFBr_2$ based upon its absorption spectrum, Merer and Travis did not observe its presence. Other unidentified absorption bands were seen, but none of these were attributed to CF. Simons and Yarwood had no quantitative data for CF production, and since the absorption spectrum of CHF was not known at that time, its presence would not have been observed (28). It is likely that CF is only a minor product of this dissociation.

Both the visible-UV and the IR absorption spectra of CHF (as well as CF and CH_2F) were obtained from the VUV photolysis of CH_3F in an argon or nitrogen matrix at 14°K by Jacox and Milligan (7). They also found that the reaction of carbon atoms, produced by the photolysis of cyanogen azide isolated in an argon matrix, with HF trapped in the matrix, led to the formation of sufficient amounts of CHF for IR spectroscopic analysis. Their results confirmed Merer and Travis' finding that the ground electronic state of the CHF radical is singlet.

Tang and Rowland produced CTF radicals by the reaction of energetic tritium atoms from nuclear recoil with fluorinated compounds such as CH_3FCl, CHF_3, and CH_2F_2 (29,30), and investigated the reaction of CTF from CH_2F_2 with olefins in the gas phase (30):

$$T^* + CH_2F_2 \rightarrow CHTF_2^* + H$$

$$CHTF_2^* \rightarrow CTF + HF$$

$CTF + C_2H_4 \rightarrow$  TF^*

The reaction of CTF with olefins was found to be completely stereospecific; the alkyl groups in the fluorocyclopropane products maintained the same orientations to each other as were present in the reacting olefins. In all cases where the formation of isomers was possible by the single-step addition of CTF, the syn and anti isomers were found in equal quantities. This complete stereospecificity, along with the insensitivity of the presence of O_2 and a large amount of inert gas on the reaction, was taken as evidence that the reacting species was singlet CTF. From an experiment in which CTF reacted with C_2H_4 in the presence of excess O_2, it was found that the reactivity of O_2 with CTF is $<$ 0.2 times that of C_2H_4.

Rowland et al. (31,32) also prepared $CH^{18}F$, along with $CF^{18}F$, by the reaction of energetic ^{18}F atoms from nuclear recoil with CH_2F_2, and studied its reaction with hydrogen halides (31). The formation of the carbenes proceeded via a mechanism similar to the aforementioned T-atom reaction. The reaction of $CH^{18}F$

with HX was found to take place by a direct insertion mechanism:

$$CH^{18}F + HX \rightarrow CH_2{}^{18}FX \qquad (3)$$

Evidence for the direct insertion was largely based on the fact that the yield of $CH_2{}^{18}FCl$ was independent of the presence of O_2, while the yields of products arising from monoradicals were strongly influenced by its presence.

It was estimated that the exothermicity of reaction (3) is about 74 kcal/mole when X = I, and about 69 kcal/mole when X = Cl. These values were based upon the estimation of $\Delta H_f°$ for CHF, which is not available, by taking the arithmetic mean of the heats of formation of CH_2 and CF_2: 25 kcal/mole. This value is probably too low, as will be discussed later. A comparison of these excitation energies with the assumed C-I and C-Cl bond energies indicated that CH_2FCl should be completely stable against C-Cl bond breakage, but that the excitation energy of CH_2FI was about 19 kcal/mole in excess of the activation energy for C-I bond dissociation. This was found to be consistent with the experimental results based on the production of $CH_3{}^{18}F$ (31).

The reaction of the CHF radical with $O(^3P)$ (33), O_2 (34), and NO (35) has been shown to generate stimulated HF emissions. In these studies, the CHF radical was produced by the successive photodetachment of Cl atoms from $CHFCl_2$ in a Suprasil tube (33) or Br atoms from $CHFBr_2$ in a quartz tube (10):

$$CHFCl_2 + h\nu(\lambda \geq 165\ nm) \rightarrow CHF + 2Cl$$

$$CHFBr_2 + h\nu(\lambda \geq 200\ nm) \rightarrow CHF + 2Br$$

The vibrationally excited HF molecule is believed to be formed primarily by the following four-centered elimination processes involving chemically activated intermediates:

$$O(^3P) + CHF \rightarrow HFCO^* \rightarrow HF^\dagger + CO^\dagger \qquad (4)$$

$$\Delta H_4° \cong -190\ kcal/mole$$

$$CHF + O_2 \rightarrow HF\dot{C}O\dot{O} \rightarrow FCOOH \qquad (5)$$

$$\rightarrow HF^\dagger + CO_2$$

$$\Delta H_5° \cong -198\ kcal/mole$$

$$CHF + NO \rightarrow HFCNO \rightarrow HF^\dagger + CNO \qquad (6)$$

$$\Delta H_6° = ?$$

where "†" stands for vibrational and "*" for electronic excitation.

Neither the kinetics nor the mechanisms of these reactions are known. The assumption that these reactions take place via long-lived intermediates was based solely on the observed HF and/or CO vibrational energy distributions, with the aid of product gas analysis. The exothermicity of reaction (6) is not available due to lack of information on the CNO radical. According to the observed HF laser intensity and the vibrational population of HF (see Fig. 2), $\Delta H_6°$ is probably very close to $\Delta H_5°$, which was estimated to be about -198 kcal/mole. The exothermicities of reactions (4) and (5) were calculated using $\Delta Hf°(CHF) = 39 \pm 4$ kcal/mole, estimated from the theoretical bond dissociation energies, De(CH-F) = 120 and De(CF-H) = 83 kcal/mole by Staemmler (36). These values are probably reliable because a similar estimate with Staemmler's calculated value for De(CF-F) gave rise to $\Delta Hf°(CF) = 61$ kcal/mole, which agrees closely with the experimental value, 59 ± 2 kcal/mole given earlier. A lower value, $\Delta Hf°(CHF) \cong 25$ kcal/mole, evaluated by taking the simple arithmetic mean of the heats of formation of ground state CH_2 and CF_2 (31), is questionable inasmuch as the ground electronic state of CH_2 is triplet, whereas CHF and CF_2 are both singlet. The use of the singlet energy, which has recently been established to be 8 ± 1 kcal/mole higher (37), would increase the value of $\Delta Hf°(CHF)$ by about 4 kcal/mole. It should be pointed out that the use of either value, $\Delta Hf°(CHF) =$ 25 or 39 kcal/mole, varies the overall exothermicities of reactions (4) and (5) by less than 10%, which does not affect our conclusion on the dynamics of these very exothermic reactions.

The HF vibrational energy distribution was measured by means of the appearance time of each vibrational-rotational laser line selectively oscillating within a cavity consisting of a grating and a highly reflective concave mirror (38,39). Assuming that v-v relaxation before the appearance of laser pulses is insignificant, the threshold times of various laser lines can be correlated with the gains and thus the relative vibrational population of the levels involved (39,40). The measured HF vibrational populations for reactions (4)-(6) are presented in Fig. 2, employing $CHFBr_2$ as the CHF radical source.

For reactions (4) and (5), whose exothermicities could be estimated, statistical models based on Eqs. (II) and (III) predict HF vibrational population distributions that agree closely with the experimental ones. This result strongly supports the proposed reaction mechanisms involving long-lived intermediates. It should be emphasized that the results of these simple calculations are not sensitive to the structure or vibrational frequencies of the intermediate assumed because the exothermicities of these reactions are so large that $E_{tot} \gg aE_z$. This is indicated by the close agreement between the value predicted by Eq. (II) and that by Eq. (III). The number of active vibrations for these non-linear intermediates was taken to be 3N - 8, with one deleted for the reaction coordinate and the other for the HF product vibrational excitation. On the basis of these statistical calculations, the HF molecule carries vibrationally about 18% of the energy available in reaction (4) and 10% in

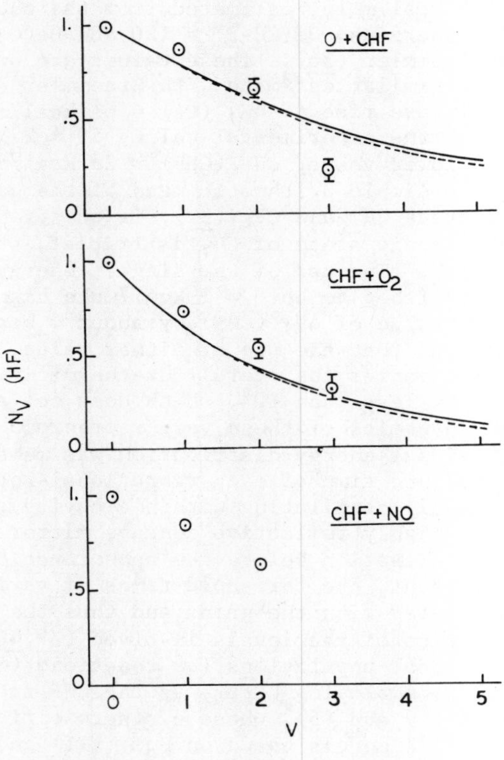

Figure 2. Vibrational distributions of the HF observed in reactions (4), (5), and (6)

reaction (5). These are, of course, much less than the values (>60%) observed in simple abstraction reactions such as F + CH_4 (41). A complex-forming reaction is therefore a less efficient laser-pumping reaction (42-45).

In addition to the major primary product HF, small amounts of F atoms were also believed to be produced from reactions (4)-(6) (34,35,46). This was suggested by the observation of weak DF stimulated emissions when D_2 was added to these systems. The production of F atoms from these reactions is not unexpected in view of the large excess of vibrational energies carried by these intermediates. In reaction (4), for example, the HFCO intermediate possesses as much as 185 kcal/mole of internal energy, which is more than sufficient to dissociate either the CH (~90 kcal/mole) or the C-F (~120 kcal/mole) bond. On account of the weakness of the DF emissions and the only slight attenuation of HF emissions when D_2 was added (46), the contribution of the secondary, but rather effective reaction, F + RH → $HF^{\dagger}$ + R (R = $CFCl_2$ or $CFBr_2$), to the HF emission is therefore not significant. This is manifested by the observed near-statistical vibrational distributions shown in Fig. 2.

The vibrational energy content of the CO formed in reaction (4) has also been measured with the CO laser resonance absorption method discussed earlier. The results obtained from photolyzing 10 torr of 1:1:98/SO_2:$CHFBr_2$:He and 1:1:20:78/SO_2:$CHFBr_2$:H_2:He mixtures in a Suprasil tube are shown in Fig. 3. In the second mixture, H_2 was added to test the possibility of the production of CO($a^3\pi$) from reaction (4). H_2 has been shown to relax CO($a^3\pi$,v=0) very effectively (47). If CO($a^3\pi$) is produced initially in significant amount, the addition of H_2 should be expected to change the population of the initial ground electronic state CO noticeably. The results presented in Fig. 3, however, do not show such an effect; this rules out the possibility that CO($a^3\pi$) is formed to any significant extent in reaction (4).

The CO formed in the flash-photolysis of both mixtures mentioned above was found to be vibrationally excited up to v = 24, which corresponds to about 127 kcal/mole of vibrational energy. A similar statistical calculation based on Eq. (II) shows that the observed vibrational population distribution is considerably colder than the statistical. Although this seems to be inconsistent with the observed near statistical HF vibrational population shown in Fig. 2, it is, however, not unexpected in view of the presence of the other reaction channel leading to the production of F atoms discussed previously. Since the ground electronic state of the CHF radical is singlet, the spin-conservation rule (if it is valid here) suggests the following two possible reaction paths for the O(^{3}P) + CHF reaction:

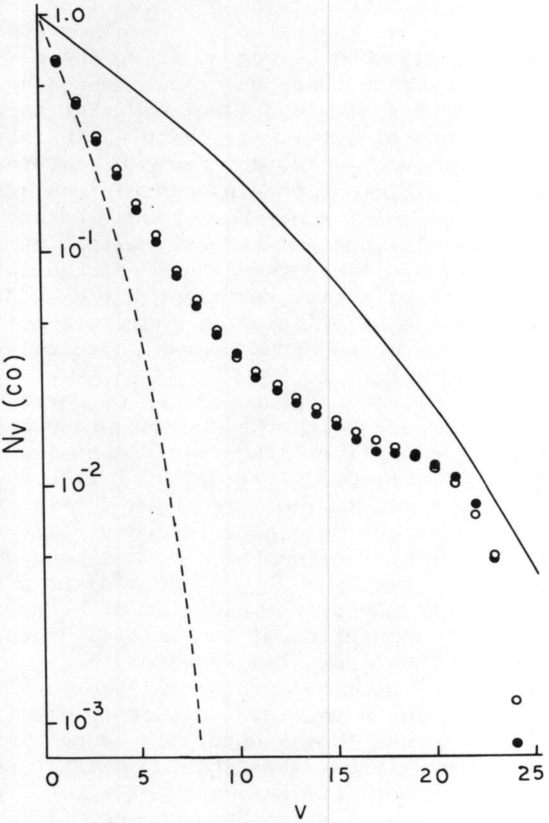

Figure 3. Vibrational distribution of CO produced from Reaction 4. Open circles = 1:1:98/$CHFBr_2$:SO_2:He in a Suprasil tube, filled circles = 1:1:20:78/$CHFBr_2$:SO_2:H_2:He flashed in the same tube; solid and broken curves = statistical distributions based on Equation (III) for Reaction 4a and Reaction 4b, respectively.

$$O(^3P) + {}^1CHF \xrightarrow{4} {}^3HFCO^{*\dagger}$$

$${}^3HFCO^{*\dagger} \xrightarrow{a} {}^1HFCO^{\dagger} \rightarrow HF^{\dagger} + CO^{\dagger} \quad (4a)$$

$$\Delta H_{4a}^{\circ} \cong -190 \text{ kcal/mole}$$

$${}^3HFCO^{*\dagger} \xrightarrow{b} H + FCO^{\dagger} \rightarrow H + F + CO^{\dagger} \quad (4b)$$

$$\Delta H_{4b}^{\circ} \cong -55 \text{ kcal/mole}$$

Because of the lack of information on the exact identity of the HFCO molecule at such a high energy, we denote the triplet vibronically excited HFCO molecule by ${}^3HFCO^{*\dagger}$ and the ground electronic vibrationally excited HFCO molecule by ${}^1HFCO^{\dagger}$.

If we assume that the O + CHF reaction takes place exclusively via (4b), then Eq. (III) predicts a much colder distribution with v_{max} = 9, compared with the observed v_{max} = 24. Evidently, the observed distribution, which lies between the two statistical distributions attributed to reactions (4a) and (4b) as noted in Fig. 3, may actually result from the simultaneous occurrence of these two reaction channels. This is consistent with the presence of F atoms in the flash-initiated SO_2-$CHFX_2$ (X=Cl,Br) systems. If the production of HF from the secondary abstraction reaction, F + $CHFX_2$, does not take place significantly before lasing, the measured HF population using the laser emission method should represent primarily that of reaction (4a). This is believed to be the case, inasmuch as the rate of F + $CHF_3 \rightarrow HF + CF_3$, which is probably comparable to that of the analogous F + $CHFX_2 \rightarrow HF + CFX_2$ reaction, was recently reported to be only about 10^{-2} times that of F + H_2 (48).

III. CF_2 Radical Reactions

Formation of CF_2. The CF_2 radical was initially detected by means of its emission (49) and absorption spectra (50-52) when fluorocarbons were passed through electrical discharges or high temperature furnaces (53). It is also formed by the pyrolysis at 600°C (54) or the flash photolysis (55,56) of CF_3COCF_3, and its transient absorption spectrum has been observed in the flash photolysis of $CF_2ClCOCF_2Cl$, $CF_2ClCOCFCl_2$, and CF_3COOH (56-58). The flash photolysis in quartz of the difluorohalomethanes: CF_2Br_2 (56, 58-60), CF_2HBr (56,58), CF_2Cl_2 (58,59) and CF_2HCl (59) all lead to the elimination of CF_2. Of this series, CF_2 is produced most efficiently from CF_2Br_2 (58,59).

The first step in the pyrolysis of CF_2HCl (61-64) and CF_2HBr (65) involves the elimination of HX to yield CF_2. The overall reaction proceeds as follows:

$$CF_2HX \rightleftarrows CF_2 + HX \quad (7,-7)$$

$$2\,CF_2 \rightarrow C_2F_4 \quad (8)$$

The kinetics of the reaction have been measured by product analyses in both flow and static systems, and the Arrhenius parameters are listed in Table I, along with those from a shock tube study of CF_3H decomposition (66). The values listed in the table from Reference (62) are probably too large (67,68).

Table I. Arrhenius Parameters for the Reaction:
$CF_2HX \rightarrow CF_2 + HX$

X	T (°K)	log A (sec^{-1})	E (kcal/mole)	Ref.
Cl	806-1023	13.84	55.8	62
Cl	943-1023	12.36 ± 0.56	51.4 ± 2.5	63
Cl	727-796	12.6 ± 0.44	52.8 ± 1.5	64
Br	710-796	14.33 ± 0.32	55.6 ± 1.1	65
F	1200-1600	11.42	47.0	66

Tetrafluoroethylene dissociates reversibly into CF_2 at elevated temperatures. The kinetics of the dissociation, as well as the equilibrium constant, have been measured in shock tubes by Modica and LaGraff (69,70) and by Carlson (71). Kushina et al. (67) investigated the pyrolysis of C_2F_4 at 550° - 670°C and obtained an expression for the rate constant. The equilibrium constant has also been determined by mass spectroscopy (72), but there is considerable disagreement among the reported values (71). The Hg-sensitized photolysis (73-76) of C_2F_4 as well as its unsensitized photolysis in quartz (56,59,68) is an efficient means of producing CF_2. Absorption by CF_2 has also been observed in the flash photolysis in quartz of CF_2CFCl and CF_2CCl_2 (77).

Higher fluoroalkenes and difluorocyclopropanes are also sources of CF_2. Dalby (59) obtained this radical by the flash photolysis of perfluoropropylene. The kinetics of CF_2 formation from the thermal decomposition of perfluorocyclopropane at 526° - 549°K were reported by Atkinson and McKeagan (78). Birchall, Haszeldine and Roberts (79) investigated the thermal decomposition of a series of gem-difluorohalocyclopropanes as sources of CF_2. The cyclopropanes studied were placed in the following approximate order of decreasing stability toward CF_2 elimination:

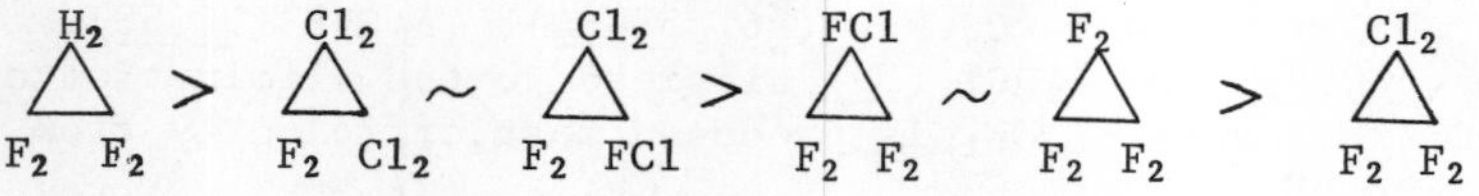

although the differences between them were not great.

The reaction of $O(^3P)$ with C_2F_4 produces CF_2 in the primary step of the reaction (77, 80-86):

$$O(^3P) + C_2F_4 \rightarrow CF_2 + CF_2O \quad (9)$$

In addition to chemical evidence for its presence as an intermediate in this reaction, both its absorption (77,85) and mass spectra (86) have been detected. Absorption by CF_2 was also observed in the reaction of $O(^3P)$ with CF_2CH_2 (s), CF_2CHCl (w), and CF_2CFCl (vw) (85). The symbols following the formulas refer to the relative strength of CF_2 absorption.

Mahler (87) found that CF_2 can be produced conveniently by the dissociation of trifluoromethylfluorophosphoranes at about 100° - 120°C. The elision of CF_2 occurs stepwise and reversibly:

$$(CF_3)_3PF_2 \rightleftarrows (CF_3)_2PF_3 + CF_2$$
$$(CF_3)_2PF_3 \rightleftarrows CF_3PF_4 + CF_2$$
$$CF_3PF_4 \rightleftarrows PF_5 + CF_2$$

These phosphoranes could be decomposed in the presence of other reagents to study the reactions of CF_2.

Another very convenient and exceptionally clean source of CF_2 is the photolysis or pyrolysis of difluorodiazirine, reported by Mitsch (88,89):

$$CF_2N_2 \xrightarrow[\text{or } \Delta]{h\nu} CF_2 + N_2$$

This process is very suitable for producing CF_2 for the study of its reactions or for synthetic purposes, as the only other major product of the decomposition is N_2.

Cavell, Dobbie and Tyerman (90) found that the single step elimination of CF_2 occurs in the flash photolysis of trifluoromethylphosphino compounds. The compounds $(CF_3)_3P$, $(CF_3)_2P$-$P(CF_3)_2$, $(CF_3)_2PX$ and CF_3PX_2 (X = F, Cl or H) all yielded CF_2, based upon their observation of its absorption spectrum.

Rowland and co-workers (31,32,91) prepared CF_2 at radioactive tracer levels by the reaction of energetic ^{18}F atoms from nuclear recoil with various molecules, including CF_4, CHF_3, CH_2F_2, and C_2F_4, followed by secondary decomposition of the vibrationally excited ^{18}F-labeled precursors. The reaction with CH_2F_2, which was useful for the production of both CF_2 and CHF, proceeds as follows:

$$^{18}F + CH_2F_2 \rightarrow CH_2F^{18}F^{\dagger} + F$$
$$CH_2F^{18}F^{\dagger} \rightarrow CH^{18}F + HF$$
$$^{18}F + CH_2F_2 \rightarrow CHF_2{}^{18}F^{\dagger} + H$$
$$CHF_2{}^{18}F^{\dagger} \rightarrow CF^{18}F + HF$$

Electronic State of CF_2. The CF_2 radical in its ground state exhibits an unusually low reactivity compared to other carbon diradicals, such as CH_2 and CCl_2, due to the strong interaction of the electronegative F atoms with the free electon pair. It has been shown spectroscopically (92,93) to have a singlet electronic ground state, which is consistent with the stereospecificity of its addition reactions with olefins (89). Simons has estimated that the first excited triplet should lie about 45 kcal/mole above the ground state (94). Staemmler deduced a value of 47 kcal/mole by ab initio calculations (36).

No direct evidence for the formation of triplet CF_2 has been reported. All the previously mentioned processes producing CF_2 appear to give rise to the singlet, with the probable exception of the $O(^3P) + C_2F_4$ reaction and perhaps CF_2 production in a glow discharge containing CF_4 (95). Heicklen and coworkers (80,83,84,95) have obtained indirect evidence for the participation of triplet CF_2 in the $O(^3P) + C_2F_4$ reaction, which will be discussed in more detail later. The Hg-sensitized photolysis of C_2F_4 apparently gives rise to singlet CF_2 only (74,76,95). This is consistent with the energetics of the system. Thus the following reactions reported for CF_2, with the exception noted, are believed to be those of the ground state singlet.

Reactions of CF_2. It was noted early in the investigation of the chemistry of CF_2 that the recombination of this diradical took place extremely slowly (50). The kinetics of the recombination reaction have been investigated by several workers. Dalby (59) and Tyerman (68) measured the recombination rates at 300° - 600°K by following the decrease in absorption of CF_2 produced by flash photolysis as a function of time. Dalby obtained the rate equation:

$$k_8 = (7.5)10^9 T^{1/2} \exp(-1200/RT) \text{ ml mole}^{-1} \text{ sec}^{-1},$$

and Tyerman:

$$k_8 = (5.0 \pm 1.0)10^9 T^{1/2} \exp(-400 \pm 100/RT) \text{ ml mole}^{-1} \text{sec}^{-1}.$$

Their results were generally in good agreement, although different values for the activation energy were obtained. The disagreement is apparently due to differences in the evaluation of the extinction coefficient (ϵ) for CF_2. Tyerman attributed the higher activation energy obtained by Dalby to his neglect to take into account the variation of ϵ with temperature. Edwards and Small (62), by the analysis of products in the CHF_2Cl pyrolysis, reported a rate constant of 1.74×10^{13} ml mole^{-1} sec^{-1}, with an assumed zero activation energy. This constant appears to be much too large in comparison with other values.

Modica and LaGraff studied the C_2F_4 - CF_2 dissociation in excess N_2 behind incident shock waves over the temperature range of 1200° to 1600°K (70). They found little temperature dependence for the recombination reaction and obtained the rate law:

$$k_8 = (4.1)10^{38} T^{6.36} \exp(-18,400/RT) \text{ ml}^2 \text{ mole}^{-2} \text{ sec}^{-1}.$$

The reaction of CF_2 with C_2F_4:

$$CF_2 + C_2F_4 \rightarrow \text{cyclo-}C_3F_6 \qquad (10)$$

was investigated by Atkinson (73) and by Cohen and Heicklen (75,82) in the Hg-sensitized photolysis of C_2F_4, by Lenzi and Mele (96) in the pyrolysis of C_2F_4O, and by Atkinson and McKeagan (78) in the pyrolysis of cyclo-C_3F_6. In these studies, the rate of CF_2 addition to C_2F_4 was compared with its rate of recombination, and values of k_{10}/k_8 were obtained which agreed within a factor of 2. These values have been compared by Tyerman (68), who, by the use of his value for k_8, obtained the following expression for k_{10}:

$$k_{10} = 8.7 \times 10^7 T^{1/2} \exp(-6400 \pm 1300/RT) \text{ ml mole}^{-1} \text{ sec}^{-1}.$$

Mahler noted that CF_2 adds to HCl to give CHF_2Cl (87). The kinetics of the reaction were measured in studies of the pyrolysis of CHF_2Cl (61-64), and the Arrhenius parameters obtained are listed in Table II. Also listed are values found for the analogous HBr reaction, obtained from the pyrolysis of CHF_2Br (65). The temperature ranges investigated

Table II. Arrhenius Parameters for the Reaction: $CF_2 + HX \rightarrow CF_2HX$

X	log A (ml/mole · sec)	E (kcal/mole)	Ref.
Cl	11.35	6.2	62
Cl		15±5	63
Cl	11.33	12.1±2.7	64
Br	11.33	9.6±1.3	65

are the same as those given in Table I. Smail and Rowland (91) studied the reactions of hydrogen halides with $CF^{18}F$ produced by reactions of energetic ^{18}F atoms. They found that at 10° - 15°C, HI was about 70 times as efficient as HBr, which in turn was about 50 times more efficient than HCl in scavenging $CF^{18}F$ from the reaction system.

The kinetics and mechanism of the reaction of CF_2 with NO was investigated by Modica (97) in shock tube experiments. He reported that the reversible reaction:

$$CF_2 + NO \rightleftarrows CF_2NO \qquad (11)$$

took place below 2500°K, and above this temperature, the reaction proceeded farther by:

$$2\ CF_2NO \rightarrow 2\ CF_2O + N_2$$

$$CF_2NO + NO \rightarrow CF_2O + N_2O$$

An expression for the rate constant for reaction (11) was reported. However, Burks and Lin pointed out that the activation energy reported was unreasonably high, and that the relative magnitude of the A factors for the forward and the reverse reactions were not consistent (98). They proposed the following mechanism, based upon a HF laser emission study and mass spectral analysis of the products:

$$CF_2 + NO \rightarrow CF_2O + N$$

$$CF_2 + N \rightarrow FCN + F$$

$$CF_2 + O \rightarrow CO + 2F$$

with O being generated by the well-known N + NO reaction.

Biordi, Lazzara, and Papp studied the reactions of CF_2 in CF_3Br-inhibited methane flames, and reported some kinetic data for radical reactions at flame temperatures (99).

The reaction of ground state CF_2 with O_2,

$$CF_2 + O_2 \rightarrow CF_2O_2 \qquad (12)$$

is slow. Dalby (59) reported an upper limit for k_{12} of 10^7 ml mole^{-1} sec^{-1}, since the lifetime of CF_2 in his experiments was independent of O_2 pressures to 110 torr. Tyerman deduced a value of ~1.3×10^4 ml mole^{-1} sec^{-1} based upon the competition of reactions (10) and (12) for CF_2 (77). Modica (69) studied the CF_2 oxidation reaction in a shock tube and found it to be first order in both CF_2 and O_2. The rate constant was given by:

$$k_{12} = 2.92 \times 10^{10} T^{1/2} \exp(-13,300/RT)\ \text{ml mole}^{-1}\ \text{sec}^{-1}.$$

Heicklen and co-workers in their investigation of the C_2F_4 + $O(^3P)$ (produced by Hg-sensitized N_2O dissociation) reaction, observed that the yield of CF_2O was enhanced when O_2 was added to the reacting system (80), and that the yield of cyclo-C_3F_6 increased under conditions in which it should have decreased if singlet CF_2 ($^1CF_2$) were produced (95). They attributed both these facts to the formation of triplet CF_2 ($^3CF_2$) in the reaction, as should be predicted by the spin-conservation rule. However, it was also shown that the reactivity toward C_2F_4 of the CF_2 obtained was identical with that of $^1CF_2$ (82). It was also postulated that $^3CF_2$ could undergo self-annihilation:

$$2\,^3CF_2 \rightarrow C_2F_4{}^* \rightarrow 2\,^1CF_2. \qquad (13)$$

Mitchell and Simons (85) studied the $O(^3P) + C_2F_4$ reaction by producing the O atoms by the flash photolysis of NO_2 in Pyrex and measuring the CF_2 absorption. They found only bands of $^1CF_2$ in the ground vibrational state. No other transient species were detected in the spectral range of 210-600 nm. There was no delay in the appearance time of the CF_2 bands as compared with its unsensitized production from C_2F_4 with light of ≥ 170 nm, which indicated that if $^3CF_2$ is a primary product, its lifetime must be not greater than 10^{-6} sec.

Tyerman (77) obtained results for this reaction that supported those of Mitchell and Simons. He also observed that in the flash photolysis experiments, the singlet CF_2 appeared more rapidly than triplet-triplet annihilation could account for. It was proposed that the rapid formation of singlet CF_2 could be explained if $^3CF_2$ is relaxed by NO_2 and NO at a rate greater than 2×10^{13} ml $mole^{-1}$ sec^{-1} (which is equivalent to 1/10 the collision rate). Experiments with 700 torr O_2 added gave an upper limit for the reaction rate of $^3CF_2$ with O_2 of 6×10^8 ml $mole^{-1}$ sec^{-1}, if less than 5% of the $^3CF_2$ was scavenged in processes not returning CF_2 to the system. Johnston and Heicklen deduced an upper limit of 6×10^9 ml $mole^{-1}$ sec^{-1} on the assumption that reaction (13) occurred at every collision (84), and the true rate constant may be much smaller (77).

We have investigated the reactions of $O(^3P)$ atoms with C_2F_4 by means of the CO laser resonance absorption method in an attempt to elucidate the mechanism of this reaction. Although CO has not been reported to be an important product of this reaction, we believed, on the basis of the energetics and the appearance of strong HF laser emission from the flash-initiated SO_2-CF_2Br_2-H_2 system, that the O + CF_2 reaction should generate F atoms, accompanied by the production of CO. Since the following reactions:

$$O(^3P) + {}^1CF_2 \rightarrow CO + 2F \qquad (14a)$$

$$\Delta H^\circ_{14a} = -4 \text{ kcal/mole}$$

$$\rightarrow CO + F_2 \qquad (14b)$$

$$\Delta H^\circ_{14b} = -42 \text{ kcal/mole}$$

$$O(^3P) + {}^3CF_2 \rightarrow CO + 2F \qquad (14c)$$

$$\Delta H^\circ_{14c} = -51 \text{ kcal/mole}$$

$$\rightarrow CO + F_2 \qquad (14d)$$

$$\Delta H^\circ_{14d} = -89 \text{ kcal/mole}$$

have largely different exothermicities, the vibrational energy content of the CO molecule formed in the O + CF_2 reaction may provide information on the identity of the CF_2 radical involved. The exothermicities of the above reactions were calculated by taking ΔH_f° ($^1CF_2$) = -44.5 kcal/mole (71) and ΔE ($^3CF_2$ - $^1CF_2$) = 47 kcal/mole (36).

In our experiments, 10-torr samples of a 2:2.7:45.3/C_2F_4:NO_2:He mixture were flash-photolyzed in a Pyrex reaction tube. CO absorption was detected up to v = 12, corresponding to as much as 69 kcal/mole of vibrational energy. The measured CO vibrational distribution is presented in Fig. 4. The detection of such a highly excited CO from the O + C_2F_4 reaction, although energetically tenable, is surprising in view of Mitchell and Simons' failure to detect vibrationally excited CF_2 in the same system (85). If the CF_2 is not vibrationally excited, only one reaction given above, namely (14d), has sufficient energy to excite CO up to v = 12. To examine this possibility, we compared the observed CO vibrational distribution with those predicted by several statistical models.

Assuming reaction (14d) occurs via a CF_2O complex, with $E_{tot} = -\Delta H_{14d}^\circ + 2.5RT$ = 90 kcal/mole, Eq. (III) predicts a distribution which is colder than the observed one, as indicated by the dotted curve I in Fig. 4. In this calculation, however, $^3CF_2$ was assumed to be vibrationally cold. If one assumes that the 53 kcal/mole of reaction energy from the primary reaction:

$$O(^3P) + C_2F_4 \rightarrow C_2F_4O^* \rightarrow CF_2O^\dagger + {}^3CF_2{}^\dagger$$

is statistically distributed among all vibrational modes of the C_2F_4O complex, then the CF_2 radical should carry 53 x $(3N_{CF_2}-6)/(3N_{C_2F_4O}-6)$ or 11 kcal/mole of vibrational energy. This increases the total available energy in reaction (14d) to a maximum of 100 kcal/mole. The use of this value for the total available energy in statistical calculations based on Eqs. (II) and (III) gave rise to CO vibrational distributions that agree closely with the experimental data as shown by the solid and dashed curves, respectively, in Fig. 4. A similar calculation assuming that vibrationally excited $^1CF_2$ is involved in reaction (14b), with E_{tot} = 63 kcal/mole, led to a distribution that is much colder than that observed, as indicated by the dotted curve II in Fig. 4. In this case, the $^1CF_2$ radical carried an additional 100 x $(3N_{CF_2}-6)/(3N_{C_2F_4O}-6)$ = 20 kcal/mole of vibrational energy from reaction (9).

On the basis of these model calculations, the reaction of vibronically excited CF_2 (i.e. $^3CF_2{}^\dagger$) with $O(^3P)$ atoms seems to be the only process which has sufficient energy to account for the extent of the CO vibrational excitation observed. In hindsight, this process also seems to be the most reasonable one if the rate of this reaction and that of electronic relaxation ($^3CF_2 + M \rightarrow {}^1CF_2 + M$ by NO or NO_2) are comparable and much faster than that of $O(^3P) + {}^1CF_2$. In this way only can the vibrational as well as the electronic energy of the $^3CF_2$ radical formed in the primary O + C_2F_4 reaction be effectively channeled into the product CO. Our present results thus appear to support the

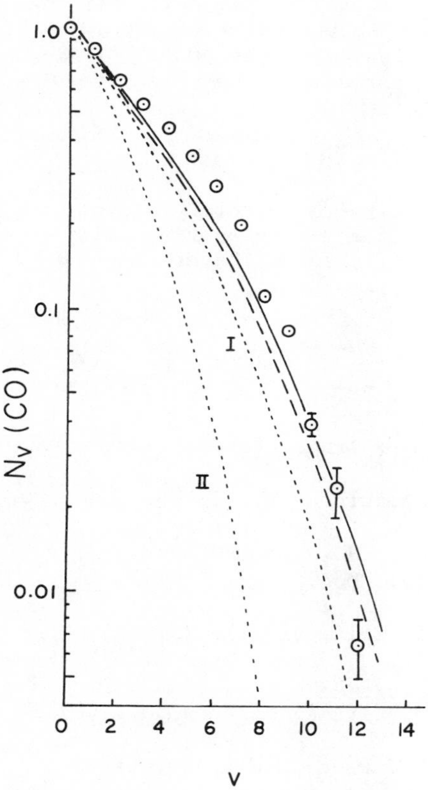

Figure 4. Vibrational population of CO observed in the $O(^3P)$ + C_2F_4 reaction. Open circles = experimental data; solid and dashed curves = statistical distributions based on Equations (II) and (III), respectively, using E_{tot} = 100 kcal/mol ; dotted curves (I) and (II) = statistical distributions based on Equation (III) using E_{tot} = 90 and 63 kcal/mol , respectively. Models employed in these calculations are given in the text.

assumption that triplet CF_2 radicals are indeed present in the $O(^3P)$ - C_2F_4 system. The $^1CF_2$ radical, which may also be present or formed initially in addition to $^3CF_2$, does not react rapidly to produce CO, according to our laser absorption results. The production of F_2 in the O + $^3CF_2$ reaction does not preclude the presence of F atoms because, energetically, the amount of available energy (100 kcal/mole) is more than sufficient to dissociate F_2. The molecular elimination of F_2 in this reaction, though rather uncommon, is not surprising. The photodissociation of Cl_2CO at 125 nm, which corresponds closely to the amount of internal energy carried by the F_2CO intermediate, has been demonstrated to eliminate Cl_2 (100).

Acknowledgement

The authors are grateful to Miss Laura Colcord (Federal Junior Fellow) for her valuable assistance in preparing this review. One of us (D.S.Y.H.) gratefully acknowledges the award of an NRC/NRL Resident Research Associateship.

Literature Cited

1. Andrews, E.B. and Barrow, R.F., Proc. Phys. Soc. A (1951), 64, 481.
2. Thrush, B.A. and Zwolenik, J.J., Trans. Faraday Soc. (1963), 59, 582.
3. Porter, T.L., Mann, D.E. and Acquista, N., J. Mol. Spectroscopy (1965), 16, 228.
4. Carroll, P.K. and Grennan, T.P., J. Phys. B (1970), 3, 865.
5. Carrington, A. and Howard, B.J., Mol. Phys. (1970), 18, 225.
6. Hall, J.A. and Richards, W.G., ibid. (1972), 23, 331.
7. Jacox, M.E. and Milligan, D.E., J. Chem. Phys. (1969), 50, 3252.
8. Simons, J.P. and Yarwood, A.J., Trans. Faraday Soc. (1963), 59, 90.
9. Yarwood, A.J. and Simons, J.P., Proc. Chem. Soc. (1962), 62.
10. Merer, A.J. and Travis, D.N., Can. J. Phys. (1966), 44, (1541).
11. Modica, A.P. and Sillers, S.J., J. Chem. Phys. (1968), 48, 3283.
12. Schatz, G. and Kaufman, M., J. Phys. Chem. (1972), 76, 3586.
13. "JANAF Thermochemical Tables", Stull, D.R., Ed., the DOW Chemical Co., Midland, Mich., 1965.
14. Hildenbrand, D.L., Chem. Phys. Letters (1975), 32, 523.
15. Kuzyakov, Yu. Ya., Vestn. Mosk. Univ. Khim. (1968), 23, 21.
16. Farber, M., Frisch, M.A. and Ko, H.C., Trans. Faraday Soc. (1969), 65, 3202.

17. Lin, M.C.,Paper No. 142, 166th Annual ACS Meeting, Chicago, Ill., Aug. 1973.
18. Lin, M.C., Int. J. Chem. Kinet. (1974), 6, 1.
19. Lin, M.C., and Shortridge, R.G., Chem. Phys. Letters (1974), 29, 42.
20. Shortridge, R.G. and Lin, M.C., J. Chem. Phys. (1976), 64, 4076.
21. Henrici, H., Lin, M.C. and Bauer, S.H., ibid. (1970), 52, 5834.
22. Gangloff, H.J., Milks, D., Maloney, K.L., Adams, T.N., and Matula, R.A., ibid. (1975), 63, 4915.
23. Hancock, G., Morley, C. and Smith, I.W.M., Chem. Phys. Letters (1971), 12, 193.
24. Djeu, N., J. Chem. Phys. (1974), 60, 4100.
25. Hudgens, J.W., Gleaves, J.T., and McDonald, J.D., ibid. (1976), 64, 2528
26. Shapiro, M. and Halavee, U., Chem. Phys. Letters (1976), 40, 387.
27. Whitten, G.Z., and Rabinovitch, B.S., J. Chem. Phys. (1964), 41, 1883.
28. Simons, J.P., private communication.
29. Tang, Y.-N. and Rowland, F.S., J. Am. Chem. Soc. (1966), 88, 626.
30. Tang, Y.-N. and Rowland, F.S., ibid. (1967) 89, 6420.
31. Tang, Y.-N., and Smail, T., and Rowland, F.S., ibid. (1969), 91, 2130.
32. Smail, T. and Rowland, F.S., J. Phys. Chem., (1970), 74, 1866.
33. Lin, M.C., Int. J. Chem. Kinet. (1973), 5, 173.
34. Gordon, R.J. and Lin, M.C., Chem. Phys. Letters (1973), 22, 107.
35. Shortridge, R.G. and Lin, M.C., IEEE J. Quantum Electron. (1974), QE-10, 873.
36. Staemmler, V., Theoret. Chim. Acta (1974), 35, 309.
37. Simons, J.W. and Curry, R., Chem. Phys. Letters (1976), 38, 171.
38. Green, W.H. and Lin, M.C., J. Chem. Phys. (1971), 54, 3222.
39. Berry, M.J., ibid. (1973), 59, 6229.
40. Parker, J.H. and Pimentel, G.C., ibid. (1969), 51, 91; (1971), 55, 857.
41. Bogan, D.J. and Setser, D.W., ibid. (1976), 64, 586 and papers cited therein.
42. Berry, M.J. and Pimentel, G.C., ibid. (1968), 49, 5190.
43. Padrick, T.D. and Pimentel, G.C., J. Chem. Phys. (1971), 54, 720., J. Phys. Chem. (1972), 76, 3125.
44. Lin, M.C., J. Phys. Chem. (1971), 75, 3642; (1972), 76, 811, 1425.
45. Clough, P.N., Polanyi, J.C. and Taguchi, R.T., Can. J. Chem. (1970), 48, 2919.
46. Hsu, D.S.Y., Shortridge, R.G., and Lin, M.C., to be published.

47. Taylor, G.W. and Setser, D.W., J. Chem. Phys. (1973), 58, 4840.
48. Goldberg, I.B. and Schneider, G.R., ibid. (1976), 65, 147.
49. Venkateswarlu, P., Phys. Rev. (1950), 77, 676.
50. Laird, R.K., Andrews, E.B. and Barrow, R.F., Trans. Faraday Soc. (1950), 46, 803.
51. Bass, A.M. and Mann, D.E., J. Chem. Phys. (1962), 36, 3501.
52. Thrush, B.A. and Zwolenik, J.J., Trans. Faraday Soc. (1963), 59, 582.
53. Margrave, J.L. and Wieland, K., J. Chem. Phys. (1953), 21, 1552.
54. Batey, W. and Trenwith, A.B., J. Chem. Soc. (1961), 1388.
55. Hertzberg, G., Chemical Institute of Canada Symposium on Free Radicals, Laval University, Quebec (1956).
56. Simons, J.P. and Yarwood, A.J., Nature (1961), 192, 943.
57. Simons, J.P. and Yarwood, A.J., ibid. (1960), 187, 316.
58. Simons, J.P. and Yarwood, A.J., Trans. Faraday Soc. (1961), 57, 2167.
59. Dalby, F.W., J. Chem. Phys. (1964), 41, 2297.
60. Mann, D.E. and Thrush, B.A., J. Chem. Phys. (1960), 33, 1732.
61. Edwards, J.W. and Small, P.A., Nature (1964), 202, 1329.
62. Edwards, J.W. and Small, P.A., Ind. Eng. Chem. (Fundamentals) (1965), 4, 396.
63. Gozzo, F. and Patrick, C.R., Tetrahedron (1966), 22, 3329.
64. Barnes, G.R., Cox, R.A. and Simmons, R.F., J. Chem. Soc. (B) (1971), 1176.
65. Cox, R.A. and Simmons, R.F., J. Chem. Soc. (B) (1971), 1625.
66. Tschuikow-Roux, E. and Marte, J.E., J. Chem. Phys. (1965), 42, 2049.
67. Kushina, D., Politanskii, S.F., Sheuchuk, V.U., Gutor, I.M., Ivashenko, A.A., and Nefedov, O.M., Izvest. Akad. Nauk SSSR, Ser. Khim. (1974), 946.
68. Tyerman, W.J.R., Trans. Faraday Soc. (1969), 65, 1188.
69. Modica, A.P. and LaGraff, J.E., J. Chem. Phys. (1966), 43, 3383.
70. Modica, A.P. and LaGraff, J.E., ibid. (1966), 45, 4729.
71. Carlson, G.A., J. Phys. Chem. (1971), 75, 1625.
72. Zmbov, K.F., Uy, O.M. and Margrave, J.L., J. Am. Chem. Soc. (1968), 90, 5090.
73. Atkinson, B., J. Chem. Soc. (1952), 2684.
74. Heicklen, J., Knight, V. and Greene, S.A., J. Chem. Phys. (1965), 42, 221.
75. Cohen, N. and Heicklen, J., ibid. (1965), 43, 871.
76. Heicklen, J. and Knight, V., J. Phys. Chem. (1966), 70, 3901.
77. Tyerman, W.J.R., Trans. Faraday Soc. (1969), 65, 163.
78. Atkinson, B. and McKeagan, D., Chem. Comm. (1966), 189.

79. Birchall, J.M., Haszeldine, R.N. and Roberts, D.W., J.C.S. Perkin I (1973), 1071.
80. Saunders, D. and Heicklen, J., J. Am. Chem. Soc. (1965), 87, 2088.
81. Saunders, D. and Heicklen, J., J. Phys. Chem. (1966), 70, 1950.
82. Cohen, N. and Heicklen, J., ibid. (1966), 70, 3082.
83. Heicklen, J. and Knight, V., ibid. (1966), 70, 3893.
84. Johnson, T. and Heicklen, J., J. Chem. Phys. (1967), 47, 475.
85. Mitchell, R.C. and Simons, J.P., J. Chem. Soc. (B), (1968), 1005.
86. Gilbert, J.R., Slagle, I.R., Graham, R.E. and Gutman, D. J. Phys. Chem. (1976), 80, 14.
87. Mahler, W., Inorg. Chem. (1963), 2, 230.
88. Mitsch, R.A., J. Heterocyclic Chem. (1964), 1, 59, 223.
89. Mitsch, R.A., J. Am. Chem. Soc. (1965), 87, 758.
90. Cavell, R.G., Dobbie, R.C. and Tyerman, W.J.R., Can. J. Chem. (1967), 45, 2849.
91. Smail, T., Miller, G.E., and Rowland, F.S., J. Phys. Chem. (1970), 74, 3464.
92. Powell, F.X. and Lide, D.R., J. Chem. Phys. (1966), 45, 1067.
93. Mathews, C.W., ibid. (1966), 45, 1068.
94. Simons, J.P., J. Chem. Soc. (1965), 5406.
95. Heicklen, J., Cohen, N. and Saunders, D., J. Phys. Chem. (1965), 69, 1774.
96. Lenzi, M. and Mele, A., J. Chem. Phys. (1965), 43, 1974.
97. Modica, A.P., ibid. (1967), 46, 3663.
98. Burks, T.L. and Lin, M.C., ibid. (1976), 64, 4235.
99. Biordi, J.C., Lazarra, C.P. and Papp, J.F., J. Phys. Chem. (1976), 80, 1040.
100. Okabe, H., Laufler, A.H. and Ball, J.J., J. Chem. Phys. (1971), 55, 373.

6

Ion-Molecule Reactions Involving Fluorine-Containing Organic Compounds

SHARON G. LIAS

Physical Chemistry Division, National Bureau of Standards, Washington, DC 20234

Since the mid-1950's the elementary chemical reactions of charged species with molecules - so-called ion-molecule reactions - have been extensively studied. Most of these investigations have been carried out in the gas phase, in the ion sources of various types of mass spectrometers where the ions are generated by electron bombardment or by absorption of high energy photons. Through the use of such mass spectrometric techniques as well as other approaches, a large body of information about the chemistry of ions has been generated during the last quarter of a century, so that it is now possible to understand the factors which influence the chemistry of ions (1,2,3).

In classical organic chemistry, a common approach to interpreting patterns of chemical reactivity has always been to examine the changes brought about in reaction rates and mechanisms when an atom or group of atoms in a reactant species is replaced by some other substituent. Within the past five years, a number of investigators have followed this approach in order to examine the factors influencing the mechanisms of the reactions of ions. Of particular interest have been the results obtained in fluorine-substituted organic compounds. Such substitution changes the electron density distribution within a reactant molecule, and therefore, as organic chemistry would predict, can often direct the site of attack of an approaching ion. Furthermore, it is seen that in an ion-molecule complex, fluorine atoms are much more reluctant to indulge in the rearrangements which are characteristic of hydrogen atoms in analogous situations. For these reasons, fluorine "labelling" can lead to the observation of reactions whose mechanisms are very specific, and therefore, are particularly interesting for ascertaining the effects of molecular structure, reaction enthalpy, or other factors on the course of a reaction. In addition, it has been shown that fluorinated molecular ions have a tendency to be formed with excess internal energy and to dissociate giving vibrationally excited fragment ions; the elucidation of the reactions and collisional deactivation of such species is also of interest.

© 0-8412-0399-7/78/47-066-152$10.00/0

This chapter presents a brief review of the reactions of positively-charged polyatomic ions in fluorinated systems. The results of many such studies have also provided information about the thermochemistry of fluorinated ions which is also reviewed here.

Reactions in Perfluorinated Alkanes

The ion-molecule reactions which occur in perfluorinated alkanes have been investigated using ion cyclotron resonance spectrometry (4,5), tandem mass spectrometry (6,7), high-pressure mass spectrometry (8,9), time-of-flight mass spectrometry (6,7), and radiation chemistry techniques (10,11,12).

Table I lists reactions observed between fragment ions and parent molecules in CF_4(9), C_2F_6 (4,5,6,8,), C_3F_8 (5,7), and c-C_4F_8 (5), and the rate constants measured for these reactions. The results show that except for CF^+ ions, which undergo condensation reactions, F^- or F transfer reactions predominate in these systems.

In Table I, we also estimate the probabilities that the collisions between these reactant pairs lead to reaction. Except for the reactions of CF_2^+, we see that the F^- or F transfer reactions observed in perfluorinated alkanes are extremely inefficient. This low reaction probability is especially striking when compared with the efficiencies of the corresponding reactions in hydrocarbon systems (3). The methyl ion reaction with ethane corresponding to reaction 4 occurs at every collision between the reactant pair; the hydride transfer reactions from propane to vinyl ions or ethyl ions, corresponding to reactions 6 and 7, occur with efficiencies of about 0.5. These differences between the efficiencies of ion-molecule reactions in fluoroalkanes and alkanes can be explained in terms of the thermochemistry of these systems, remembering that endothermic or thermoneutral ion-molecule reactions are quite inefficient, and often can not even be seen on the time scale of ion collection in a mass spectrometer. The highly efficient hydride transfer reactions observed in hydrocarbon systems are all exothermic. In the fluorocarbon systems, on the other hand, the F^- transfer reactions listed for CF_3^+, $C_2F_3^+$, and $C_2F_5^+$ are actually slightly endothermic, as demonstrated by the ion cyclotron resonance technique known as double resonance (5). This is a surprising result for those used to thinking in terms of hydride transfer reactions in hydrocarbon systems, where fragment carbonium ions can usually be assumed to undergo efficient exothermic reactions with the parent molecule. The thermochemistry of the fluorocarbon ions will be discussed in more detail below.

Table I. Ion-Molecule Reactions in Perfluoroalkanes.

F^- and F Transfer Reactions	k x 10^{10} cm^3/molecule-second	$k_{Rn}/k_{Collision}$[a]
(1) $CF_2^+ + CF_4 \rightarrow CF_3^+ + CF_3$	2.7[b]	0.36
(2) $CF_2^+ + C_2F_6 \rightarrow C_2F_5^+ + CF_3$	14.5±3[c]	1.7
(3) $CF_2^+ + C_3F_8 \rightarrow C_3F_7^+ + CF_3$	(1.9)[d]	0.27
$\rightarrow CF_3^+ + C_3F_7$	(0.86)[d]	
(4) $CF_3^+ + C_2F_6 \rightarrow C_2F_5^+ + CF_4$	(0.4)[d], 0.34[c]	(0.06), 0.05
(5) $CF_3^+ + C_3F_8 \rightarrow C_3F_7^+ + CF_4$	(0.3)[d]	0.04
(6) $C_2F_3^+ + C_3F_8 \rightarrow C_3F_7^+ + C_2F_4$	(0.8)[d]	0.10
(7) $C_2F_5^+ + C_3F_8 \rightarrow C_3F_7^+ + C_2F_6$	(0.086)[d]	0.01
(8) $C_3F_5^+ + c\text{-}C_4F_8 \rightarrow C_4F_7^+ + C_3F_6$	e	
Condensation Reactions		
(9) $CF^+ + CF_4 \rightarrow C_2F_5^+$	3.3[b]	0.40
(10) $CF^+ + C_2F_6 \rightarrow [C_3F_7^+]^* \rightarrow CF_3^+ + C_2F_4$	(0.13)[d], 1.9[c]	(0.01), 0.20
(11) $CF^+ + C_3F_8 \rightarrow [C_4F_9^+]^* \rightarrow CF_3^+ + C_3F_6$	0.32[d]	0.03

a. k (collision) estimated with the Langevin-Gioumousis-Stevenson equation, reference 13.
b. High pressure mass spectrometer, reference 9. Rate constant for thermal ions.
c. High pressure photoionization mass spectrometer, reference 8. Rate constant for thermal ions.
d. Tandem mass spectrometer, reference 7. These rate constants correspond to those for ions having 0.3 eV kinetic energy.
e. Observed in an ion cyclotron resonance spectrometer, reference 5.

In several of the studies of fluorinated alkanes (4,5,6,7) endothermic collision-induced dissociations of the reactant ions were also observed. It was suggested (4,5,6,7,8) that when a fluorinated alkane is ionized by electron impact, it undergoes dissociation to give fragment ions containing internal excitation energy. The observation of these endothermic reactions under conditions such that the reacting ions have nominally thermal energies, also indicates that the fragment ions, if formed with excess energy, are capable of retaining this energy for times as long as milliseconds.

If indeed the fluorinated carbonium ions observed in these investigations are reacting from an excited state, it would be expected that the average rate constant of reaction would change with pressure as the reactant ion is collisionally deactivated. The effect of pressure on the rate constant for reaction 5 was examined qualitatively in an ion cyclotron resonance study (5), and it was found that the average rate constant decreases as the pressure is increased, as one would expect for an endothermic reaction.

In a high pressure mass spectrometric study (8) of reaction 4, it was shown that at pressures above about 0.5 torr, essentially all of the CF_3^+ ions had reacted to form $C_2F_5^+$. On the other hand, the rate constants which have been reported (7,8) for this reaction may be too high for collisionally deactivated ions, since no reaction was seen between CF_3^+ and C_2F_6 under the conditions of an ion cyclotron resonance experiment (10^{-5} torr, 0.5 sec. observation time) (14). From this latter result, it can be estimated that for thermal reactant ions, the rate constant for reaction 4 can not be greater than 5 x 10^{-12} cm^3/molecule-second. This corresponds to a collision efficiency of 0.0066. If the inefficiency of this reaction can be attributed to an energy barrier, then the maximum endothermicity of reaction 4 is 3.0 kcal/mole (0.13 eV).

In agreement with the high pressure mass spectrometric result (8), there is also evidence that reaction 4 occurs in the radiolysis of C_2F_6 at pressures as high as two atmospheres (9,10), or in the liquid phase (11), since CF_4 is observed as a product under these conditions, even in the presence of a free radical scavenger. The formation of this product from the dissociation of neutral excited C_2F_6 in the radiolysis:

$$C_2F_6^* \rightarrow CF_4 + CF_2 \qquad (12)$$

can be excluded since CF_4 is not formed when bromine is added to the system, and there is no reason to expect that small concentrations of bromine would have any effect on the primary decomposition of an electronically excited molecule. We can tentatively conclude (8) therefore, that ionic fragmentation of $C_2F_6^+$ to form CF_3^+ does occur at pressures as high as two atmospheres, as well as in the liquid phase, and that the fragment ions undergo reaction 4 under these conditions.

Reactions in Perfluorinated Olefins

Table II shows the ion-molecule reactions which have been observed in C_2F_4 (15,16) and C_3F_6 (5).

Table II. Ion-Molecule Reactions in Perfluorinated Olefins.

Reaction	
$C_2F_4^+ + C_2F_4 \rightarrow C_3F_5^+ + CF_3$	(13)
$C_2F_3^+ + C_2F_4 \rightarrow C_3F_5^+ + CF_2$	(14)
$CF_2^+ + C_2F_4 \rightarrow C_2F_4^+ + CF_2$	(15)
$CF^+ + C_2F_4 \rightarrow CF_3^+ + C_2F_2$	(16)
$CF_3^+ + C_3F_6 \rightarrow C_3F_5^+ + CF_4$	(17)
$CF^+ + C_3F_6 \rightarrow C_3F_5^+ + CF_2$	(18)

Results from references 5, 15, and 16.

In C_2F_4, it has been shown (16) that when ionization is effected by bombardment with 13 eV electrons, only the parent $C_2F_4^+$ ions are produced initially at pressures of about 10^{-6} torr. In the ion cyclotron resonance spectrometer (16) the parent $C_2F_4^+$ ion undergoes reaction 13. The rate constant for this reaction increases from a limiting value of 3.0×10^{-12} cm^3/molecule-second at low pressures (10^{-5} torr) to a value of 1.0 (16) or 1.5 (17) x 10^{-11} cm^3/molecule-second at higher pressures (10^{-4} torr). This was interpreted in terms of the reaction scheme:

$$C_2F_4 + e^- \rightarrow (C_2F_4^+)^* + 2\ e^- \qquad (19)$$

$$(C_2F_4^+)^* + C_2F_4 \rightarrow (C_4F_8^+)^* \rightarrow C_3F_5^+ + CF_3 \qquad (20)$$

$$(C_2F_4^{+})^* + C_2F_4 \rightarrow C_2F_4^{+} + C_2F_4^{*} \quad (21)$$

$$C_2F_4^{+} + C_2F_4 \rightarrow C_4F_8^{+} \rightarrow C_3F_5^{+} + CF_3 \quad (22)$$

The molecular ions formed initially by electron impact are considered to contain internal excitation energy. Some of these ions react to form product $C_3F_5^+$ ions, while the others undergo non-reactive collisions or resonant charge exchange reactions with the neutral C_2F_4 molecules in the system, producing molecular ions of lower internal energy. These moderated ions then react to form products. It was found (16) that the rate constant observed for formation of $C_3F_5^+$ reaches a constant value after the parent ion has undergone an average of six collisions.

The bimolecular rate constant for the condensation reaction 13 is low relative to the rate constant for collision between $C_2F_4^+$ and C_2F_4. Approximately one collision in fifty leads to the formation of products different from the reactants. In the discussion below of condensation reactions in partially fluorinated ethylenes, it will be shown that the rearrangement which is necessary for methyl radical elimination from a $C_4(H,F)_8^+$ complex is substantially inhibited by the presence of F-atoms in the complex. This would explain the fact that the rate constants of these reaction are considerably lower, and also more dependent on the internal energy content of the reactants as compared to the $C_2H_4^+$ - C_2H_4 reaction pair (18,19).

The decrease in reaction rate with increase in internal energy can be rationalized in terms of the lifetime of the collision complex. Rearrangements such as those involved in reaction 13 may be expected to increase in probability when the collision partners remain together for a longer period of time. Because the lifetime of the complex will be shortened by increased internal or translational energy the rate coefficient of such an exothermic reaction will also be reduced.

In the next section we will present evidence that the intermediate condensation ion ($C_4F_8^+$ in this case) formed in fluorinated ethylenes has the cyclobutane structure. It has been shown (20) that cyclobutane ions formed with a large amount of internal energy have a high probability of dissociating to give an ethylene ion and ethylene molecule, while those formed with only a small amount of excess energy dissociate by losing a methyl radical, i.e. through a process analogous to reactions 20 and 22.

The results given in Table II show that in perfluorinated propylene, fragment CF^+ and CF_3^+ ions undergo exothermic F^- transfer reactions with the parent molecule. The only other reactions observed in C_3F_6 as well as in 2-C_4F_8 (5) are endothermic collision-induced dissociation reactions.

Reactions in Partially Fluorinated Ethylenes

A number of studies have been carried out exploring the effects of fluorine substitution on the ion-molecule chemistry of ethylenes (21,22,23,24,25,26,27). From results obtained in C_2H_4, C_2H_3F, CH_2CF_2, cis-CHFCHF, trans-CHFCHF, C_2F_3H, and C_2F_4, and mixtures thereof, a consistent picture of the reactions of fluorinated ethylene parent ions with fluorinated ethylene emerges.

In general, when charge transfer is exothermic for a given $C_2(H,F)_4$ reactant pair, it predominates. The only other reaction channel which is observed in nearly all systems is condensation:

$$C_2(H,F)_4^+ + C_2(H,F)_4 \rightarrow C_4(H,F)_8^+ \qquad (23)$$

(where the two reactant species may or may not have the same empirical formula and/or structure). The lifetime of the $C_4(H,F)_8^+$ condensation ion is dependent on its internal energy content. When the pressure is increased sufficiently, the $C_4(H,F)_8^+$ adducts are collisionally stabilized. The collisional stabilization of the $C_4H_8^+$ ion formed in ethylene, at pressures above about 0.1 torr, has been well documented (28,29,30). The $C_4H_6F_2^+$ ion formed in vinyl fluoride:

$$C_2H_3F^+ + C_2H_3F \rightarrow C_4H_6F_2^+ \qquad (24)$$

has been observed in a mass spectrometer at pressures in the range 0.01-0.3 torr (26), and the $C_4H_4F_4^+$ ion formed in CF_2CH_2:

$$CH_2CF_2^+ + CH_2CF_2 \rightarrow C_4H_4F_4^+ \qquad (25)$$

is stabilized at the pressures typical of ion cyclotron resonance experiments (10^{-6} to 10^{-5} torr).

Two modes of dissociation are observed for the $C_4(H,F)_8^+$ condensation ions. One of these leads to the formation of a substituted ethylene ion and ethylene molecule having identities different from those of the original reactant species. Because in this reaction there is never any scrambling of H (or in deuterium labelling experiments, D) and/or F species, it is called a "methlyene switching reaction" and is assumed to occur from a condensation ion having the structure of a cyclobutane ion. Examples of methylene switching reactions which have been observed (17) include:

$$C_2H_3F^+ + CDFCDF \rightarrow \left[\begin{array}{c} CH_2-CDF \\ |\;\;+\;\;| \\ CHF-CDF \end{array}\right]^* \rightarrow CHFCDF^+ + CH_2CDF \qquad (26)$$

$$C_2H_3F^+ + CDFCF_2 \rightarrow \left[\begin{array}{c} CH_2-CDF \\ |\;\;+\;\;| \\ CHF-CF_2 \end{array}\right]^* \rightarrow CHFCF_2^+ + CH_2CDF \qquad (27)$$

$$C_2H_3F^+ + CF_2CF_2 \rightarrow \left[\begin{array}{c} CH_2-CF_2 \\ |\;\;+\;\;| \\ CHF-CF_2 \end{array}\right]^* \rightarrow CHFCF_2^+ + CH_2CF_2 \qquad (28)$$

An examination of these reactions as written here shows that a symmetrical cleavage along the vertical axis returns the complex to reactants or results in a charge transfer, while a cleavage along the horizontal axis produces the "methylene switching" products. An isotopically mixed product is never produced, even in small yield.

The other mode of dissociation of the $C_4(H,F)_8^+$ complexes is through the elimination of a methyl radical. Some examples of methyl radical elimination reactions which have been observed in fluorinated ethylene systems (17) are:

$$CH_2CF_2^+ + CDFCF_2 \rightarrow \left[\begin{array}{c} CH_2-CF_2 \\ |\;\;+\;\;| \\ CDF-CF_2 \end{array}\right]^* \rightarrow C_3H_2DF_2^+ + CF_3 \qquad (29)$$

$$CHFCHF^+ + CDFCF_2 \rightarrow \left[\begin{array}{c} CHF-CHF \\ |\;\;+\;\;| \\ CDF-CF_2 \end{array}\right]^* \rightarrow C_3H_2DF_2^+ + CF_3 \qquad (30)$$

$$cis\text{-}CHFCHF^+ + tr\text{-}CDFCDF \rightarrow \left[\begin{array}{c} CHF-CHF \\ |\;\;+\;\;| \\ CDF-CDF \end{array}\right]^* \begin{array}{l} \rightarrow C_3H_2DF_2^+ + CDF_2\,(76\%) \\ \rightarrow C_3D_2HF_2^+ + CHF_2\,(24\%) \end{array} \qquad (31)$$

As in the methylene switching reactions, there seems to be little scrambling of the H(D) and F atoms within the ion-molecule complex, since in every case the observed products can be explained by the shift of a single H(D) or F atom to another methylene group followed by loss of the resulting methyl group from the ionic complex. It has been suggested that the mechanism of this reaction is a 1,4-F shift in a linear tetramethylene ionic inter- mediate (17) or that the mechanism predominantly involves a 1,2-H(D) shift in a cyclic ion intermediate (22). Assuming either one of these mechanisms, it is possible to calculate the statistical probability for transfer of H or F to and from $C(H,F)_2$ groups in the $(C_2D_4-C_2(H,F)_4)^+$ complexes studied by Ferrer-Correia and Jennings (22). Comparing the observed and statistical probabilities of H(D) or F migration to CHF, CD_2, or CF_2 groups in these complexes, it can be seen that the probability for F-migration to a $CH_2(CD_2)$ group is always very low, not more than 20-30% of the statistical probability. The probability of F-migration from one CHF group to an adjacent CHF group is slightly greater than the statistical prediction, and the prob- ability of F-migration to a CF_2 group is three times greater than one would predict from statistical considerations (26). The products formed in reactions 29, 30, and 31 are in agreement with these predicted trends. It would seem from these results that the migration of an F-atom in an ionic complex is activated by the presence of the polarized C-F bonds on an adjacent carbon.

It is somewhat more difficult to generalize about the mecha- nism of H migration in such complexes, but certain patterns do emerge from the data reported by Ferrer-Correia and Jennings (22). In a $C_4(H,F)_8^+$ complex in which two or more CH_2 groups are ad- jacent, the probability of CH_3 elimination is greater than would be statistically predicted. On the other hand, the H-species have a very low probability of migrating to a CHF or CF_2 group.

It is interesting to examine the chemistry seen in the CH_2CF_2 system in the light of these predicted trends for prob- abilities of migration of H and F atoms in $C_4(H,F)_8^+$ complexes. As mentioned above, the $C_4H_4F_4^+$ complex formed in reaction 25 is stabilized at pressures in the range 10^{-6} to 10^{-5} torr. This com- plex does not undergo any methylene switching or methyl radical elimination reactions. If this complex were ever formed with the structure:

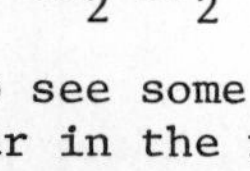

$$\begin{matrix} CH_2-CH_2 \\ \left\lceil \quad + \quad \right\rceil \\ CF_2-CF_2 \end{matrix} \qquad\qquad I$$

one would expect to see some methyl radical elimination, since such processes occur in the reaction of $C_2F_4^+$ with C_2H_4 where the complex must necessarily have structure I. Therefore, the complex formed in reaction 25 must have the structure:

$$\begin{matrix} CH_2-CF_2 \\ \left\lceil \quad + \quad \right\rceil \\ CF_2-CH_2 \end{matrix} \qquad\qquad II$$

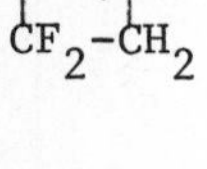

as has been suggested before (17,22,26). This is consistent with the observation that an H-species has a low probability of migrating to a CF_2 group and that an F-atom has a low probability for migrating to a CH_2 group. That is, the stability of the $C_4H_4F_4^+$ complex formed in CH_2CF_2 may result from the reluctance of the H and F atoms to rearrange in a structure such as II. Methylene switching reactions could not be seen if the intermediate has this structure.

Reactions in Partially Fluorinated Alkanes

Investigations of the ion chemistry occurring in fluorine-substituted alkanes have demonstrated that protonated parent ions and carbonium ions attack at the fluorine substituent to form a dialkyl fluoronium ion intermediate:

$$R^+ + R'F \rightarrow R\text{-}F\text{-}R'^+ \qquad (32)$$

$$RFH^+ + R'F \rightarrow R\text{-}F\text{-}R'^+ + HF \qquad (33)$$

Fluoromethanes. In methyl fluoride (9,31,32,33,34,35,36) the parent ion reacts with the parent molecule mainly by transferring a proton, although a small fraction of the ions undergo a condensation reaction in which HF and H are displaced:

$$CH_3F^+ + CH_3F \rightarrow CH_3FH^+ + CH_2F \qquad (34)$$

$$\rightarrow C_2FH_4^+ + HF + H \qquad (35)$$

The protonated parent ion formed in reaction 34 reacts further with methyl fluoride to generate a dimethylfluoronium ion which does not react further:

$$CH_3FH^+ + CH_3F \rightarrow CH_3FCH_3^+ + HF \qquad (36)$$

There is evidence that the structure of the protonated parent ion is as written, that is, that the protonation is on the fluorine atom. When deuterium labelling is utilized, it is seen that a deuteronated parent ion undergoes reaction 36 with no incorporation of deuterium in the dimethyl fluoronium ion product (31):

$$CH_3FD^+ + CH_3F \rightarrow CH_3FCH_3^+ + DF \qquad (37)$$

The fragment ions formed in methyl fluoride are CH_2F^+ and CH_3^+. Deuterium labelling experiments (32) have demonstrated that the methyl ion initially formed by fragmentation reacts with the parent CH_3F molecule by F^- transfer:

$$CD_3^+ + CH_3F \rightarrow (CD_3FCH_3^+)^* \rightarrow CD_3F + CH_3^+ \qquad (38)$$

The less energetic methyl ions formed as products in reaction 38 react with the parent molecule to form CH_2F^+. The CD_3^+ ion reacts to form nearly equal amounts of CH_2F^+ and CD_2F^+:

$$CD_3^+ + CH_3F \rightarrow (CD_3FCH_3^+)^* \rightarrow CH_2F^+ + CD_3H \quad (53\%)$$
$$\rightarrow CD_2F^+ + CH_3D \quad (47\%) \qquad (39)$$

again suggesting a dimethyl fluoronium ion intermediate. The CH_2F^+ ion is not seen to react with CH_3F under the low pressure conditions of an ion cyclotron resonance experiment, but in a high pressure mass spectrometer (9) CH_2F^+ disappears slowly. The suggested (9) reaction mechanism for this ion is:

$$CH_2F^+ + CH_3F \rightarrow (CH_3FCH_2F^+)^* \xrightarrow{CH_3F} (CH_3)_2F^+ + CH_2F_2 \qquad (40)$$

The ion-molecule chemistry observed (9,33,34,37) in CF_2H_2 and CF_3H follows the same pattern as described above for methyl fluoride. The CF_2H_2 parent ion reacts with the parent molecule to transfer a proton:

$$CF_2H_2^+ + CF_2H_2 \rightarrow CH_2F_2H^+ + CF_2H \qquad (41)$$

and the protonated parent molecule in turn reacts with the parent molecule to form a difluoromethyl fluoronium product ion:

$$CH_2F_2H^+ + CH_2F_2 \rightarrow (CH_2F)_2F^+ + HF \qquad (42)$$

The abundance of the parent ion formed in CF_3H by electron impact is so small that it is not observed in the mass spectrum of this compound. However, when a proton is transferred to CF_3H from some othe protonating agent (such as CH_5^+ generated in excess methane, for instance), the protonated parent ion is seen to undergo a condensation reaction with CF_3H analogous to reactions 36 and 42 observed in CH_3F and CH_2F_2:

$$CHF_3H^+ + CHF_3 \rightarrow (CHF_2)_2F^+ + HF \qquad (43)$$

In a mixture of CH_3F with CF_2H_2, it was seen (33) that initially the three halonium ions $(CH_3)_2F^+$, $(CH_2F)_2F^+$ (reactions 36 and 42), and $CH_3FCH_2F^+$ are formed:

$$CH_3FH^+ + CF_2H_2 \rightarrow CH_3FCH_2F^+ + HF \qquad (44)$$

$$CH_2F_2H^+ + CH_3F \rightarrow CH_3FCH_2F^+ + HF \qquad (45)$$

At sufficiently high pressures, however, the halonium ions $(CH_2F)_2F^+$ and $CH_3FCH_2F^+$ disappear as a result of reactions with CH_3F, such as the second step of reaction 40 and:

$$(CH_2F)_2F^+ + CH_3F \rightarrow CH_3FCH_2F^+ + CF_2H_2 \qquad (46)$$

That is, reaction occurs to reduce the number of fluorine atoms in the fluoronium ion products.

In attempts to establish information about the relative heats of formation of various fluorinated methyl ions, studies (33,34, 38,39) have been carried out examining the directions of fluoride and hydride transfer reactions between fluoromethyl ions and various fluorinated methanes. Some studies (38,39) have also included ions and molecules containing chlorine atoms. Although the more quantitative aspects of this work will be covered in detail later in our discussion of the thermochemistry of fluorinated ions, it is interesting to examine here the relative importances of competing reaction channels, shown in Table III. It should be pointed out that at sufficiently high pressures, stabilization of the dimethyl fluoronium ion intermediates of the reactions is observed (9).

In reactions of various ions with CH_3F, CF_2H_2, and CF_3H, it is found that exothermic fluoride transfer reactions occur to the virtual exclusion of hydride transfer, even when the latter reaction channel is significantly exothermic. In reactions with $CFCl_3$, CF_2Cl_2, $CHFCl_2$, or CH_2FCl it is seen that where chloride transfer is exothermic, it strongly predominates even when alternate channels are more exothermic. Fluoride transfer does compete with chloride transfer, but hydride transfer is never obser-observed when the reactant molecule contains chlorine. Apparently it is not the strength of the bond broken in the reaction which determines the preferred channel since the C-F bonds are stronger than the C-H bonds in some of these compounds. A likely interpretation is that since these reactions all proceed via a dialkylhalonium ion intermediate the electrophilic reactant ions preferentially attack high electron density chlorine atoms in the reactant molecules. Attack at fluorine atoms occurs with a significantly lower probability, and attack at hydrogen atoms occurs with a negligible probability. This is consistent with the picture given below of the ion chemistry occurring in alkyl fluorides, where reaction nearly always involves attack of the ion at the fluorine atom.

Table III. Relative Importances of Competing Cl^-, F^-, or H^- Abstraction from the Halomethanes.

	Mode of Reaction and Enthalpy (eV,kcal/mole)		
Reactant Pair	Cl^- Transfer	F^- Transfer	H^- Transfer
$CFCl_3$			
$CF_3^+ + CFCl_3$	90% (-1.60, -37)	10% (-1.93, -44.5)	
$CF_2H^+ + CFCl_3$	90% (-1.37, -37.5)	10% (-1.55, -35.8)	
$CF_2Cl^+ + CFCl_3$	99% (-0.93, -21.5)	1% (-1.17, -26.9)	
CF_2Cl_2			
$CF_3^+ + CF_2Cl_2$	80% (-0.67, -15.5)	20% (-1.83, -42.3)	
$CF_2H^+ + CF_2Cl_2$	86% (-0.43, -10.0)	14% (-1.46, -33.6)	
CF_2H_2			
$CF_2Cl^+ + CHFCl_2$	95% (-0.03, -0.8)	5% (-0.88, -20.3)	0% (-0.88, -20.4)
$CHFCl^+ + CHFCl_2$	0% (0,0)	100% (-0.66, -15.2)	0% (-0.73, -16.9)
CH_2FCl			
$CF_3^+ + CH_2FCl$	95% (-0.004, -0.1)	5% (-0.81, -18.7)	0% (-0.78, -17.9)
$CH_2F^+ + CH_2FCl$	-- (0,0)	Major channel (-0.45, -10.4)	Minor channel (-0.67, -15.5)
CH_3F			
$CF_3^+ + CH_3F$		100% (-0.01, -0.3)	0% (-2.61, -60.6)
CF_2H_2			
$CH_3^+ + CF_2H_2$		Major channel (-0.36, -8.3)	Minor channel (-1.2, -27.3)

Table III (cont'd.)

Reactant Pair	Mode of Reaction and Enthalpy (eV,kcal/mole) Cl^- Transfer	F^- Transfer	H^- Transfer
CF_2H_2			
$CF_2H^+ + CF_2H_2$		Occurs (+0.02, +0.4)	-- (0,0)
$CF_3^+ + CF_2H_2$		100% (-0.37, -8.6)	0% (-0.40, -9.2)
CF_3H			
$CH_3^+ + CF_3H$		100% (-0.38, -8.7)	0% (-0.78, -18.1)
$CH_2F^+ + CF_3H$		100% (-0.02, -0.40)	0% (+0.12, +2.7)

Results from references 38 and 39.

Alkyl Fluorides. The ion-molecule reaction mechanism observed in ethyl fluoride (31) is similar to that described above for the CH_3F system. The parent ion transfers a proton to the parent molecule:

$$C_2H_5F^+ + C_2H_5F \rightarrow C_2H_5FH^+ + C_2H_4F \quad (47)$$

and the protonated parent ion reacts with the ethyl fluoride molecule to displace HF:

$$C_2H_5FH^+ + C_2H_5F \rightarrow C_2H_5FC_2H_5^+ + HF \quad (48)$$

and form the diethyl fluoronium ion. The fragment $C_2H_4F^+$ ion undergoes the reactions:

$$C_2H_4F^+ + C_2H_5F \rightarrow C_2H_5^+ + C_2F_2H_4 \text{ (F}^- \text{ transfer)} \quad (49)$$

$$\rightarrow C_2H_5F_2^+ + C_2H_4 \text{ (F}^+ \text{ transfer)} \quad (50)$$

$$\rightarrow C_4H_7^+ + 2\ HF \text{ (Condensation)} \quad (51)$$

Of the ions generated in reactions 49, 50, and 51, only $C_2H_5^+$ reacts further with C_2H_5F, giving $C_2H_5FH^+$, $C_3H_5^+$, and $C_4H_9^+$ as product ions.

A somewhat different pattern of reactivity is observed in CH_3CHFCH_3 (40). The parent $CH_3CHFCH_3^+$ ion fragments to yield $CH_3CFCH_3^+$, CH_3CHF^+, and CH_2CHF^+ as the abundant primary ions, all of which react with the parent 2-fluoropropane molecule by F^- transfer to yield sec-$C_3H_7^+$ as a product. The sec-$C_3H_7^+$ ion, in turn, reacts with the 2-fluoropropane molecule to regenerate $CH_3CFCH_3^+$, probably through a dipropyl fluoronium ion intermediate:

$$\text{sec-}C_3H_7^+ + CH_3CHFCH_3 \rightarrow (CH_3)_2CHFCH(CH_3)_2^+ \rightarrow (CH_3)_2CF^+ + C_3H_8 \quad (52)$$

$$\rightarrow C_6H_{13}^+ + HF \quad (53)$$

Since the $CH_3CFCH_3^+$ ion produced in reaction 52 will react to regenerate sec-$C_3H_7^+$, a chain reaction occurs in this system. The concentration of the chain carrier is attenuated, however, by the occurrence of the alternate reaction channel 53.

Reactions of Fluorinated Ions with Other Organic Compounds

Reactions with Alkanes. Figure 1 shows the probability that a hydride transfer reaction will occur when CF_3^+, CF_2H^+, CF_2Cl^+, or CCl_2F^+ undergoes a collision with the C_2-C_9 linear alkanes (41).

$$CX_3^+ + C_nH_{2n+2} \rightarrow CX_3H + C_nH_{2n+1}^+ \quad (54)$$

Hydride transfer is the only available reaction channel for these reactants.

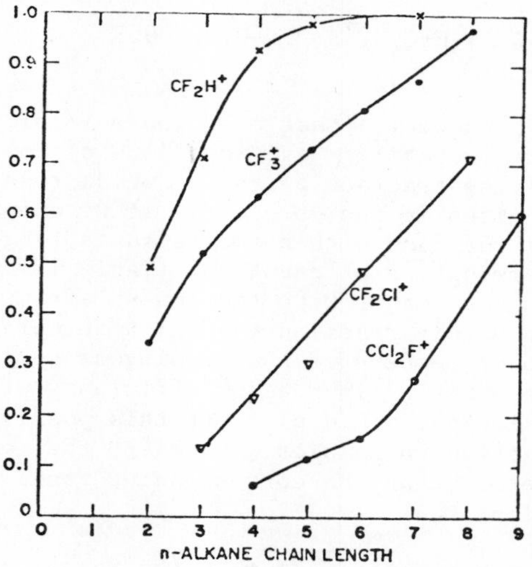

International Journal of Mass Spectrometry and Ion Physics

Figure 1. Probabilities of a reactive collision between CF_2H^+(×), CF_3^+(●), CF_2Cl^+(▽), CCl_2F^+(○), and the linear alkanes. Probabilities are estimated by dividing measured rate coefficients by the rate coefficient for collision of the particular ion–molecule pair.

It follows from Figure 1 that for each of these ions, the probability of a reactive collision increases with increasing chain length of the n-alkane. Table IV lists the enthalpies of reaction for each of these ions at a secondary site and a primary site in n-butane; these values will give an indication of the ordering of reaction enthalpies for hydride transfer reactions of these ions.

Table IV. Enthalpies of Hydride Transfer Reactions of Halomethyl Ions with n-Butane.

Reaction	Enthalpy (e,V kcal/mole) Secondary Site	Primary Site
$CF_3^+ + n\text{-}C_4H_{10} \rightarrow CF_3H + C_4H_9^+$	-2.0, -46.4	-1.2, -28.4
$CF_2H^+ + n\text{-}C_4H_{10} \rightarrow CF_2H_2 + C_4H_9^+$	-1.6, -36.8	-0.8, -18.8
$CF_2Cl^+ + n\text{-}C_4H_{10} \rightarrow CF_2ClH + C_4H_9^+$	-1.4, -31.7	-0.6, -13.7
$CFCl_2^+ + n\text{-}C_4H_{10} \rightarrow CFCl_2H + C_4H_9^+$	-0.4, -9.3	-0.4, +8.7

It has been shown (3) that many ion-molecule reactions which occur with a low probability are reactions of low exothermicity or reactions involving reactant pairs for which reaction is endothermic at certain sites in the molecule. If we compare the relative efficiencies of reaction with normal alkanes (Fig. 1) and the relative exothermicities of reaction, (Table IV) we see that indeed, the ion which reacts with the lowest efficiency, $CFCl_2^+$ does not have an exothermic reaction channel available at the primary sites in the lower members of the homologous series. Furthermore, the next ion in order of increasing efficiency of reaction, CF_2Cl^+, is the ion one would pick for this position on the basis of relative reaction enthalpies. However, the reactions of CF_2H^+ are less exothermic than the corresponding reactions of CF_3^+ but are more efficient.

It has been suggested (41) that the reaction efficiency of an ion-molecule collision will increase with an increase in the lifetime of the ion-molecule complex; there is evidence that the ordering of efficiencies of reactions of CF_2H^+, CF_3^+, and CF_2Cl^+ is to a large degree, determined by the relative lifetimes of the collision complexes. The selectivity of reaction site in deuterium labelled n-butane and propane of these three ions, as shown in Table V throws some light on this matter.

Table V. Positional Selectivity of Hydride Transfer Reactions of Fluorinated Methyl Ions with Alkanes.

Reactant Molecule	Percent Reaction			
	Statistical	CF_2Cl^+	CF_3^+	CF_2H^+
$CD_3CH_2CD_3$				
Secondary sites	25	83		52
Primary sites	75	17		48
$CD_3CH_2CH_2CD_3$				
Secondary sites	40	84	64	56
Primary sites	60	16	36	44

Results from reference 41.

If collision duration is indeed a factor in determining the reaction efficiency, one would expect that the reactions associated with longer-lived ion-molecule complexes would exhibit less selectivity of reaction site. From the results given in Figure 1, we would predict that the three ions CF_3^+, CF_2H^+, and CF_2Cl^+ (all of which undergo highly exothermic reactions at every site in the alkanes) form complexes with alkanes whose lifetimes are ordered $CF_2Cl^+ < CF_3^+ < CF_2H^+$. This is also the ordering one would predict from the positional selectivity of these ions with $CD_3CH_2CD_3$ or $CD_3CH_2CH_2CD_3$, shown in Table V. That is, CF_2Cl^+ shows the most selectivity (i.e. the shortest lifetime), followed by CF_3^+ and CF_2H^+ in that order.

Reactions with Alkenes, Cycloalkanes, and Alkynes. Tables VI and VII list the rate constants and modes of reaction observed for CF_3^+ and $C_2F_5^+$, respectively with alkenes, cycloalkanes, and alkynes (42).

The results given in Table VI show that 50-90% of the reactive collisions between CF_3^+ and C_2H_4 or C_3H_4 consist of the HF-elimination processes:

$$CF_3^+ + H_2C{=}CH_2 \rightarrow (C_3F_3H_4^+)^* \rightarrow C_3F_2H_3^+ + HF \qquad (55)$$

$$CF_3^+ + CH_3C{\equiv}CH \rightarrow (C_4F_3H_4^+)^* \rightarrow C_4F_2H_3^+ + HF \qquad (56)$$

$$CF_3^+ + CH_2{=}C{=}CH_2 \rightarrow (C_4F_3H_4^+)^* \rightarrow C_4F_2H_3^+ + HF \qquad (57)$$

Table VI. Rate Constants and Modes of Reaction of CF_3^+ with Alkenes, Cycloalkanes, and Alkynes.

Reactant Molecule	$k_{Rn} \times 10^{10}$ cm^3/molecule-second	Condensation-Dissociation	Hydride Transfer	Charge Transfer
C_2H_2	1.2	100%		
C_2H_4	7.2	85%	15%	
CH_3CCH	7.5	78%	22%	
CH_2CCH_2	7.5	79%	21%	
CH_3CHCH_2	6.7	82%	18%	
c-C_3H_6	4.7	85%	15%	
2-C_4H_8	6.7	9%	45%	46%
c-$C_3H_5CH_3$	7.6	36%	60%	4%
c-C_4H_8	8.2	3%	84%	13%

Results from reference 42.

At pressures around 10^{-6} to 10^{-5} torr, the abundances of the intermediate adduct ions were negligibly small (<1%). In each case, the only other reaction observed for these reactant pairs was hydride transfer to CF_3^+.

It is interesting to contrast reactions 55, 56, and 57 and the condensation reaction:

$$CF_3^+ + C_2H_2 \rightarrow C_3F_3H_2^+ \qquad (58)$$

which accounts for about 95% of the reactive encounters between CF_3^+ and C_2H_2 at 10^{-5} torr. However, only 10% of the CF_3^+-acetylene collisions lead to reaction. Thus, when CF_3^+ and acetylene collide, the adduct has a high probability of simply falling apart to regenerate the original reactants. Depending on the structure one assumes the $C_3F_3H_2^+$ condensation ion to have, it can be estimated that its internal energy content will be 50-65 kcal/mole. In the reactions of CF_3^+ with ethylene, propyne, and propadiene, condensation ions with considerable excess energy are also formed, but this energy is largely disposed of by elimination of HF (reactions 55-57). It can be estimated that the overall reaction:

$$CF_3^+ + C_2H_2 \rightarrow (C_3F_3H_2^+)^* \rightarrow HF + C_3F_2H^+ \qquad (59)$$

is as much as 50 kcal/mole exothermic, and indeed, about 5% of the product ions formed in the reaction between CF_3^+ and acetylene are $C_3F_2H^+$. However, the low probability of this channel indicates that the dissociation leading to HF elimination is not readily available to the $C_3F_3H_2^+$ condensation ion. It is generally accepted that the structure of a $C_3H_3^+$ ion is cyclic, and it is reasonable to assume that the $C_3F_2H^+$ analogue would also have as its most stable structure, a cyclic structure. The $C_3F_3H_2^+$ condensation ion would most likely initially be formed with either of the two structures:

CF_3CHCH^+

III

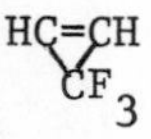

IV

Either of these intermediates would have to undergo some rearrangement in order to eliminate HF and form a product ion of the structure:

In the case of reactions 55, 56, and 57, product ions of a stable structure can be formed by elimination of HF from condensation ions without the necessity for much rearrangement.

When CF_3^+ reacts with propylene or cyclopropane, 82-85% of the reactive encounters result in a dissociation of the intermediate ion to eliminate CH_2CF_2:

$$CF_3^+ + C_3H_6 \rightarrow (C_4F_3H_6^+)^* \rightarrow C_2FH_4^+ + CH_2CF_2 \quad (60)$$

The remainder of the reaction between these pairs is hydride transfer:

$$CF_3^+ + C_3H_6 \rightarrow CF_3H + C_3H_5^+ \quad (61)$$

Essentially, the same pattern of reactivity is seen for the next higher homologue of propylene, 1-butene.

Insight into the mechanism of the CH_2CF_2 elimination, has been obtained by using cyclopropane-1, 1-d_2 and CH_3CHCD_2. If the H's and D's are statistically distributed throughout the $C_4F_3D_2H_4^+$ intermediates, one would expect to see $C_2FH_4^+$, $C_2FH_3D^+$, and $C_2FH_2D_2^+$ product ions in the ratio 0.07:0.53:0.40, while a mechanism involving transfer of an unrearranged $CH_2(CD_2)$ group would result in a 0.33:0:0.67 distribution. In the propylene experiment, the ratio $C_2FH_4^+$:$C_2FH_3D^+$: $C_2FH_2D_2^+$ was 0.08:0.49:0.43, which corresponds to a statistical distribution, within experimental error. In cyclopropane, the ratio of these ionic products was 15:35:49. This means that about 30% of the CF_3^+-cyclopropane condensation ions undergo dissociation process 60 through a clean transfer of a methylene group, without any rearrangement.

When CF_3^+ reacts with 2-butene, 90% of the reactive encounters are accounted for by the exothermic hydride- and charge transfer reactions:

$$CF_3^+ + C_4H_8 \rightarrow C_4H_7^+ + CF_3H \quad (62)$$

$$CF_3^+ + C_4H_8 \rightarrow C_4H_8^+ + CF_3 \quad (63)$$

which are of approximately equal importance (Table VI). The only other product ion observed in a 2-butene-CF_4 mixture is $C_3F_2H_3^+$. This is the ion which would be formed if the product of an HF-elimination reaction had enough excess energy to dissociate further by losing C_2H_4:

$$CF_3^+ + C_4H_8 \rightarrow (C_5F_3H_8^+)^* \rightarrow C_3F_2H_3^+ + HF + C_2H_4 \quad (64)$$

The isomeric cyclic alkanes, cyclobutane and methylcyclopropane also undergo reactions 62, 63, and 64, although for these compounds hydride transfer reaction 62 is of major importance.

The observation of $C_4H_8^+$ ions in mixtures containing cyclobutane or methylcyclopropane is somewhat surprising, since charge transfer from CF_3^+ is respectively, 38 and 22 kcal/mole (1.6 and 0.95 eV) endothermic. To explain the formation of $C_4H_8^+$ ions, we must postulate that ring opening and rearrangement to the 2-butene or isobutene structure occurs in the ion-molecule complex:

$$CF_3^+ + (c\text{-}C_4H_8 \text{ or } c\text{-}C_3H_5CH_3) \rightarrow (C_5H_8F_3^+)^* \rightarrow (2\text{-}C_4H_8^+ \text{ or } i\text{-}C_4H_8^+) + CF_3 \quad (65)$$

which would make this reaction exothermic.

Table VII gives the reactions occurring between $C_2F_5^+$ and C_2 or C_3 alkenes and alkynes, as well as the observed rates of reaction. Again, as in the case of CF_3^+, hydride transfer is observed when it is exothermic. The $C_2F_5^+$ ion also undergoes a CF_2 elimination reaction with these compounds:

$$C_2F_5^+ + M \rightarrow CF_3M^+ + CF_2 \quad (66)$$

which leads to the formation of a product ion of the same indentity as the condensation ions of the CF_3^+ reactions. However, unlike the CF_3M^+ ions formed in condensation reactions of CF_3^+, the ions formed in CF_2 elimination reactions of $C_2F_5^+$ do not undergo further dissociation, because excess energy is carried away by the departing CF_2 species.

This is substantiated by the reaction between $C_2F_5^+$ and C_2H_2, where the rate constant for formation of $C_3F_3H_2^+$ is nearly an order of magnitude greater when the reactant ion is $C_2F_5^+$ (Table VII) than when it is CF_3^+. In the reaction of $C_2F_5^+$:

$$C_2F_5^+ + C_2H_2 \rightarrow CF_3C_2H_2^+ + CF_2 \quad (67)$$

the $C_3F_3H_2^+$ ion is formed with a maximum of about 13 kcal/mole excess energy, while in reaction 58 the ion has about 50-65 kcal/mole excess energy, and dissociates with a high probability to give back the original reactants:

$$C_3F_3H_2^+ \rightarrow CF_3^+ + C_2H_2 \quad (68)$$

This dissociation is endothermic for $C_3F_3H_2^+$ ions formed in reaction 67.

Table VII. Rate Constants and Modes of Reaction of $C_2F_5^+$ with Alkenes and Alkynes.

Reactant Molecule	$k_{Rn} \times 10^{10}$ cm^3/molecule-second	Products
C_2H_2	5.0	$C_3F_3H_2^+ + CF_2$
C_2H_4	3.8	$C_2H_3^+ + C_2F_5H$
		$C_2H_3F_2^+ + C_2F_3H$
		$C_3H_4F_3^+ + CF_2$
CH_3CCH	6.5	$C_3H_3^+ + C_2F_5H$
		$C_4H_4F_3^+ + CF_2$
		$C_5H_3F_4^+ + HF$
CH_2CCH_2	7.0	$C_3H_3^+ + C_2F_5H$
		$C_4H_4F_3^+ + CF_2$

Results from reference 42.

Reactions with Aromatic Compounds. In the reaction of fluorinated methyl ions with benzene and toluene, the most important process is the HF elimination reaction (42,43). For instance, CF_3^+ reacts with these compounds as follows:

$$CF_3^+ + C_6H_6 \rightarrow (CF_3C_6H_6^+)^* \rightarrow c\text{-}C_6H_5CF_2^+ + HF \qquad (69)$$

$$CF_3^+ + c\text{-}C_6H_5CH_3 \rightarrow (CF_3C_6H_5CH_3^+)^* \rightarrow CH_3C_6H_4CF_2^+ + HF \qquad (70)$$

These reactions apparently proceed without any rearrangements occuring in the ion-molecule complex, since in the reaction of CH_2F^+ with C_6D_6 only DF loss is observed (43):

$$CFH_2^+ + C_6D_6 \rightarrow C_6D_5CH_2^+ + DF \qquad (71)$$

An examination of the reaction between CF_3^+ and $c\text{-}C_6H_5CD_3$ confirmed that the HF elimination process 2 involved only ring hydrogens rather than hydrogens from the methyl group (42).

For these aromatic compounds, the major competing reaction channel is charge transfer, which accounts for 8 and 15% of the reactive encounters between CF_3^+ and benzene or toluene, respectively. In toluene, the highly exothermic hydride transfer reaction:

$$CF_3^+ + c\text{-}C_6H_5CH_3 \rightarrow CF_3H + c\text{-}C_6H_5CH_2^+ \qquad (72)$$

accounts for only about 3% of the reactive encounters.

The $C_2F_5^+$ ion also reacts with benzene to form a $C_6H_5CF_2^+$ product ion. This reaction can be written:

$$C_2F_5^+ + C_6H_6 \rightarrow C_6H_5CF_2^+ + CF_3H \qquad (73)$$

When chlorofluoromethyl ions react with benzene or toluene, it is seen that HCl elimination strongly predominates over HF elimination in reactions analogous to 69-71.

Reactions with Oxygenated Compounds. A recent study (14,44) has reported on the reactions of CF_3^+, CF_2Cl^+, CCl_2F^+, and $C_2F_5^+$ with a series of organic molecules containing a carbonyl functional group. These ions undergo four-center reactions with such compounds, which in the case of aldehydes and ketones lead to a product in which the carbonyl oxygen of the reactant compound has been replaced by a F^+ (or Cl^+). For instance, CF_3^+ reacts with acetone as follows:

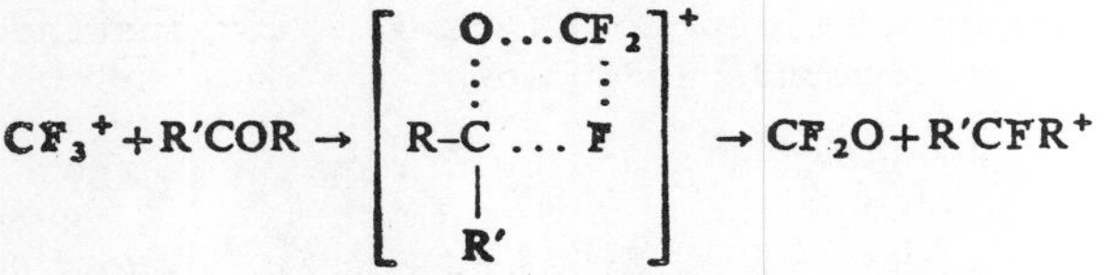

$$CF_3^+ + R'COR \rightarrow \left[\begin{array}{c} O \cdots CF_2 \\ \vdots \quad\quad \vdots \\ R-C \cdots F \\ | \\ R' \end{array} \right]^+ \rightarrow CF_2O + R'CFR^+ \qquad (74)$$

The $C_2F_5^+$ ion undergoes an analogous reaction:

$$C_2F_5^+ + CH_3COCH_3 \rightarrow CH_3CFCH_3^+ + C_2F_4O \qquad (75)$$

but in the chlorine-containing ions, CF_2Cl^+ and $CFCl_2^+$, the product ion contains Cl rather than F. When larger aldehydes or ketones are allowed to react with CF_3^+, the monofluorinated carbonium ions homologous to that formed in reaction 74 are observed only in very small abundances. Instead, the major product ions are usually the ions corresponding to the loss of HF from the monofluorinated ions:

$$CF_3^+ + C_2H_5COCH_3 \rightarrow CF_2O + (C_4H_8F^+)^* \rightarrow CF_2O + HF + C_4H_7^+ (76)$$

$$CF_3^+ + c\text{-}C_5H_8O \rightarrow CF_2O + (C_5H_8F^+)^* \rightarrow CF_2O + HF + C_5H_7^+ \qquad (77)$$

The extent of this dissociation is, not unexpectedly, governed by the energetics of the reaction in which the fluorinated ion is formed. It is interesting that the $C_5H_8F^+$ ion formed in the reaction between $C_2F_5^+$ and cyclopentanone:

$$C_2F_5^+ + c\text{-}C_5H_8O \rightarrow C_2F_4O + C_5H_8F^+ \qquad (78)$$

does not undergo dissociation. It can be estimated that the four-center reaction 78 is as much as 4 eV less exothermic than the first step of reaction 77.

The reactions of these ions with acids and esters were also examined. Reaction 79, a displacement involving cleavage of the acyl-oxygen bond and the formation of acylium ion products:

$$RCOOR' + CX_3^+ \rightarrow \left[\begin{array}{c} O \\ \| \\ R-C \\ | \\ O \cdots CX_3 \\ R' \end{array} \right]^+ \rightarrow RCO^+ + R'OCX_3 \qquad (79)$$

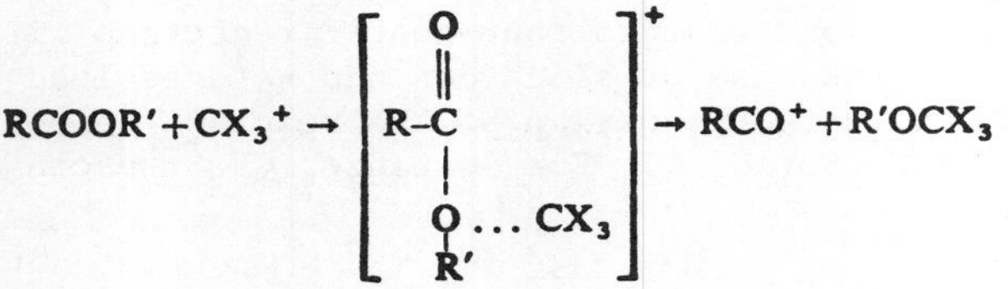

(where R and R' are H or alkyl groups) is important for many such reaction pairs. In fact, when R is CH_3 (i.e. acetic acid, acetic anhydride, methyl acetate), reaction 79 is the only channel observed.

However, formic acid and all of the formates (all compounds where R=H) which were investigated undergo alternate reactions to a greater or lesser extent. An alternate reaction which is generally important is another four-center reaction, this one involving breakage of the acyl-oxygen bond:

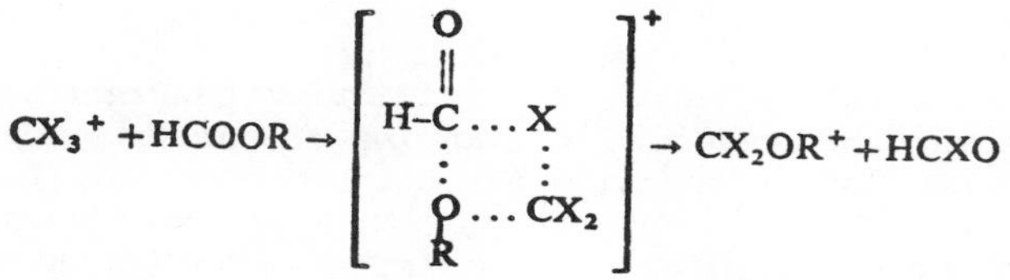

$$CX_3^+ + HCOOR \rightarrow \begin{bmatrix} \overset{\displaystyle O}{\underset{\displaystyle }{\overset{\|}{H-C}}} \cdots X \\[6pt] \underset{\displaystyle R}{O} \cdots CX_2 \end{bmatrix}^+ \rightarrow CX_2OR^+ + HCXO \qquad (80)$$

(where R is H or an alkyl group). The prevalence of one channel or another can be explained in terms of relative exothermicities. In the case of the R=CH_3 compounds, reaction 79 is about 2.7 eV more exothermic than the four-center reaction analogous to reaction 80, but when R=H, the four-center reaction is more exothermic. The occurrence of reaction 79 and reaction 80 indicate that the halomethyl ions attack acids and esters at the acyl oxygen rather than the carbonyl oxygen.

Reaction of fluoromethyl cations with ROH (R=H, CH_3, C_2H_5) have been examined (42, 43), and can be represented in a general fashion as:

$$C(H,F)_3^+ + ROH \rightarrow C(H,F)_2{=}OR^+ + HF \qquad (81)$$

$$\rightarrow C(H,F)_3OH + R^+ \qquad (82)$$

When R=C_2H_5, only the latter channel is observed.

Thermochemical Information about Fluorinated Species Derived from Ion-Molecule Reactions

Under the usual conditions used to investigate ion-molecule reactions, reactions which are strongly endothermic can not be observed, and reactions which are close to thermoneutral, ($\Delta H = 0 \pm \sim$ 2-3 kcal/mole) occur with efficiencies such that less than one collision in 10 is reactive. On the other hand, it can usually be safely assumed that any reaction which is observed to occur with a high efficiency is exothermic. Within the framework of these general observations, it is possible to derive upper and/or lower limits for the heats of formation of particular ions or molecules from the observation or non-observation of particular ion-molecule reactions. More quantitative information about ionic heats of formation can be obtained for systems in which an ion-molecule equilibrium:

$$A^+ + B \rightleftarrows C^+ + D \qquad (83)$$

can be observed. The equilibrium constant observed leads to a value for $\Delta G°$:

$$-RT\ln K_{eq} = \Delta G° = \Delta H - T\Delta S \qquad (84)$$

If the temperature variation of the equilibrium constant is determined, or if some other means of estimating ΔS can be employed, a value for the enthalpy of reaction is obtained.

In several recent studies, new or improved estimates of the heats of formation of several fluorinated ions or radicals have been obtained from such experiments involving ion-molecule reactions. These results are summarized in Table VIII. In this table and Table IX, for the sake of consistency with other discussions of the subject in the literature, we shall use as the unit of energy kcal/mole; 1 kcal/mole corresponds to 4.18 kjoule or 0.04 eV.

In the case of uncharged radicals and molecules, it is well known that the substitution of a fluorine atom for a hydrogen atom results in a "stabilization" of the species, i.e. a lowering of the heat of formation of the radical or molecule. This is illustrated in Table IX, where the heats of formation of CH_3, CH_2F, CHF_2, and CF_3 are listed (33,34,50,51). In the case of the corresponding cations, it is interesting to examine whether the decreased electronegativity difference between C^+ and F (as compared to C and F), and the increased double bond character of the C-F bond will cause any differences in the "stabilization" effected by fluorine substitution in the ion as compared to the neutral species. A detailed LCAO-SCF calculation in the INDO approximation (52) predicts that in these fluorine-substituted ions, there is substantial polarization of the carbon-fluorine π-bond toward the positively charged carbon-atom, balanced by a polarization in the opposite direction of the C-F σ bond. Thus, although the magnitude of both the destabilizing σ-polarization and the stabilizing π back-donation increase in going from CF_3^+ to CF_2H^+ to CFH_2^+, the net total charge density on fluorine varies by only 0.089 in this series. The data given in Table IX (39,47,53,54,55) indeed confirm that the energy required to effect ionization of the CFH_2, CF_2H, and CF_3 radicals is within the limits of error, the same for the three fluorine-substituted species. That is, the degree of stabilization effected by substitution of one fluorine atom is, within the limits of error of presently available data, not strongly influenced by the presence of the positive charge on the carbon except for the first substitution. Significantly greater stabilization is effected in going from CH_3^+ to CH_2F^+ than from CH_3 to CH_2F. Some authors (55,56) have defined the "stabilization energy" of fluorine-substitution in these methyl ions as the difference in the CH_3^+-H^- bond energy and the corresponding CX_3^+-H^- bond energy (where X is H or F). These bond energies are listed in Table IX, along with the CX_3^+-F^- and CX_3^+-Cl^- bond energies. Here it can be seen that although considerable "stabilization" is effected by the presence of F atoms in the methyl group, the number of F-atoms present in the ion has, within the limits of error, little effect upon the amount of stabilization. (This is just another way of saying that the degree of stabilization per F atom varies in the same way in the ion as in the corresponding RH, RF, and RCl molecules.) Earlier considerations of this subject (33,34,55,56) have usually led to the conclusion that the substitution of F for H in CF_2H^+ resulted in significantly less stabilization than the corresponding substitution in CFH_2^+, partly because until recently, the heat of formation of CF_3^+ was taken as 99 kcal/mole (46), rather than 94.1 kcal/mole (45) or 91.5 kcal/mole (47). The data listed in Table IX show that the the presently available limits of error on the heats of formation of these ions are such that small differences in the degree of "stabilization" effected by consecutive fluorine-substitution can not be discerned with certainty.

Anicich and Bowers (17) have reached a similar conclusion about the stabilization brought about by fluorine substitution in the fluorinated ethylene ions. They point out that a plot of the experimentally determined heats of formation of $C_2H_4^+$, $C_2H_3F^+$, $C_2H_2F_2^+$, $C_2HF_3^+$, and $C_2F_4^+$ fall on a straight line, as shown in Figure 2.

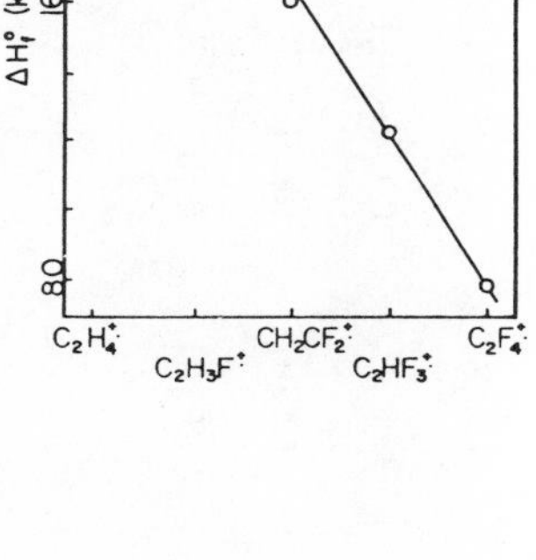

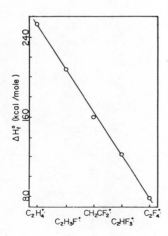

Figure 2. Plot of the heat of formation (49) vs. the number of fluorine substituents in the ion series $C_2(H,F)_4^+$. Figure from Reference 17.

Table VIII. Thermochemical Information about Fluorinated Ions from Ion-Molecule Reactions.

Ion	Reaction	Thermochemical Information	ΔH_f(Ion) kcal/mole From Ion-Molecule Reactions	ΔH_f(Ion) kcal/mole Other Methods
CF_3^+	$CH_3^+ + CF_4 \rightleftarrows CF_3^+ + CH_3F$[a,b]	$\Delta H_{Rn} = 0.0 \pm 1$ kcal/mole	95.5 ± 4[a]	94.1 ± 0.8[c]
				99[d]
				91.5 ± 3[e]
$C_2F_5^+$	$CX_3^+ + C_2F_6 \rightarrow C_2F_5^+ + CF_4$[f]	$k_{Rn}/k_{collision} \leq 0.0067$;[g]		
		$\Delta H_{Rn} \geq +3.0$ kcal/mole	≥ -1.2	
sec-$C_3F_7^+$	$C_2F_5^+ + C_3F_8 \rightarrow C_2F_6 +$ sec-$C_3F_7^+$	Endothermic	> -108	
CH_2F^+	$CF_3^+ + CH_2FCl \rightarrow CH_2F^+ + CF_3Cl$[a]	$\Delta H_f(CH_2F^+) \leq 200 \pm 6$	199.5 ± 6[a]	200.3 ± 2[h]
	$CD_5^+ + CH_2FCl \rightarrow CH_2F^+ + DCl + CD_4$	$\Delta H_f(CH_2F^+) \geq 199 \pm 3$		
CF_2H^+	$CF_2H^+ + CH_2F_2 \rightleftarrows CH_2F^+ + CF_3H$[a,b,i]	$\Delta H_{Rn} = 0.3$ kcal/mole	142 ± 5[a,b]	
CF_2Cl^+	$C_2H_5^+ + CF_2Cl_2 \rightarrow CF_2Cl^+ + C_2H_5Cl$[a]	$k_{Rn}/k_{collision} \sim 0.008$	130 ± 5[a]	130 ± 2[c]
		$128 \pm 5 < \Delta H_f < 131 \pm 5$		132.4 ± 4[e]
CCl_2F^+	sec-$C_3H_7^+ + CFCl_3 \nrightarrow CFCl_2^+ + C_3H_7Cl$[a]	$\Delta H_f > 157 \pm 1.5$	155 ± 5	168.6 ± 1.5[c]
				181 ± 1.5[j]

Table VIII. Continued

Ion	Reaction	Thermochemical Information	ΔH_f(Ion) kcal/mole: From Ion-Molecule Reactions	ΔH_f(Ion) kcal/mole: Other Methods
$CHFCl^+$	$C_2H_5^+ + CHFCl_2 \rightarrow CHFCl^+ + C_2H_5Cl$ [a]	$\Delta H_f(CHFCl^+) > 177 \pm 3$	179 ± 5 [a]	194.7 [j]
	$CF_2Cl^+ + CHFCl_2 \rightarrow CHFCl^+ + CF_2Cl_2$ [a,i]	$\Delta H_f(CHFCl^+) \leq 181 \pm 3$		
$C_3H_4F^+$	$C_2H_4^+ + C_2H_3F \rightarrow C_3H_4F^+ + CH_3$ [k]		<185	
$C_3H_3F_2^+$	$C_2H_4^+ + CH_2CF_2 \rightarrow C_3H_3F_2^+ + CH_3$ [k]		<137	
$C_3H_2F_3^+$	$C_2H_3F^+ + CH_2CF_2 \rightarrow C_3H_2F_3^+ + CH_3$ [k]		<92	
$C_3HF_4^+$	$C_2H_4^+ + C_2F_4 \rightarrow C_3HF_4^+ + CH_3$ [k]		<56	
$C_3F_5^+$	$C_2HF_3^+ + C_2F_4 \rightarrow C_3F_5^+ + CHF_2$ [k]		<30	

References

a. Reference 39. b. References 33 and 34. c. Reference 45. d. Reference 46. e. Reference 47.
f. Reference 8. g. See discussion. h. Reference 48. i. Reference 38. j. Reference 49.
k. Reference 17.

Table IX. Thermochemical Properties of Fluoromethyl Cations.

R	$\Delta H_f(R)$	Stabilization[a]	$\Delta H_f(R^+)$	Stabilization[a]	IP eV	$D[R^+-H^-]$	$D[R^+-F^-]$	$D[R^+-Cl^-]$
CH_3	34.82 ± 0.2[b]	---	260.9 ± 1[e]	---	9.80[h]	311 ± 0.1	255 ± 2	224 ± 0.2
					9.84[i]			
CH_2F	-4.9[c(1)] ±2	39.7 ± 2	200.3 ± 2[c,f]	60.6 ± 4	8.90[j]	290 ± 2	246 ± 2	206 ± 5
	$-7.8\pm$?[d]	$42.6\pm$?						
CHF_2	-59.2[d] $\pm$?	$54.3\pm$?	142 ± 5[c,f]	58.3 ± 7	8.74[c(2)] ±0.2	283 ± 5	246 ± 6	201 ± 8
CF_3	-112.4 ± 1[b]	$53.2\pm$?	94.1 ± 2[f,g]	47.9 ± 7	8.95[l]	293 ± 3	255 ± 2	207 ± 3
			91.5[h]	50.5 ± 7	±0.1			

a. Lowering of ΔH_f effected by replacement of one H-atom by one F-atom, i.e. the difference between the heat of formation of R^+ and that of the ion containing one less F-atom.

b. Reference 50.

c. References 33 and 34. (1) Calculated from $\Delta H_f(CH_2F^+)$ and $IP(CH_2F)$. (2) Calculated from $\Delta H_f(CHF_2^+) - \Delta H_f(CHF_2)$.

d. Reference 51.

e. Reference 53.

f. Reference 39.

g. Reference 45. h. Reference 47. i. Calculated from $\Delta H_f(CH_3^+) - \Delta H_f(CH_3)$. j. Reference 49.

k. Reference 54. l. Calculated from $\Delta H_f(CF_3^+) - \Delta H_f(CF_3)$, taking $\Delta H_f(CF_3^+) = 94.1$ kcal/mole.

Each fluorine atom substituted for a hydrogen atom decreases the heat of formation of the ion by ca. 44 kcal/mole.

Quantitative information about the heats of formation of the $C_3(H,F)_5^+$ ions is not available, although a detailed theoretical calculation of the geometries and charge densities has appeared (52). Anicich and Bowers (17), have reported maximum values for the heats of formation of the ions in the series based on the observation of the reactions listed in Table VIII. They point out that these upper limits correlate well with the trends in the heats of formation one would predict taking $\Delta H_f(C_3H_5^+)$ as 216 kcal/mole (49), and assuming that the stabilization effected by F-atom substitution is -48 $\pm$ 5 kcal/mole, as illustrated in Figure 3.

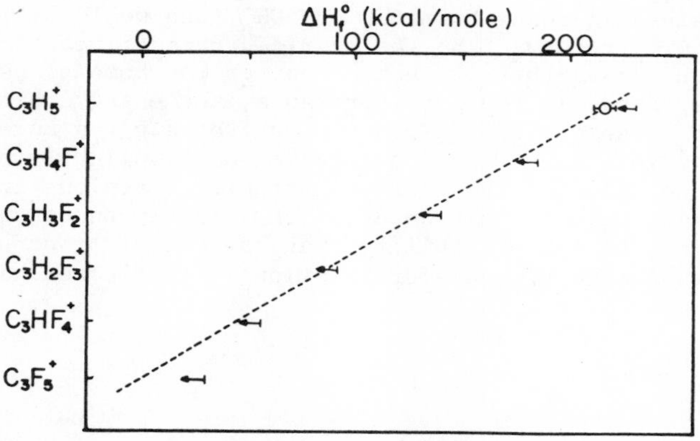

Figure 3. Plot of the heat of formation vs. the number of fluorine substituents in the ion series $C_3(H,F)_5^+$. Arrows and bars note the upper limits to the heats of formation set by reactions listed in Table VIII. Heat of formation of $C_3H_5^+$ was taken as 216 kcal/mol (49). The dashed line represents the expected change in heat of formation with fluorine substitution as discussed in the text.

In the earlier discussion of F^- transfer reactions in perfluorinated alkanes, we remarked upon the contrast between the ion-molecule chemistry in fluorocarbon and hydrocarbon systems. That is, hydrocarbon carbonium ions will nearly always undergo exothermic hydride transfer reactions with alkanes having a larger number of carbon atoms, but this is apparently not the case in fluorocarbon systems. An examination of the thermochemistry of these reactions explains these differing trends. Table X contrasts the energy gained in going from R^+ to RH for CH_3^+, $C_2H_5^+$, and sec-$C_3H_7^+$, with the energy gained in going from R^+ to RF in the corresponding fluorinated systems.

Table X. Energy Differences Associated with F^- Transfer and H^- Transfer in Corresponding Fluorocarbon and Hydrocarbon Systems.

Reactant-Product Pair	ΔH kcal/mole	Reactant-Product Pair	ΔH kcal/mole
CH_3^+-CH_4	- 279	CF_3^+-CF_4	-317
$C_2H_5^+$-C_2H_6	- 239	$C_2F_5^+$-C_2F_6	-319
sec-$C_3H_7^+$-C_3H_8	- 217	sec-$C_3F_7^+$-C_3F_8	<-317

These numbers show that in hydrocarbon systems, more energy would be gained in going from CH_3^+ to CH_4 than would be lost in going from C_2H_6 to $C_2H_5^+$, so the hydride transfer reaction between CH_3^+ and C_2H_6 is exothermic, and so on, up the homologous series. On the other hand, in the perfluorokanes, the energy change associated with F^- transfer to a given ion or from a given molecule does not change significantly or possibly increases slightly over the homologous series, and therefore F^- transfer reactions in these systems will always be thermoneutral or slightly endothermic. This explains the low probability that thermal fluorocarbon carbonium ions undergo F^- transfer reactions with fluoroalkanes (Table I).

Literature Cited

1. For reviews, cf. Franklin, J. L., ed., "Ion-Molecule Reactions", Plenum Press, New York, 1972.
2. For reviews, cf. Ausloos, P., ed., "Interactions Between Ions and Molecules", Plenum Press, New York, 1975.
3. For review, cf. Lias, S. G. and Ausloos, P., "Ion-Molecule Reactions; Their Role in Radiation Chemistry", Amer. Chem. Soc. and ERDA Monograph, Washington, D.C., 1975.
4. King, J., and Elleman, D. D., J. Chem. Phys. (1968), 48, 412.
5. Su, T., and Kevan, L., J. Phys. Chem. (1973), 77, 148.
6. Marcotte, R. E., and Tiernan, T. O., J. Chem. Phys. (1971), 54, 3385.
7. Su, T., Kevan, L., and Tiernan, T. O., J. Chem. Phys. (1971), 54, 4871.
8. Sieck, L. W., Gorden, R., Jr. and Ausloos, P., J. Res. NBS (1974), 78A, 151.
9. Pabst, M. J. K., Tan, H. S., and Franklin, J. L., Int. J. Mass Spectrom. Ion Phys. (1976), 20, 191.
10. Kevan, L., and Hamlet, P., J. Chem. Phys. (1965), 42, 2255.
11. Cooper, R., and Haysom, H. R., J. Chem. Soc. Faraday Trans. II (1973), 69, 904.
12. Sokolowska, A., and Kevan, L., J. Phys. Chem. (1967), 71, 2220.
13. Gioumousis, G., and Stevenson, D. P., J. Chem. Phys. (1958), 29, 294.
14. Eyler, J. R., Ausloos, P., and Lias, S. G., J. Am. Chem. Soc. (1974), 96, 3673.
15. Derwish, G. A. W., Galli, A., Giardini-Guidoni, A., and Volpi, G. G., J. Am. Chem. Soc. (1964), 86, 4563.
16. Anicich, V. G., Bowers, M. T., O'Malley, R. M., and Jennings, K. R., Int. J. Mass Spectrom. Ion Phys. (1973), 11, 99.
17. Anicich, V. G., and Bowers, M. T., Int. J. Mass Spectrom. Ion Phys. (1974), 13, 359.
18. Sieck, L. W., and Ausloos, P. J., J. Res. NBS (1972), 76A, 253.
19. LeBreton, P. R., Williamson, A. D., and Beauchamp, J. L., J. Chem. Phys. (1975), 62, 1623.
20. Lifshitz, C., and Tiernan, T. O., J. Chem. Phys. (1971), 55, 3555.
21. O'Malley, R. M., Jennings, K. R., Bowers, M. T., and Anicich, V. G., Int. J. Mass Spectrom. Ion Phys. (1973), 11, 89.
22. Ferrer-Correia, A. J., and Jennings, K. R., Int. J. Mass Spectrom. Ion Phys. (1973), 11, 111.
23. Anicich, V. G., and Bowers, M. T., Int. J. Mass Spectrom. Ion Phys. (1973), 12, 231.
24. Anicich, V. G., and Bowers, M. T., Int. J. Mass Spectrom. Ion Phys. (1974), 13, 351.
25. Anicich, V. G., and Bowers, M. T., J. Am. Chem. Soc. (1974), 96, 1279.
26. Sieck, L. W., Gorden, R., Jr., Lias, S. G., and Ausloos, P., Int. J. Mass Spectrom. Ion Phys. (1974), 15, 181.

27. Williamson, A. D., and Beauchamp, J. L. J. Chem. Phys., (1976), in press.
28. Field, F. H., J. Am. Chem. Soc. (1961), 83, 1523.
29. Meisels, G. G., J. Chem. Phys. (1965), 42,, 3237.
30. Gorden, R., Jr., and Ausloos, P., J. Chem. Phys. (1967), 47, 1799.
31. Beauchamp, J. L., Holtz, D., Woodgate, S. D., and Patt, S. L., J. Am. Chem. Soc. (1972), 94, 2798.
32. Henis, J. M. S., Loberg, M. D., and Welch, M. J., J. Am. Chem. Soc. (1974), 96, 1665.
33. Blint, R. J., McMahon, T. B., and Beauchamp, J. L., J. Am. Chem. Soc. (1974), 96, 1269.
34. McMahon, J. B., Blint, R. J., Ridge, D. P., and Beauchamp, J. L., J. Am. Chem. Soc. (1972), 94, 8934.
35. Marshall, A. G., and Buttrill, S. E., J. Chem. Phys. (1970), 52, 2752.
36. McAskill, N. A., Aust. J. Chem. (1970), 23, 2301.
37. Harrison, A. G., and McAskill, N. A., Aust. J. Chem. (1971), 24, 1611.
38. Dawson, J. H. J., Henderson, W. G., O'Malley, R. M., and Jennings, K. R., Int. J. Mass Spectrom. Ion Phys. (1973), 11, 61.
39. Lias, S. G., and Ausloos, P., Int. J. Mass Spectrom. Ion Phys., submitted for publication.
40. Beauchamp, J. L., and Park, J. Y., J. Phys. Chem. (1976), 80, 575.
41. Lias, S. G., Eyler, J. R., and Ausloos, P., Int. J. Mass Spectrom. Ion Phys. (1976), 19, 219.
42. Ausloos, P., and Lias, S. G., unpublished results.
43. Beauchamp, J. L., "Interactions Between Ions and Molecules" (Ausloos, P., ed.), p. 413, Plenum Press, New York, 1975.
44. Ausloos, P., Lias, S. G., and Eyler, J. R., Int. J. Mass Spectrom. Ion Phys. (1975), 18, 261.
45. Ajello, J. M., Huntress, W. T., and Rayerman, P., J. Chem. Phys. (1976), 64, 4746.
46. Walker, T. A., Lifshitz, C., Chupka, W. A. and Berkowitz, J., J. Chem. Phys. (1969), 51, 3531.
47. Jochims, H. W., Lohr, W., and Baumgartel, H., Ber. Bunsengesellschaft. (1976), 80, 130.
48. Krauss, M., Walker, J. A., and Dibeler, V. H., J. Res. NBS (1968), A72, 281.
49. Franklin, J. L., Dillard, J. G., Rosenstock, H. M., Herron, J. T., Draxl, K., and Field, F. H., NSRDS-NBS 26 (1969).
50. JANAF Thermochemical Tables, NSRDS-NBS 37 (1971).
51. Kerr, J. A., and Timlin, D. M., Int. J. Chem. Kinetics (1971), 3, 427.
52. Kispert, L. D., Pittman, C. U., Jr., Allison, D. L., Patterson, T. B., Jr., Gilbert, C. W., Jr., Hains, C. F., and Prather, J., J. Am. Chem. Soc. (1972), 94, 5979.
53. Herzberg, G., and Shoosmith, J., Can J. Phys., (1956), 34, 523.

54. Lossing, F. P., Bull. Soc. Chim. Belg. (1972), 81, 125.
55. Martin, R. H., Lampe, F. W., and Taft, R. W., J. Am. Chem. Soc. (1966), 88, 1353.
56. Olah, G. A., and Mo. Y. K., Adv. in Fluorine Chem. (1973), 7, 69.

Non-Boltzmann Kinetics

7

Crossed Molecular Beam Studies of Fluorine Chemistry

J. M. FARRAR and Y. T. LEE

Materials and Molecular Research Division, Lawrence Berkeley Laboratory, and Department of Chemistry, University of California, Berkeley, CA 94720

Fluorine atom chemistry has received significant attention from chemical kineticists in recent years. Indeed, the present volume bears testimony to the proliferation of activity in this field. The molecular beam technique, which has come of age during the past several years, has been applied with great success to the chemistry of fluorine atoms and it is this subject which we will address in the present article. The crossed molecular beam technique (1) applied to problems in chemical kinetics involves colliding well-defined beams of atoms and molecules in a vacuum chamber; from the direct measurements of results of two-body collisions, in particular, angular and energy distributions of reaction products, one attempts to visualize in great detail how the reaction occurs. These dynamical features of chemical reactions, the favored orientation of molecules, reaction intermediate lifetimes, and product energy distributions are among the major concerns of a crossed molecular beam experiment.

The experimental results which we will describe are primarily those obtained in this laboratory but a few experimental data exist which have been collected elsewhere. Our experimental program in fluorine atom chemistry has been motivated primarily by two facts which have also been important to studies performed by other methods: (1) atomic fluorine abstraction of hydrogen atoms from appropriate molecules has been demonstrated to be an important class of reactions for chemical lasers (2). In particular, the reactions of $F + H_2 \rightarrow HF + H$ and $F + D_2 \rightarrow DF + D$ have been investigated in great detail by various theoretical and experimental approaches (3-11); the latter reaction provides us with an example from the general class of reactions of fluorine atoms with diatomic molecules. (2) Substitution reactions of fluorine atoms with unsaturated hydrocarbons involving the formation of C-F bonds frequently are observed to proceed through a "complex" which

© 0-8412-0399-7/78/47-066-191$10.00/0

lives many rotational periods (12-19). Since the C-F bond is the strongest single bond involving a carbon atom, these reactions are generally quite exoergic with cleavage of C-C, C-H, and C-Cl bonds occurring readily. Since many of these reactions do proceed through intermediates, the competition among various modes of bond cleavage in the long-lived intermediate yields valuable information on intramolecular energy transfer in the complex prior to decomposition to products. Various studies of reactions of fluorine atoms with hydrocarbons, both aliphatic and aromatic, as a function of substituent placement and identity, chain length, and initial kinetic energy of the reactants thus make the fluorine atom chemistry observed a very interesting probe of unimolecular decay. Many of the first studies of unimolecular decompositions from this laboratory were concerned with the determination of product branching ratios and recoil velocity distributions in olefins, aromatics, and heterocyclics at a fixed collision energy. Certain characteristic features of these reactions began to emerge in terms of the nature of the exit channel, that is, whether the critical configuration corresponded to a potential energy maximum or merely resulted from angular momentum, and the nature of the group emitted. Additional work on selected systems as a function of the initial collision energy has led to a further understanding of the role of the potential surface in unimolecular decompositions.

Experimental Method

The molecular beam technique, in general, is ideally suited to the study of reactions of highly reactive species such as fluorine atoms; since beam intensities and detector sensitivities are limited even under the most favorable experimental conditions, only for reactions with large cross sections and favorable kinematic relations can one hope to measure both energy and angular distributions of product molecules by crossing two well-defined reactant beams at a given collision energy. Since the internal states or the velocities of the reactants can be selected, the initial relative velocity vector can be well-defined in magnitude and direction and thus the measured quantities in a molecular beam experiment, the angular and energy distributions of scattered products, can be related to the mechanics of single collisions. The differential cross section for chemical reaction, which is the quantity most easily related to classical and quantum mechanical models based on collision theory and from a knowledge of potential energy hypersurfaces, is determined rather directly in a beam experiment, although present experimental data yield cross sections averaged to a great extent over initial impact parameter and final internal state distributions. This situation contrasts, however, with the much more highly averaged rate constant of classical kinetics which is an average over the initial Boltzmann distribution of the reaction cross section at a particular relative velocity weighted by the relative velocity. In principle the beam technique provides the means for studying state-to-state chemical kinetics on favorable systems, and highly refined experiments employing two crossed beams with specified internal states and relative velocities are definitely possible in the near future.

In molecular beam experiments, all measurements are carried out in a laboratory system of coordinates, but one wishes to interpret data in terms of a coordinate system which moves with the center of mass (c.m.) of the colliding system. The manner in which this transformation is made is well-understood (20-23), but the Jacobian involved in the transformation distorts certain portions of the lab data; in addition, the lab-c.m. transformation is often not single-valued. In order to circumvent this problem, data fitting routines have been developed which assume c.m. angular and velocity distributions, averaging them and transforming them back to the lab for comparison with the data. When the initial velocity distributions are quite narrow, this technique allows one to recover the c.m. distributions reliably. A velocity vector diagram shown in Figure 1 illustrates the nature of the transformation. The vector $\underset{\sim}{C}$, which points in the direction of the c.m. determined by partitioning the initial relative veocity vector according to conservation of linear momentum, allows us to relate laboratory velocity, $\underset{\sim}{v}$, and scattering angle, Θ, to c.m. velocity, $\underset{\sim}{u}$, and angle, θ, through the vector addition $\underset{\sim}{u} + \underset{\sim}{C} = \underset{\sim}{v}$. Since we determine $\underset{\sim}{v}$ and we desire $\underset{\sim}{u}$, we must deconvolute $\underset{\sim}{C}$. If the initial conditions create a substantial spread in vectors $\underset{\sim}{C}$, then the set of vectors $\underset{\sim}{u}$ which we determine to fit a measured $\underset{\sim}{v}$ are not unique. In fact, some experimental data have been interpreted incorrectly by failing to account correctly for the spread in initial conditions and also failing to use the correct Jacobian to relate lab and c.m. fluxes. We refer the reader to an earlier review (24) for further discussion of this point.

The development of a crossed molecular beam apparatus capable of the detection of arbitrary molecular fragments resulting from chemical reaction has been central to the advancement of molecular beam kinetics. The first reliable and sensitive apparatus of this sort employing electron-bombardment ionization of reaction products in an ultrahigh vacuum chamber followed by mass spectrometric detection became operational in 1968 (25) and a number of instruments of this type are now employed in various laboratories. This so-called "universal" detector together with the advent of free jet molecular beams (26) moved molecular beam chemical kinetics out of the "alkali age", so named because surface ionization detection techniques used previously limited the species to be studies to alkali atoms and alkali halides. The general experimental arrangement employed in a crossed molecular beam apparatus is shown in Figure 2. Reagent atomic species formed by thermal dissociation in a heated

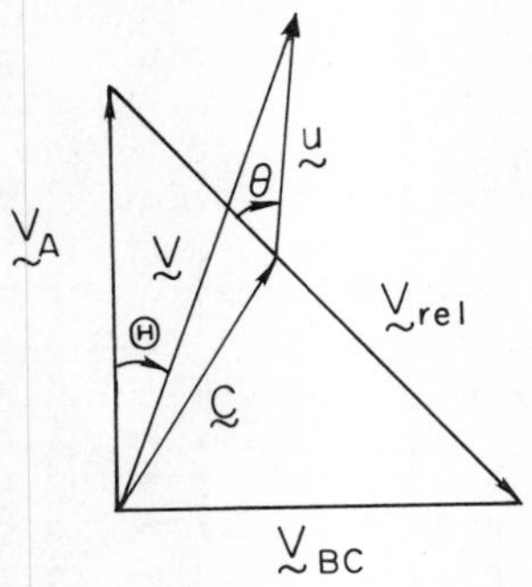

Figure 1. Velocity vector Newton diagram showing relationship among vectors ***u****,* ***v****, and* ***C***

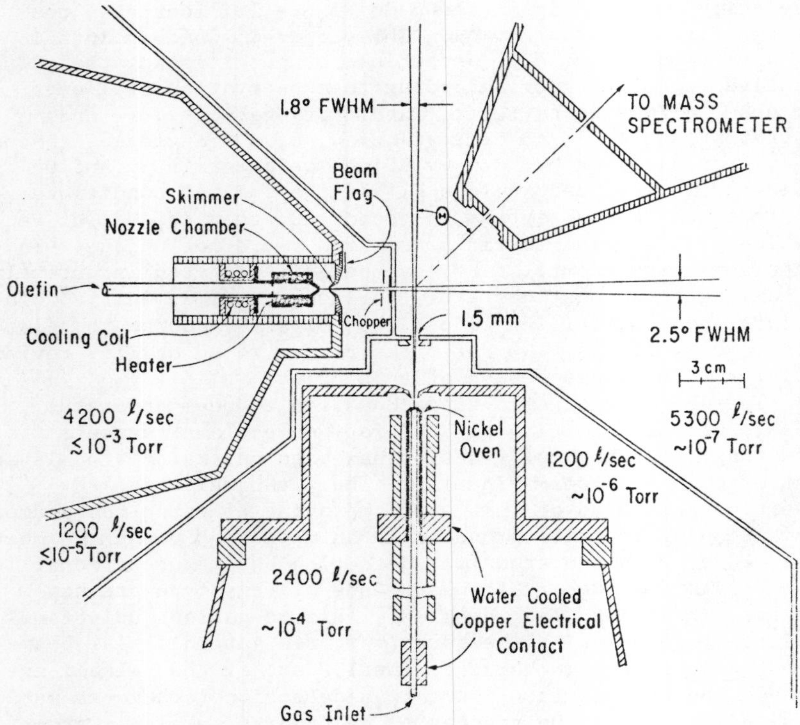

Figure 2. Experimental arrangement for molecular beam studies of fluorine atom chemistry

oven are collimated into a beam; separate beam formation, collimation, and reaction chambers are used to insure that the beams are well-defined and, depending on the nature of their formation, do not present a large gas load to the collision chamber. The details of halogen-atom beam formation are discussed in greater detail below. The secondary beam of reactant molecules is formed in a similar manner and intersects the atomic halogen beam at 90° in a collision chamber maintained at 10^{-7} torr by an oil diffusion pump. Collisions then occur at the intersection zone of the beams; for measurements of angular distributions, reactively scattered products are detected in the plane defined by the beams as a function of laboratory scattering angle, Θ, by a rotatable mass spectrometer detector. This detector is comprised of three nested vacuum chambers, each pumped with a separate ion pump (and cryogenic pumping in the innermost chamber) to achieve a pressure of less than 10^{-10} torr in the ionization region. An electron bombardment ionizer of approximately 0.1% efficiency produces ions which are then injected into a quadrupole mass filter. While the pressure in the ionization region is ~10^{-10} torr, the partial pressure of reaction products may be as low as 10^{-15} to 10^{-16} torr when they reach ionization region. At such low signal levels, very careful work must be done to extract meaningful signals, including mechanical design to eliminate sources of background molecules in the detector, elaborate cryogenic pumping, and beam modulation techniques as discussed below. After the ions pass through the mass filter, they are accelerated and collimated into a beam which is directed to an aluminum coated cathode held at -30 kV. Several secondary electrons are emitted per incident ion; these electrons are accelerated in the same 30 kV potential to a Al coated plastic scintillator which emits several photons per secondary electron. The photon pulses are amplified by a photomultiplier tube and counted using conventional techniques. As indicated above, beam modulation is used for low-level signal recovery and to effect this, the secondary (molecular) beam is chopped with a tuning fork at 150 Hz. A signal from the chopper is processed by a digital logic circuit which gates a dual scaler in synchronization with the beam modulation. In this configuration, one scaler counts signal plus background corresponding to the beam open period; the other scaler collects background only. In this manner, reactive scattering signals which generally range from a few counts per second to several hundred per second can be recovered.

The total energy of the system under study consists of the initial relative and internal energies plus the exoergicity of the reaction. This energy can be divided among internal degrees of freedom plus relative translational energy of the products. In our molecular beam experiments, we measure velocity distributions of scattered products to infer internal energy distributions by conservation of energy. On the other hand, detailed state distributions for some simple diatomic product molecules could be measured by using the molecular beam resonance method (27,28) or laser-induced fluorescence (29,30).

Velocity distributions of the reactively scattered products can be measured by time of flight (TOF) methods in which the velocity of a particle is determined by measuring the time required for the particle to traverse a known distance in the detector. Although not indicated in Figure 2, a chopper with very narrow slits chops the reactively scattered products as they enter the detector at a frequency (1600 Hz, typically) which is high compared to the secondary beam modulation frequency. The burst of molecules then enters the detector at time t_o and the pulse spreads in time during transit through the detector in accordance with the velocity distribution of molecules in the pulse. A multi-channel scaler then records the distribution of flight times of the products; signal averaging is accomplished by minicomputer control of the TOF unit. Flight times of 100-400 μsec through our detector (pathlength 17 cm) are measured quite readily using this technique.

Having discussed in general terms how angular and velocity distributions of reactively scattered products are measured using a "universal" molecular beam apparatus, we now focus attention on methods of beam production, in particular, techniques for the production of stable, intense beams of fluorine atoms. Since the bond energy of F_2 is 36 kcal mole^{-1}, thermal dissociation of F_2 in a heated nickel oven has been found to be a successful method for fluorine atom production. The high reactivity of fluorine atoms with various materials rules out a great number of them for oven construction and we have found that nickel affords the best readily available material. Thermal dissociation of F_2 in a nickel oven at a pressure of ~0.1 - 1.0 torr at 750°C can be accomplished with the degree of dissociation approaching 50 - 90%. At such low pressures, beam formation is determined by effusion through an orifice and the velocity distribution of the atoms is the Maxwell-Boltzmann distribution. In the absence of velocity selection, such a broad distribution of initial speeds leads to a very broad distribution of initial kinetic energies, with the relative velocity vector and the center of mass vector distributed over a large region of velocity space. A slotted disc velocity selector (31) can be used to reduce this spread in initial conditions and much of the work discussed in this paper was performed with such a selector with a full width half maximum (FWHM) velocity spread of 20%. Such a small spread in the initial relative velocity of the reactants enables the lab - c.m. transformation to be made more uniquely, thus assisting us in the interpretation of the data.

The velocity selected fluorine atom source was the workhorse for much of the work done in this laboratory, but recent developments in seeded supersonic beam technology have led to much more efficient means of halogen atom beam production. The expansion of a gas at a few hundred torr through a small orifice into a vacuum has been shown to be an effective means for the production of high intensity, narrow velocity spread atomic and molecular beams. By mixing 1 to 2% F_2 with a rare gas carrier such as argon or helium and expanding the mixture at 500 torr through an orifice (0.1 mm dia) in a nickel oven heated to 750-800°C, fluorine atom beams were produced which were one to two orders of magnitude more intense than velocity selected beams and had velocity distributions which were somewhat narrower than selected beams, generally with a 15% FWHM spread. Such a method of beam production maintains the low partial pressure of halogen to make the equilibrium favor atom production while providing a high pressure expansion under hydrodynamic flow conditions leading to a narrowing of the velocity distribution.

Under the hydrodynamic flow conditions of supersonic expansions, species of different masses in a molecular beam achieve the same terminal velocity which is related to the average mass number of the expanding species. This technique can be used to increase the speed of a heavy particle relative to its thermal speed if a small fraction of the gas mixture is the desired heavy species and the bulk of the mixture is a light carrier gas. This technique has been used very successfully to produce hyperthermal beams of Xe atoms formed by expansions of mixtures composed of 1% Xe in 99% H_2. By varying the diluent gas from argon to helium, high energy beams of fluorine atoms can be produced for reactions in the hyperthermal region (10-40 kcal $mole^{-1}$).

This technique of halogen atom production is not unique to fluorine; in fact, the first halogen atom beam produced in this laboratory was a chlorine atom beam produced by dissociation of Cl_2, in argon buffer gas in a heated graphite oven (32). The technique should be applicable to all of the halogen atoms and a vast variety of scattering experiments can be initiated using such sources.

The molecular beams in our experiments are also formed by supersonic expansion. In addition to enhanced intensity and narrow velocity distributions obtained in this way, the molecules in such beams can generally be considered to be in their lowest vibrational and rotational states. This "freezing out" of internal degrees of freedom is important in further defining the initial experimental conditions. In understanding the mechanics of collisions, one can obtain simplifications in interpretation by knowing that reactant molecules are rotationally and vibrationally "cold". This point will be discussed further in later sections of this paper.

Examples of Experimental Results

I. $F + D_2 \rightarrow DF + D$. The reaction of atomic fluorine with molecular hydrogen containing molecules has been demonstrated to be the chief mechanism responsible for the HF chemical laser and has received significant experimental and theoretical attention. In addition to chemical laser (2,11,33,34) and infrared chemiluminescence (5,6) measurements on this system which have yielded values for a number of rate constant ratios for production of specific vibrational states of HF, the molecular beam technique (3,4) has also provided significant data on this reaction. The $F + D_2$ reaction to form DF is exoergic by 31.5 kcal $mole^{-1}$, yielding enough energy in conjunction with translation to produce DF in excited vibrational states up to v' = 4. The equal gain and zero gain variations of the chemical laser technique (2) allow for the determination of rate constant ratios $k_{v'}/k_{v'-1}$ for v' = 1, 2, 3, and 4 where v' is the DF vibrational quantum number and infrared chemiluminescence measurements allow for determinations of this ratio for v' = 2-4. In contrast, the crossed molecular beam technique is not very ideal for measurements of relative state populations although, as we shall see, some information of this type can be determined under favorable conditions.

The laws of conservation of energy and angular momentum play a very important role in determining what information is contained in the molecular beam data on this system. A consideration of rate constant data (35) for this system in conjunction with the recognition that the rate constant is an ensemble average of cross section weighted by relative velocity, allows us to estimate the maximum impact parameter contributing to chemical reaction. At 300°K, the cross section is $\lesssim 1$ Å^2 so b_{max} is roughly 1 Å. For an impact parameter of 1 Å and $v_{rel} \sim 2 \times 10^5$ cm sec^{-1} (for a collision energy of 1.7 kcal $mole^{-1}$), $L = \mu vb$ or approximately $10\hbar$ in magnitude. Since at room temperature, J, the rotational angular momentum of D_2, is no more than $2\hbar$, we can use the conservation relation, $\mathcal{J} = \underset{\sim}{L} + \underset{\sim}{J} = \underset{\sim}{L}' + \underset{\sim}{J}'$, to allow us to conclude that J' is likely to be small also ($10 \sim 12\hbar$). Since the exit channel reduced mass is small, even if all of the initial orbital angular momentum appears as product rotational excitation, the rotational energy of DF will be approximately 3 - 5 kcal which is less than the vibrational level spacing of 8 kcal. Thus, by conservation of energy, only discrete regions of product translational energy space will be populated and "quantization" of the product velocity distribution should occur. In a crossed beam experiment with well-defined beam velocities, one thus expects significant structure in the product velocity spectrum. Under very favorable kinematic circumstances, particularly at low collision energies where the angle between $\underset{\sim}{V}_{rel}$ and $\underset{\sim}{V}_{D_2}$ is quite large, such structure should also be observable in the angular distributions. Figure 3 demonstrates this point clearly. Two discrete peaks are observable at both the collision energies; the lower panel of this figure shows Cartesian plots corresponding to these experiments. The significant feature of both of these contour maps is that the DF product scatters predominantly in the backward direction in the c.m. system, corresponding to net repulsive encounters; as the collision energy is increased, the sharpness of the backward scattering diminishes and products are formed scattered through angles significantly smaller than 180°.

Experiments performed at higher collision energies continue to show the trend toward more forward scattering, but the predominance of backward scattering is still in evidence. While all of the experimental data show structure arising from individual vibrational states of DF, determination of rate constant ratios is very difficult since one must integrate flux over c.m. velocity to assess these ratios. The $1/u^2$ Jacobian singularity near the c.m. is most pronounced in calculating the k_4/k_3 ratio and the centroid distribution, although narrow by contemporary standards, is broad enough to introduce prohibitively large uncertainties into calculations of relative populations.

The observation of DF predominatly scattered into the backward hemisphere at the collision energies studied here suggests that the favored orientation for reaction is collinear. Ab initio calculations of the potential energy surface for this system (10,36) suggest that collinear approach of F + H_2 is the dominating geometry leading to chemical reaction. Indeed, the backward scattered DF product is that which one expects from low impact parameter collisions. One can thus conclude that the beam data are consistent with a potential energy surface with predominant collinear character. For simple systems containing light atoms, the comparison between theoretical studies and experimental results will provide further understanding of reaction dynamics from first principle in the very near future.

II. Substitution Reactions of F Atoms with Unsaturated Hydrocarbons.

A. General Remarks on Unimolecular Decompositions of Chemically Activated Radicals. A major portion of the molecular beam studies of fluorine atom chemistry (12-19,37) has been concerned with a class of reactions characterized by the formation of a transient species from bimolecular association of the reactants and whose lifetime is long compared to its rotational or vibrational periods. The formation of such a long lived complex implies that the reactants experience a net attraction and consequently the potential energy surface for the reaction possesses a deep well; of course, the total energy of the system is greater than that required to dissociate the intermediate, either to reactants or products, so that in the absence of a third body or relatively improbable photon emission (radiative lifetime $\gtrsim 10^{-2}$ sec), the intermediate must decay prior to detection. Under certain favorable conditions where a large number of internal degrees of freedom participate in the redistribution of the excitation energy, the reactive intermediate could live long enough to travel to our detector, but none of the systems studied to date have exhibited this behavior.

The decay of a complex formed in such a bimolecular encounter provides us with an example of the general class of reactions denoted as unimolecular decompositions (38). Under circumstances in which a long-lived complex decomposes to products, a vast collection of questions can be asked regarding the dynamics of the decomposition, including the extent to which the initial excitation energy of the intermediate species is distributed throughout the entire molecule before a sufficient amount localizes in the bond to be broken as the products form. The most widely accepted model of unimolecular decay, the Rice-Ramsperger-Kassel-Marcus (RRKM) theory (38,39), states that a quasiequilibrium between the complex and the initial configuration is maintained through the reaction. This assumption is equivalent to stating that energy randomization in the complex is rapid compared to chemical reaction. The technique of chemical activation developed by Kistiakowsky (40) and extensively employed by Rabinovitch and co-workers (41) has provided kineticists with a large body of experimental data, the majority of which lend support to the assumptions of RRKM theory. In this technique, the complex is "synthesized" in a bulb reactor by a bimolecular reaction and the ratio of the concentration of complexes stabilized by collisions with background gas molecules to products of decomposition is measured as a function of pressure. Under these circumstances, the "clock" for these experiments becomes the time between collisions and if one assumes that the "strong collision" assumption is valid, that is that an activated molecule (one with enough energy to decompose) can be deactivated in a single collision with a buffer gas molecule, one can infer whether or not the product yields are characteristic of a statistical distribution of energy in the complex. Direct comparisons with RRKM theory can be made since this model is concerned with complex lifetimes and rates of decomposition, quantities which these conventional kinetic methods determine directly.

Information of a different sort is obtained in a molecular beam experiment, although the means for producing the species undergoing unimolecular decomposition is also chemical activation. Whereas the conventional kinetic studies yield reaction rates for direct comparison with RRKM lifetimes, the beam technique yields product recoil energy distribution which, in principle, contain information regarding exit channel dynamics specifically ignored in RRKM. Comparison of experimental results with RRKM theory is indirect, requiring additional assumptions whose validity must be determined. Fortunately, however, statistical theories of a different sort exist which base their predictions on asymptotic (and therefore measureable) properties of the products; these "phase space" theories (42,43) will be discussed in greater detail below.

Reactions studied using the chemical activation technique in beam or bulb experiments yield important information because the intermediate radicals have well-defined excitation energies; unimolecular decompositions involving the formation of a C-F bond have additional appeal because many competing channels for decomposition open up and studies of branching ratios for C-H, C-C, or C-Cl bond cleavage provide important chemical information. In a statistical model, the rate for a particular channel is proportional to the density of states at the critical configuration; consequently, one expects that more exoergic reactions should proceed with greater rates since more states are accessible under such circumstances.

In the discussion of specific experimental systems which follows, the bulk of our remarks will be limited to a discussion of product recoil velocity distributions and the information which such studies provide for us. An understanding of the energetics of unimolecular decomposition and their relation to product velocity distributions can be facilitated by referring to Figure 4. The excitation energy of the complex, E^*, is determined by the initial translational and internal energy of the reactants plus the well depth of the complex. Unimolecular decomposition to products can occur when energy E_o flows into the reaction coordinate such that the barrier V_a is surmounted. This barrier can include centrifugal energy plus potential energy terms; whenever terms of the latter type are present, the critical configuration is "tight", that is, incipient product rotations are not free, but correspond to bending vibrations (44). This potential energy barrier implies that the reverse association reaction has a nonzero activation energy; the nature of this barrier and its disposal as product translation and internal energy is a vital factor in interpreting measured product recoil distributions. The energy in excess of the critical configuration barrier, $E_T - V_a$, where $E_T = E_{trans} + E_{int} - \Delta D_o^o$ is the total available energy, is that quantity of energy which can be distributed among the various product oscillators plus the relative translational energy of the separating fragments. If some energy $\varepsilon^\dagger$ at the critical configuration is apportioned to the coordinate which becomes product translation, then $\varepsilon^\dagger$ should be a lower bound to the translational energy of the products. The manner in which the exit channel barrier releases its energy V_a among product degrees of freedom is a dynamical consideration not addressed by conventional RRKM theory, although recent efforts to include these dynamics in a "tight" transition state theory have been made (45). In the case where the critical configuration is tight, the product recoil distributions cannot be predicted by RRKM theory and any assumptions which one makes to connect the energy distribution at the critical configuration with that of reaction products must be assessed carefully. Certain

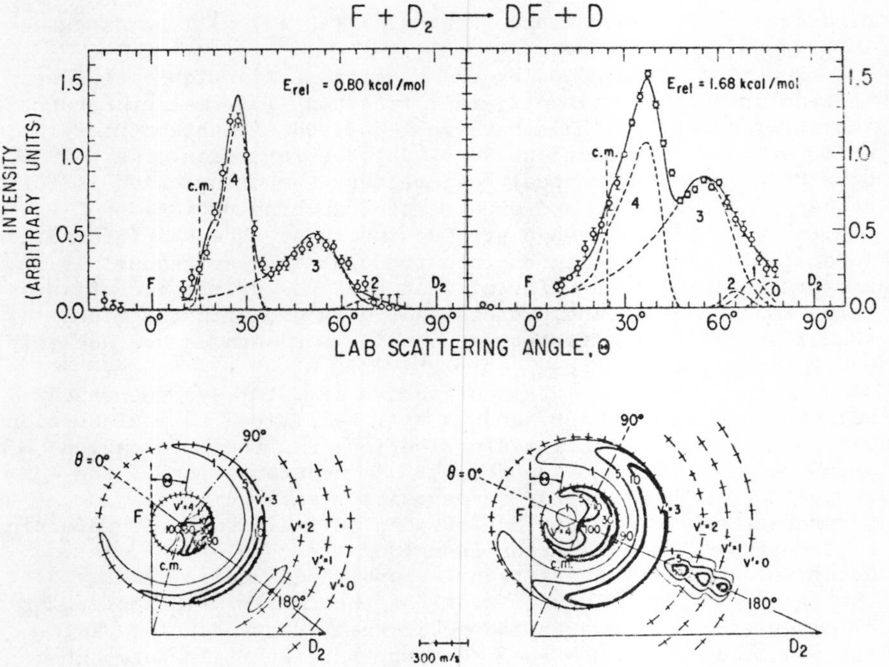

Figure 3. Upper panel: DF angular distributions at 0.80 kcal/mol and 1.68 kcal/mol collision energy. Lower panel: Cartesian flux contour maps generated from these data.

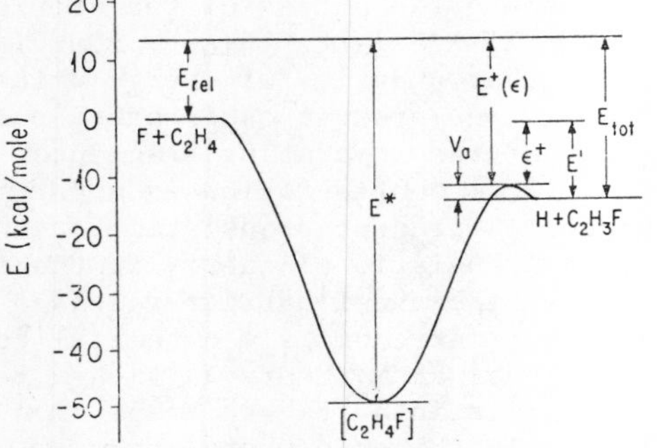

Figure 4. Schematic reaction coordinate for the system $F + C_2H_4$ relating various energies of complex and critical configuration along the reaction coordinate.

trajectory calculations (46) formulated to probe energy release from such a repulsive force in the exit channel suggest that when a light atom is emitted from a complex, such as during C-H bond cleavage, most of the exit channel barrier V_a appears in product translational energy. While such an energy transfer mechanism is valuable in interpreting our experimental data, we feel that one should avoid using RRKM theory without considering the exit barrier to predict product recoil energy distributions.

Other models have been formulated which predict product recoil energy distributions, most notably the phase space theory of Light and co-workers (42,43). This model possesses the desirable attribute that cross sections are computed by evaluating the volume in phase space which the products can sample which is accessible from a statistical complex whose conserved quantum numbers are given by the total energy, the total angular momentum and its projection on a space-fixed axis. Thus the theory is parameterized in terms of the observable properties of product molecules; however, establishment of the limits on the formation and decomposition of a strong-coupling complex frequently necessitates considerations of intermediate geometry, particularly when the rotational energy of the complex must become product angular momentum under the dictates of an exit channel barrier. A number of variations on the phase space theory have been proposed and the literature can be consulted for details (16).

A comparison of RRKM and phase space theory indicates that the degree and philosophy for parameterizing these models is quite different. In the former case, the only constant of the motion is the total energy of the system. Phase space theory is appealing because two more constants of the motion corresponding to angular momentum variables have been considered in the derivation. The recent versions of transition state theory which consider explicitly angular momentum in limiting cases where the total angular momentum becomes either product rotational or orbital angular momentum yield results in accord with phase space theory (45). The various models for product recoil distributions can be illustrated most fruitfully by consideration of several systems to which we now turn.

B. F + C_2H_4. A system which has received substantial attention in molecular beam experiments in this laboratory and which provides us with an opportunity to discuss many of the important features of unimolecular decay is the F + C_2H_4 reaction (12,19). The reaction to form C_2H_3F plus a hydrogen atom is exoergic by approximately 14 kcal mole^{-1} and the stability of the intermediate radical, C_2H_4F, is estimated to be ~50 kcal mole^{-1} with respect to reactants. A molecular beam experiment performed as indicated in the previous section, then, reveals the characteristic formation of the C_2H_3F product near the direction of the center of mass of the system, as shown in Figure 5. Velocity analysis of the reaction products confirms the symmetry of the product emission about $\theta_{c.m.} = \pi/2$ as expected for a complex which lives many rotational periods. Initial angular and velocity distribution measurements on this system yielded valuable information on the c.m. angular and energy distributions with the following notable points: the c.m. angular distribution measured at the relative collision energy of 1.98 kcal $mole^{-1}$ was symmetric about $\pi/2$ as expected, but the distribution was peaked at $\pi/2$ in the c.m. system. This peaking was explained by considering the angular momentum disposal in the chemical reaction: while the explanation has appeared in detail in previous publications (18), we repeat the salient features here. Recalling that the total angular momentum must be conserved in a collision, we note that since the C_2H_4 molecules are produced in a nozzle expansion and are therefore rotationally "cold", the initial orbital angular momentum then becomes the total angular momentum of the system and will be perpendicular to the initial relative velocity vector. During the lifetime of the complex, the total angular momentum is the rotational angular momentum of the complex, which rotates in the plane defined by the heavy atom C-C-F framework. As the C-H bond extends and breaks along the reformed p_z orbital on the sp^2 hybridized carbon atom, the emitted hydrogen atom is emitted along the rotational angular momentum vector. This vector becomes the final product rotational angular momentum since the emitted hydrogen atom cannot remove a significant amount of orbital angular momentum. We thus see that the products recoil along $\underset{\sim}{J}'$ which is perpendicular to $\underset{\sim}{V}$, thus producing the observed sideways peaking. A second notable feature of the observed c.m. distributions is found in the c.m. energy distribution: the shape of the distribution is quite wide and yields an average product kinetic energy which is ~50% of the total available energy, far in excess of statistical predictions. The interpretation of this result at a single collision energy is complicated by the fact that hydrogen addition to a halogenated olefin on the end of the molecule where the halogen resides proceeds with an activation of energy of ~3 kcal $mole^{-1}$ (47), therefore suggesting that the exit channel for the C_2H_4F decomposition to C_2H_3F + H has a barrier. The manner in which energy is released in descending this barrier then destroys the energy distribution at the critical configuration which RRKM theory predicts. A contour map of c.m. product flux as a function of polar coordinates (u,θ) is shown in Figure 6.

In order to provide further insight into the perturbation of the critical configuration energy distribution by the exit barrier, additional experiments on this system at higher collision energies have been undertaken. The seeded beam technique can be used to vary the collision energy from 2.2 kcal $mole^{-1}$ to 12.1 kcal $mole^{-1}$, thereby increasing the total energy available to the system from 16 to 28 kcal $mole^{-1}$.

The determination of the initial collision energy dependence of the unimolecular reaction dynamics can in principle, provide

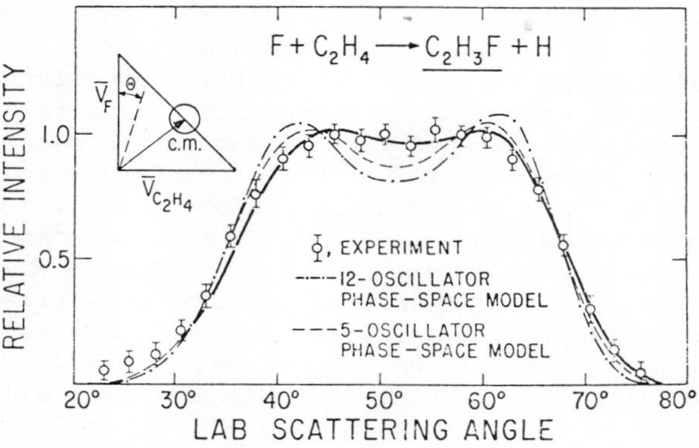

Figure 5. Experimental angular distribution for C_2H_3F from $F + C_2H_4$ $E_{rel} = 1.98$ kcal/mol

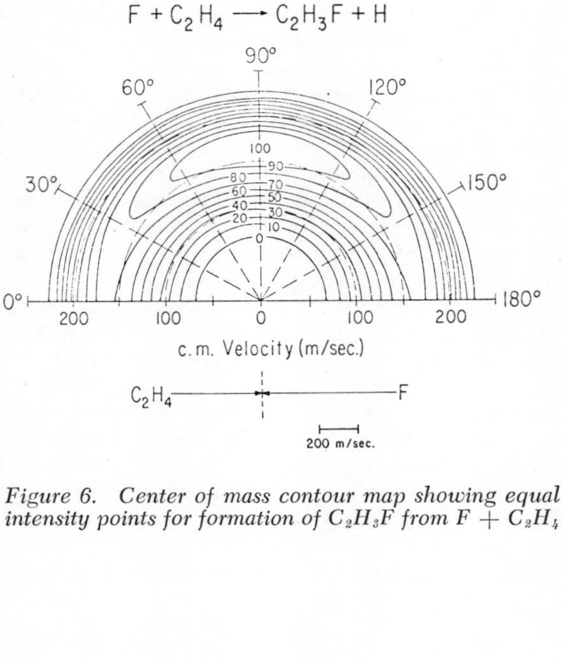

Figure 6. Center of mass contour map showing equal intensity points for formation of C_2H_3F from $F + C_2H_4$

two kinds of information. Since the energy which can be distributed amoung the various modes of the critical configuration is the total available energy in excess of the barrier, increasing this total energy should minimize the nonrandom contribution of V_a to P(E') as E_T increases. Thus, if a statistical model is correct and requires corrections arising from the specific dynamics associated with V_a, then as the collision energy is increased, P(E') should approach statistical behavior more closely. This is the information which we seek in the C_2H_4F system. In principle, information of a different sort can be obtained from the initial collision energy dependence of g(θ), the c.m. angular distribution function. When the binding energy E_o of the chemically activated intermediate becomes comparable to the collision energy, the complex lifetime should approach one rotational period and the g(θ) function should lose its symmetry about π/2. By then estimating the rotational period of the complex, one can then infer the complex lifetime. If one then knows accurately the stability, E_o, of the complex, model calculations to test statistical theories of unimolecular decay can be made. This experimental handle on the decay of a long-lived complex yields lifetimes which can be compared directly with quantities to which RRKM theory addresses itself directly. This desirable state of affairs is generally not achieved in beam experiments on chemically activated radicals; the large stability of the intermediate complex (~30-60 kcal $mole^{-1}$) compared to available collision energies in our experiments does not allow us to use the rotational period as a "clock". The C_2H_4F system, however, does allow us to use the first method for separating the role of the exit channel barrier in the measured recoil energy distributions.

The experimental data obtained over the range of collision energies 2.2 to 12.1 kcal $mole^{-1}$ are shown in Figure 7. Over this range of collision energies, the reaction appears to proceed through a long-lived complex, as evidenced by angular distributions symmetric about π/2 in the center of mass coordinate system. The sideways peaking noted in the earlier work is observed at all energies here, suggesting that the fractional partititioning of the total angular momentum into rotational and orbital angular momentum remains constant. The significant conclusion from these experimental data, however, is that the recoil energy distributions maintain a rather constant shape as a function of the initial kinetic energy, channeling ~50% of the total energy into translation of the products. A comparison of the measured average kinetic energy of the products with calculations performed with various statistical models is quite instructive. We have performed phase space calculations of the recoil energy distributions including the phase space volume associated with all of the product vibrational modes and with a selected subset of product vibrational modes; the former calculation, denoted "12-oscillator" model yields average kinetic energy releases which place a smaller fraction of the total energy into translation as the total energy increases. The latter phase space calculation in which the phase space volume associated with 5 product oscillators including only those modes expected to be active based on gross geometry changes during the reaction yields results in closer accord with experiment, but still somewhat discordant. This model, which heuristically incorporates partial activation of product oscillators, places a larger fraction of the total excess energy in translation since fewer oscillators share the energy, but again, the fraction of the total energy appearing in translation decreases at higher energies, in contrast to the 50% value observed at all energies. A third model, recently proposed by Marcus (45) accounts for the nature of the "tight" critical configuration by assuming that the quantum numbers for the bending vibrations in the transition state which become product rotations are "adiabatic on the average". This model then channels additional energy into translation since the energy of a bending vibrational quantum is greater than for a rotational quantum, the deficit accounted for by translation; calculations with this model yield results quite similar to the 12-oscillator phase space calculation. The statistical models, then, yield values of the average kinetic energy which are neither qualitatively or quantitatively correct. In all cases, the models place too little translational energy in the products and the fraction of the total energy appearing in translation decreases with increasing collision energy. The 5-oscillator phase space model yields results in closer accord with experiment, but this model certainly cannot be correct because a calculation of the complex lifetime performed using the density of states of these vibrations yields lifetimes which are smaller than one rotational period at all collision energies. Since the experimentally observed angular distributions place a lower bound on the lifetime of the complex of several rotational periods, we can be sure that the C_2H_3F recoil energy distributions cannot be explained by simple reduction of the number of oscillators participating in the decomposition.

Rowland (48) has studied the C_2H_4F system by measuring the pressure dependence of the stabilization-decomposition ratio for C_2H_4F produced in the hot atom reaction of $^{18}F + C_2H_4$; his results suggest that the lifetime of the complex is on the order of 10^{-9} sec, a result which can only be explained easily by including all of the complex's vibrations in a lifetime calculation. An explanation of the lifetime results and the recoil distributions must certainly require an explicit treatment of the dynamics of intramolecular energy transfer connecting the complex with the critical configuration. The normal modes associated with formation of the C_2H_4F complex are quite different from those required to form the products

(49); slow energy migration between those two subsets of oscillators might explain the long observed lifetime while the participation of a reduced set of vibrational modes in the decomposition may account for the disproportionately large fraction of the total energy in product translation. While the mechanical explanation for the large average kinetic energy release in this system has not yet been developed, the study of this reaction as a function of initial collision energy suggests strongly that the exit channel barrier does not account for the nonstatistical distribution of the excess energy.

C. The Effect of Potential Energy Barrier in the Exit Channel. We have indicated in the previous section that exit channel interactions can play a significant role in determining the translational energy distribution of product molecules. In the decomposition of C_2H_4F, we have demonstrated that to a large extent, the effect of the barrier can be minimized by studying the energy dependence of the product recoil distribution. A number of important points can be made regarding the systematics of the exit channel barrier by considering the decompositions of substituted fluorobenzenes. Table I and Figure 8 show the energetics associated with a number of these reactions and Figure 9 portrays some of the recoil distribution results graphically.

Table I

Energetics and Average Product Translational Energies, Fluorine + Substituted Benzenes

Reactants	Products	Exoergicity kcal mole^{-1}	Barrier kcal mole^{-1}	<E'> kcal mole^{-1}
F + C_6H_6	C_6H_5F + H	13	~4	7.7
F + toluene	C_7H_7F + H	13	~4	6.8
	C_6H_5F + CH_3	22	~7.5	9.3
F + m-xylene	C_8H_9F + H	13	~4	7.8
	C_7H_7F + CH_3	22	~7.5	11.0
F + C_6H_5Cl	C_6H_4ClF + H	13	~4	8.2
	C_6H_5F + Cl	29	~1	3.4

A number of conclusions can be reached from an examination of these data. We find that average product translational energy does not correlate well with reaction exoergicity. In

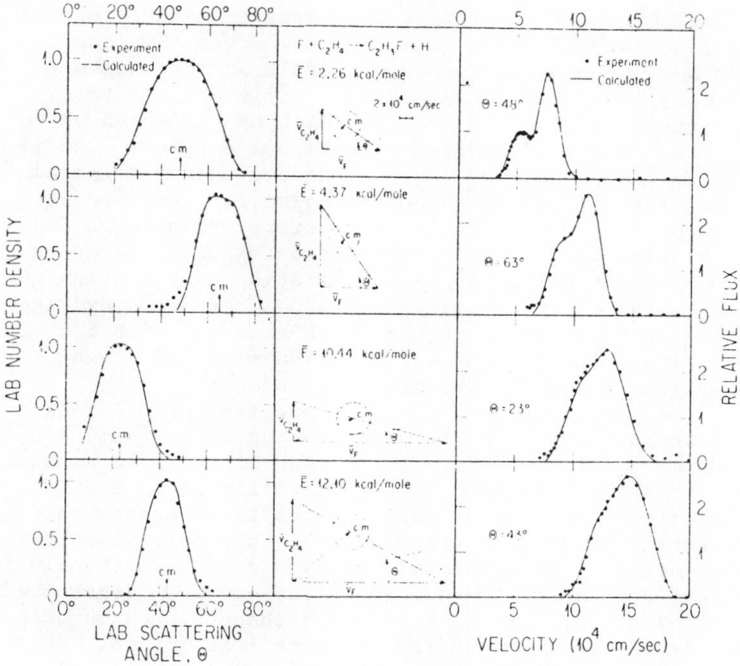

Figure 7. Experimental angular and TOF distribution data for F + $C_2H_4 \rightarrow C_2H_3F$ at 4 collision energies

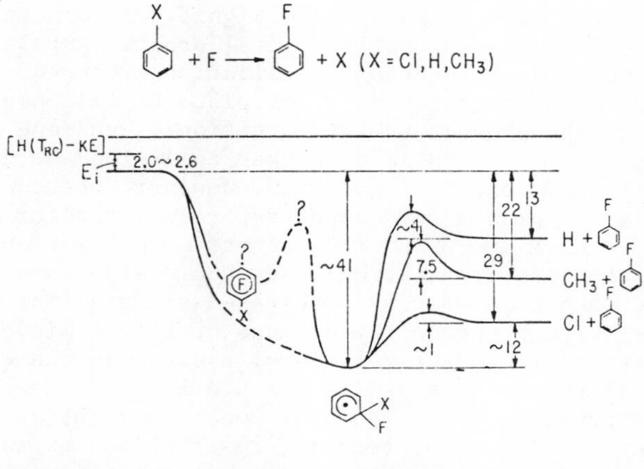

Figure 8. Schematic reaction coordinate for the reactions of F + substituted benzenes

the case of Cl emission, this very exoergic channel results in only 3.4 kcal $mole^{-1}$ in translation whereas the less exoergic H-atom emission reactions result in much higher translational energy release. Higher still are the translational energy releases associated with CH_3 emission which are accompanied by very large exit channel barrier. One is immediately led to the conclusion that the energy released in descending the barrier is channeled primarily into product translation; very high translational energy release is found in CH_3 emission where a large barrier is expected, whereas Cl emission proceeds with very low translational excitation in accordance with a very small exit channel barrier. Hydrogen atom emission provides an intermediate case in which a 4 kcal barrier releases energy into translation with $\langle E' \rangle$ lying between values found for CH_3 or Cl emission.

One can therefore conclude that predictions of product energy distributions from the decompositon of long-lived complexes must first consider the height of the exit channel barrier. Most reactions under consideration here have exoergicities of 20 kcal or less; if the total excess energy of the complex is distributed randomly, then only ~1-2 kcal can be expected in the reaction coordinate if more than 10 vibrational degrees of freedom share the energy. Consequently, a barrier of height 5 to 10 kcal could channel much more energy than the statistical amount into product translation. For different reactions with comparable exoergicities, one thus expects that the higher the exit channel barrier, the "colder" the products.

Angular momentum conservation plays an important role in many of these unimolecular decompositions, particularly at higher collision energies. In the case of decompositions involving emission of a hydrogen atom from the collision complex, the light atom cannot remove significant orbital angular momentum and consequently initial orbital angular momentum become product rotational excitation. Consequently angular momentum conservation does not place kinetic energy restrictions on the product energy partitioning and one expects this kinematic freedom to hold at higher collision energies. In the case of heavy particle emission, however, the conservation laws can play a very important role. Earlier in this section we have discussed the conversion of exit channel potential energy into translation; one should also note, however, that when this energy is released along a line which does not pass through the center of mass of the collision complex, substantial product rotational excitation can occur, particularly if the moments of inertia are large. Under these circumstances product orbital angular momentum might be substantially decreased thereby removing restrictions on the recoil energy distribution. Conversely, if significant orbital angular momentum is carried away by the products, reduced masses and interaction potential ranges might require that a large fraction of the rotational energy of the complex be carried away as product translation such that recoil velocities might be larger than one initially expects from simple statistical factors. In particular, the "hot atom preparative conditions" discussed by Bunker (50) regarding unimolecular decomposition of species formed from translationally "hot" reactants may play a very important role in decomposition lifetimes and recoil energy distributions. If an activated molecule is formed from high angular momentum reactants, its decomposition may be significantly governed by conservation laws; especially when the total angular momentum is placed into final orbital angular momentum, we expect translationally "hot" products to be formed. This situation should be particularly marked in cases where a slight activation energy for the reaction causes the total cross section to increase with collision energy. Under such circumstances, the initial orbital angular momentum increases dramatically with energy and its conversion into product orbital angular momentum requires the mean velocity with which the products retreat from one another to increase dramatically. The net effect of the conservation laws then is to force the average fraction of the total energy in translation to increase with collision energy, in opposition to the usual statistical result. This point is shown more clearly in the $F + CH_3I$ reaction discussed in the section immediately following.

III. Other Examples.

A. $F + CH_3I \rightarrow IF + CH_3$. This chemical reaction provides us with an example of an abstraction reaction from an aliphatic compound; three channels are open in this system as indicated here:

$$F + CH_3I \rightarrow IF + CH_3 \qquad \Delta H = -11 \text{ kcal}$$

$$\rightarrow HF + CH_2I \qquad \Delta H = -35 \text{ kcal}$$

$$\rightarrow CH_3F + I \qquad \Delta H = -53 \text{ kcal}$$

The last reaction, while highly exoergic, has been found by Tal'rose (51) to be of negligible importance in this system; the second reaction is known to produce vibrationally excited HF yielding laser action, but we have not studied this channel. Tal'rose has also studied the reaction to produce IF using mass spectrometric probing of fluorine flames and has measured a rate constant $k = 2 \times 10^{-10}$ cm^3 sec^{-1}.

We have studied the IF production reaction (37) at two collision energies, 2.6 and 14.1 kcal $mole^{-1}$ produced by Ar and He seeded F atom beams; the laboratory angular distribution for

IF at the lower collision energy in Figure 10 shows some symmetry about the centroid, suggesting that the reaction proceeds through a long-lived complex. The time of flight data in conjunction with the laboratory angular distribution data indeed confirm that the c.m. angular distribution is symmetric about $\theta = \pi/2$ as expected from decomposition of a complex which lives many rotational periods. The angular distribution in the center of mass system is of the form $(1 + a \cos^2\theta)$, symmetric about $\pi/2$ but anisotropic, thereby providing information about angular momentum partitioning in the decomposition of the complex. In particular, the strong forward-backward peaking at this collision energy indicates that the complex is quite prolate, which is not surprising since CH_3I is a highly prolate (cigar-shaped) symmetric top, and that the complex rotational angular momentum becomes product orbital angular momentum. This situation occurs, as pointed out by Miller, Safron, and Herschbach (52), because the creation of products with low rotational excitation in conjunction with the azimuthal symmetry of $\mathbf{L}'$ about $\mathbf{V}'$ causes the product $\mathbf{V}'$ vectors to map out a sphere and the intensity of products "piles up" at the poles. This situation is relatively common in alkali atom-alkali halide exchange reactions in which long range forces created by dipole-polarizable atom interactions make L and L' large; the F + CH_3I case, however, is unique in that the forces are relatively short range. The disposal of angular momentum in this case is governed by the highly prolate nature of CH_3IF.

The recoil distribution for this system is rather interesting; we find that the products are formed with ~30% of the total energy in translation. This result is quite close to statistical predictions and, in fact, the phase space volume associated with the high frequency H-atom motions in the CH_3 fragment is so small that the experimental data do not allow one to conclude the extent to which vibrational energy is shared among the modes of the collision complex.

At higher collision energy the angular distribution in Figure 11 shows quite dramatically the forward-backward peaking mentioned previously. Now, however, the peaks are not of equal intensity in the c.m. system; this result would be expected if the collision complex lived only a fraction of a rotational period. One can convolute the random lifetime distribution expected for the decomposition with the classical angular momentum form factors to relate the ratio of the forward-backward intensities to the rotational period of the complex. This "osculating" model (53) predicts that a forward-backward ratio of ~1.5 corresponds to a complex lifetime of half a rotational period. Based on estimates of the total reaction cross section, we can compute the maximum orbital angular momentum quantum number leading to reaction and compute a rotational period of 8×10^{-12} sec. A comparison of this lifetime with RRKM theory

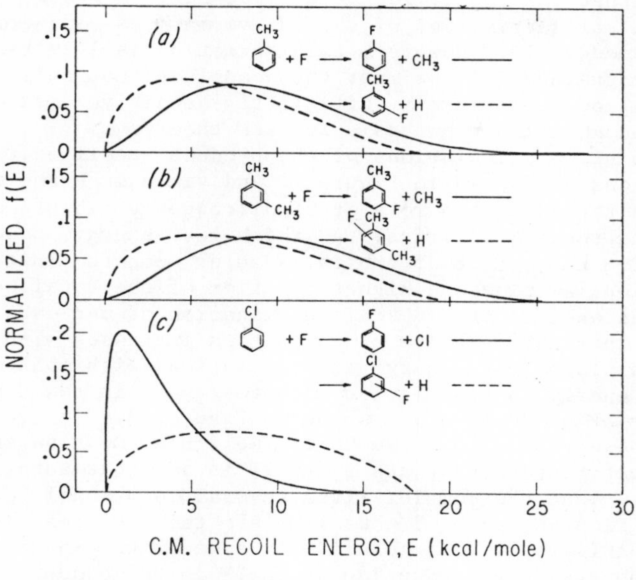

Figure 9. Product recoil energy distributions for F + substituted benzenes

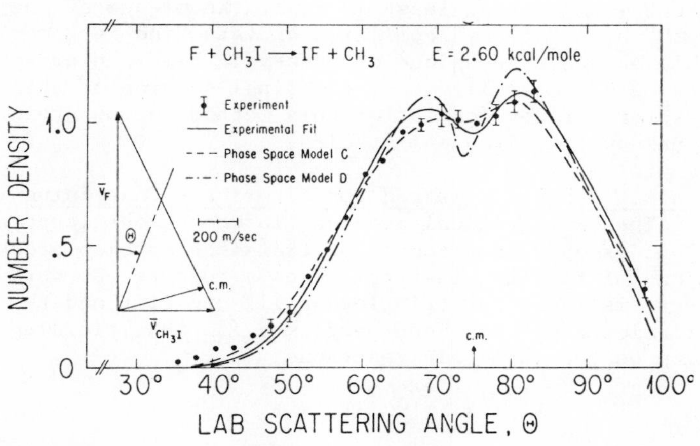

Figure 10. IF product angular distribution from the reaction F + CH_3I at $E_{rel} = 2.6$ kcal/mol

indicates that the "observed" value lies between predictions based upon full activation of the CH_3 vibrations and activation of the ν_2 mode only. However, this variation is less than an order of magnitude and one must thus conclude that this system does not allow us to make a sufficiently sensitive test of the energy randomization hypothesis in RRKM theory.

Phase space calculations of the product recoil energy distributions are shown in Figure 12 and yield much the same conclusion regarding the role of high frequency vibrations in determining these quantities. At the higher energy, however, one does observe the effect of orbital angular momentum in that the conservation law places a higher fraction of the total energy in translation than at 2.6 kcal; examination of our data indicates that the reactive cross section increases with energy and places more energy in translation, with ~55% of the total energy in translation. Contour maps in the c.m. system for IF production are shown in Figure 13.

The observation that the CH_3IF "molecule" must be stable with respect to the reactants F + CH_3I is somewhat surprising initially. That the complex lives several rotational periods at a collision energy of 2.6 kcal $mole^{-1}$ tells us that it is bound by at least 20 kcal $mole^{-1}$. In a separate experiment (54), we have observed the production of CH_3IF in the endoergic bimolecular reaction $F_2 + CH_3I \rightarrow F + CH_3IF$. The energetics of this process and the corresponding reaction of atomic fluorine are shown in Figure 14. The energy scale which locates the stability of CH_3IF with respect to both reactions has been determined by integrating the differential cross section for CH_3IF production as a function of initial kinetic energy. Such measurements indicate that the threshold for CH_3IF formation is ~11 kcal, thus suggesting that the radical is stable with respect to F + CH_3I by at least 25 kcal. Knowledge of the "well depth" of a "sticky" collision of this kind is generally unavailable and this technique of endoergic bimolecular synthesis appears to be of generality. In the final section of this paper, we discuss further development of this method for observing hitherto unknown free radical species.

B. A New Synthetic Method for Fluorine - Containing Radicals. The CH_3IF radical produced in the endoergic reaction of F_2 + CH_3I is only one example of transient species produced in reactions of F_2 with other molecules. More recent work on similar systems involving trihalogens IIF and ClIF and the pseudo-trihalogen HIF has been performed (55,56), yielding information on the endoergic reactions

$$F_2 + I_2 \rightarrow IIF + F$$

$$F_2 + ICl \rightarrow ClIF + F$$

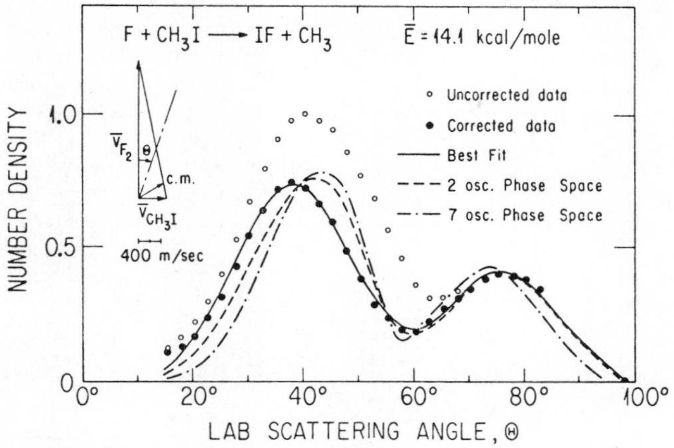

Figure 11. IF product angular distribution from the reaction $F + CH_3I$ at $E_{rel} = 14.1$ kcal/mol

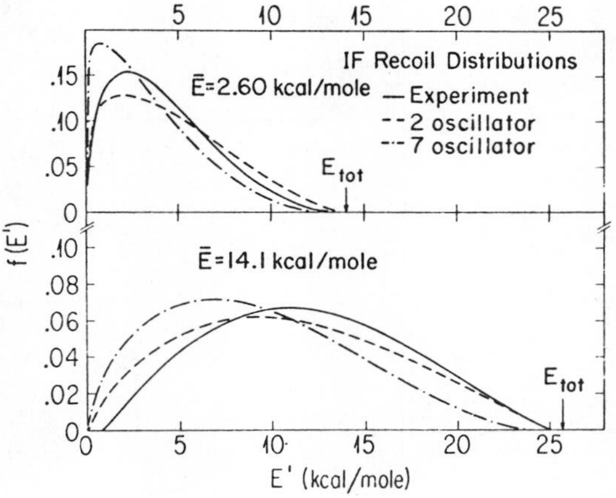

Figure 12. IF product recoil energy distributions at $E_{rel} = 2.6$ and 14.1 kcal/mol. Best fit and model calculation distributions as noted.

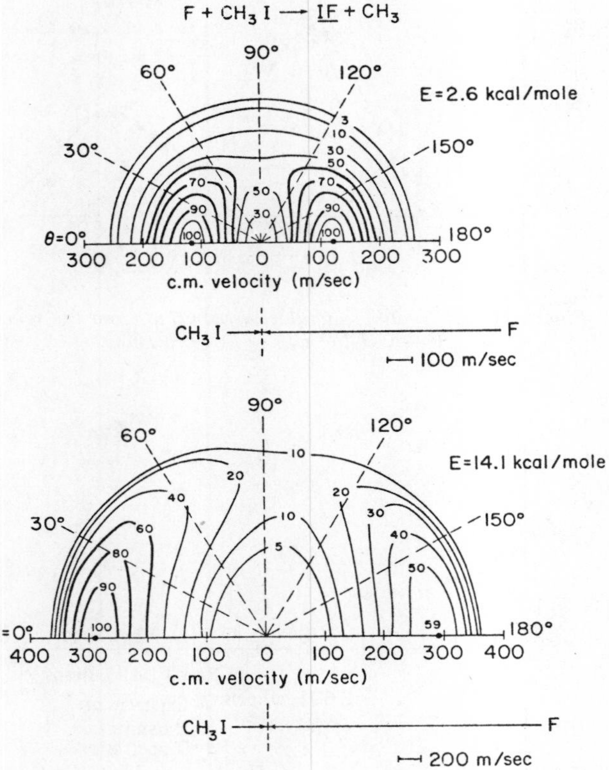

Figure 13. C.M. contour map for IF production at 2.6 and 14.1 kcal/mol

$$F_2 + HI \rightarrow HIF + F.$$

The IIF radical is of some importance in understanding the macroscopic reaction mechanism in the $F_2 + I_2$ system, particularly in view of predicted stabilities of trihalogen molecules from molecular orbital calculations (54), termolecular recombination work (58) and matrix isolation (59).

One must realize that the reaction

$$F_2 + XI \rightarrow XIF + F \quad (X = H, Cl, I)$$

can only yield observable XIF product if the decomposition of XIF into IF and X is endoergic. By measuring thresholds for the above reactions for X = H, Cl, I, we find values of 11, 6, and 4 kcal respectively and, further, the predominant products formed are XIF and F. The dissociation of XIF into IF + X at higher collision energies becomes important first for $F_2 + I_2$ at 7 kcal, and then for F_2 + ICl; the dissociation of HIF is an inaccessible channel for the collision energy range of these experiments.

In the F_2 + ICl system, no product mass peaks at m/e 56 ($Cl^{37}F$) or m/e ($Cl^{35}F$) were observed and no m/e 20 (HF) peak was observed for the F_2 + HI reaction. These mass spectral results indicate that the trihalogen and pseudo-trihalogen species formed involve bonding schemes such as ClIF and HIF with the more electropositive atom in the center of the molecule as predicted by Walsh's rules (60,61).

Figure 15 shows a contour map for the F_2 + ICl system which is typical of all the data. Very forward peaked angular distributions for the products suggest that abstraction of an F atom from F_2 by XI proceeds through a somewhat bent geometry, i.e., F-F-I or F-I-X angle less than 180° in the system F-F-I-X.

Aside from the unique chemical identity of these trihalogens, the stability of the I_2F molecule is of importance in elucidating some features of the kinetics of the F_2/I_2 system (62). First of all, it is important to note that the gas phase generation of F atoms in F_2 and I_2 mixtures only requires 4 kcal of relative kinetic energy between F_2 and I_2 through the $F_2 + I_2 \rightarrow I_2F + F$ reaction. A recent study of this system indicates that electronically excited IF is formed as evidenced by chemiluminescence observations. The four center exchange reaction $F_2 + I_2 \rightarrow 2IF$ is exoergic by 60 kcal, more than enough to produce electronically excited IF, but our crossed molecular beam experiments have shown that at low energies collisions between $F_2 + I_2$ produce I_2F + F or IF + I + F, but not 2IF. If I_2F were not stable, only the $F + I_2$ and $I + F_2$ reactions would be exoergic and then only by 30 kcal, too little to account for the chemiluminescence. However, the reaction $F + I_2F \rightarrow 2IF$ is exoergic by 64 kcal and, in view of the beam measurements of I_2F stability and the ease of production, could account for the observed chemiluminescence

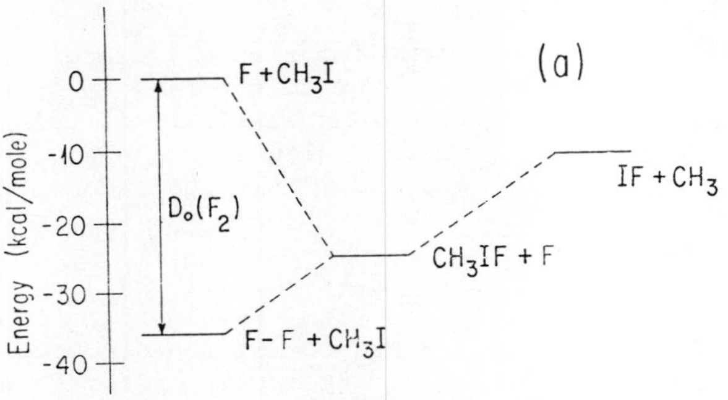

Figure 14. Schematic energy level showing the relationship between the formation of CH_3IF as a transient species via the $F + CH_3I$ path or as a stable radical via the $F_2 + CH_3I$ pathway

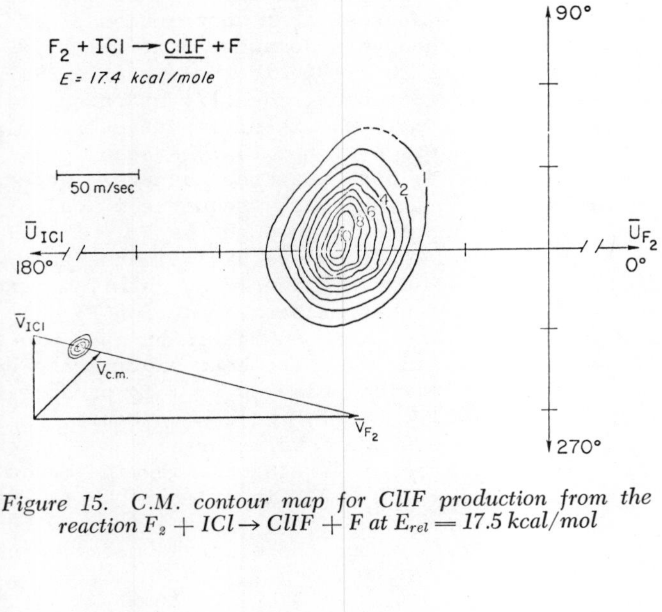

Figure 15. C.M. contour map for ClIF production from the reaction $F_2 + ICl \rightarrow ClIF + F$ at $E_{rel} = 17.5$ kcal/mol

in F_2 and I_2 mixtures.

Abstract

A summary of work in fluorine chemistry as determined from crossed molecular beam studies of reaction dynamics is presented. Special emphasis is given to studies of unimolecular decay of long-lived complexes formed from bimolecular association of fluorine atoms with unsaturated hydrocarbons. The experimental results are discussed in terms of statistical models for energy randomization in the complex as well as the role of the exit channel potential energy barrier in determining the product translational energy distribution.

Experimental results on recent studies of reactions of F_2 with a variety of species are also presented, with special attention to the utility of the method of endoergic bimolecular reactions in "synthesizing" unstable radicals such as trihalogens and pseudotrihalogens. Results are presented for $F_2 + I_2 \rightarrow I_2F + F$, $F_2 + HI \rightarrow HIF + F$, $F_2 + CH_3I \rightarrow CH_3IF$, and $F_2 + ICl \rightarrow FICl + F$.

Literature Cited

1. Lawley, K. P. ed, "Molecular Scattering", Adv. Chem. Phys. 30, Wiley, New York, 1975.
2. Molina, M. J., and Pimentel, G. C., IEEE J. Quantum Electron. (1973), QE9, 64.
3. Schafer, T. P., Siska, P. E., Parson, J. M., Tully, F. P., Wong, Y. C., and Lee, Y. T., J. Chem. Phys. (1970), 53, 3385.
4. Schafer, T. P., Ph.D. Dissertation, Univ. of Chicago (1972).
5. Polanyi, J. C., and Woodall, K. B., J. Chem. Phys. (1972), 57, 1574.
6. Jonathan, N., Melliar-Smith, C. M., and Slater, D. H., Mol. Phys. (1971), 20, 93.
7. Wilkins, R. L., J. Chem. Phys. (1972), 57, 912.
8. Muckerman, J. T., J. Chem. Phys. (1971), 54, 1155.
9. Schatz, G. C., Bowman, J. M., and Kuppermann, A., J. Chem. Phys. (1973), 58, 4023.
10. Bender, C. F., O'Neil, S. V., Pearson, P. K., and Schafer, H. F., J. Chem. Phys. (1972), 56, 4626.
11. Berry, M. J., J. Chem. Phys. (1973), 59, 6229.
12. Parson, J. M., and Lee, Y. T., J. Chem. Phys. (1972), 56, 4658.
13. Parson, J. M., Shobatake, K., Lee, Y. T., and Rice, S. A., J. Chem. Phys. (1973), 59, 1402.
14. Shobatake, K., Parson, J. M., Lee, Y. T., and Rice, S. A., J. Chem. Phys., (1973), 59, 1416.
15. Shobatake, K., Parson, J. M., Lee, Y. T., and Rice, S. A., J. Chem. Phys. (1973), 59, 1427.

16. Shobatake, K., Lee, Y. T., and Rice, S. A., J. Chem. Phys. (1973), 59, 1435.
17. Shobatake, K., Lee, Y. T., and Rice, S. A., J. Chem. Phys. (1973), 59, 6104.
18. Parson, J. M., Shobatake, K., Lee, Y. T., and Rice, S. A., Faraday Discuss. Chem. Soc. (1973), 55, 344.
19. Farrar, J. M., and Lee, Y. T., J. Chem. Phys. (1976), 65, 1414.
20. Morse, F. A., and Bernstein, R. B., J. Chem. Phys. (1962), 37, 2019.
21. Entemann, E. A., Ph.D. Dissertation, Harvard Univ. (1967).
22. Warnock, T. T., and Bernstein, R. B., J. Chem. Phys. (1968), 49, 1878.
23. Siska, P. E., Ph.D. Dissertation, Harvard Univ. (1969).
24. Farrar, J. M., and Lee, Y. T., Ann. Rev. Phys. Chem. (1974), 25, 357.
25. Lee, Y. T., McDonald, J. D., LeBreton, P. R., and Herschbach, D. R., Rev. Sci. Instrum. (1969), 40, 1402.
26. Anderson, J. B., Andres, R. P., and Fenn. J. B., Adv. Chem. Phys. (1969), 10, 289.
27. Mariella, R. P., Herschbach, D. R., and Klemperer, W., J. Chem. Phys. (1973), 58, 3785.
28. Bennewitz, H. G., Haerten, R., and Müller, G., Chem. Phys. Lett. (1971), 12, 335.
29. Schultz, A., Cruse, H. W., and Zare, R. N., J. Chem. Phys. (1972), 57, 1354.
30. Cruse, H. W., Dagdigian, P. J., and Zare, R. N., Faraday Discuss. Chem. Soc. (1973), 55, 277.
31. Hostettler, H. U., and Bernstein, R. B., Rev. Sci. Instrum. (1960), 31, 872.
32. Valentini, J. J., and Lee, Y. T., unpublished data.
33. Coombe, R. D., and Pimentel, G. C., J. Chem. Phys. (1973), 59, 251.
34. Coombe, R. D., and Pimentel, G. C., J. Chem. Phys. (1973), 59, 1535.
35. Fettis, G. C., Knox, J. H., and Trotman-Dickenson, A. F., J. Chem. Soc. London (1960), 1064.
36. Bender, C. F., O'Neil, S. V., Pearson, P. K., and Schaefer, H. F., Science (1972), 176, 1412.
37. Farrar, J. M., and Lee, Y. T., J. Chem. Phys. (1975), 63, 3639.
38. Forst, W., "Theory of Unimolecular Reactions", Academic New York, 1973.
39. Marcus, R. A., J. Chem. Phys. (1952), 20, 359.
40. Butler, J. N., and Kistiakowsky, G. B., J. Am. Chem. Soc. (1960), 82, 759.
41. Rynbrandt, J. D., and Rabinovitch, B. S., J. Chem. Phys. (1971), 54, 2275.
42. Pechukas, P., Light, J. C., and Rankin, C., J. Chem. Phys. (1966), 44, 794.

43. Lin, J., and Light, J. C., J. Chem. Phys. (1966), 45, 2545.
44. Wieder, G. M., and Marcus, R. A., J. Chem. Phys. (1966), 37, 1835.
45. Marcus, R. A., J. Chem. Phys. (1975), 62, 1372.
46. Shobatake, K., Rice, S. A., and Lee, Y. T., J. Chem. Phys. (1973), 59, 2483.
47. Jones, W. E., Macknight, S. D., Teng, L., Chem. Rev. (1973), 73, 407.
48. Williams, R. L., and Rowland, F. S., J. Chem. Phys. (1972), 76, 3509.
49. Light, J. C., private communication.
50. Bunker, D. L., J. Chem. Phys. (1972), 57, 332.
51. Leipunskii, I. O., Morozov, I. I., and Tal'rose, V. L., Dokl. Akad. Nauk. SSSR (1971), 198, 1367.
52. Miller, W. B., Safron, S. A., and Herschbach, D. R., Faraday Discuss. Chem. Soc. (1967), 44, 108.
53. McDonald, J. D., Ph.D. Dissertation, Harvard Univ. (1971).
54. Farrar, J. M., and Lee, Y. T., J. Am. Chem. Soc. (1974), 96, 7570.
55. Valentini, J. J., Coggiola, M. J., and Lee, Y. T., Faraday Discuss. Chem. Soc. (1976), to be published.
56. Valentini, J. J., Coggiola, M. J., and Lee, Y. T., J. Am. Chem. Soc. (1976), 98, 853.
57. Peyerimhoff, S. D., and Buenker, R. J., J. Chem. Phys. (1968), 49, 2473.
58. Bunker, D. L., and Davidson, N., J. Am. Chem. Soc. (1958), 80, 5090.
59. Nelson, L. Y., and Pimentel, G. C., J. Chem. Phys. (1967), 47, 3671.
60. Walsh, A. D., J. Chem. Soc. (1953), 2266.
61. Walsh, A. D., J. Chem. Soc. (1953), 2288.
62. Birks, J. W., Gabelnick, S. D., and Johnston, H. S., J. Mol. Spectr. (1975), 57, 23.

8

Molecular Beam Studies with F^+ Ions

WALTER S. KOSKI

Department of Chemistry, The Johns Hopkins University, Baltimore, MD 21218

There have been very few beam studies of reactions involving F^+ ions. In recent years, however, there has been an increasing number of publications of detailed studies of 3 atom-ion-molecule reactions of the type $A^+ + BC \rightarrow AB^+ + C$ involving hydrogen molecules and elements of the second row of the Periodic Table. The F^+ ion has been included in some of these studies. The main objectives of such studies has been to shed light on the details of the collision process and on the distribution of energy in the ionic product. The reactions are frequently classified as direct or proceeding through persistent complex formation. The rotational period of the system ($\sim 10^{-12}$ sec) offers a convenient kinematic clock. The symmetry of the velocity distribution of the ionic product in the center of mass coordinate system can be determined from the measured angular and energy distribution of the product ions. If the projectile ion interacts with the target molecule for a time longer than the rotational period of the system the velocity distribution is symmetrical about a line passing through the center of mass and perpendicular to the projectile ion direction. Such a distribution suggests that the reaction is proceeding through a persistent complex formation. On the other hand if the distribution is forward of the center of mass and consequently asymmetric relative to the $\pm 90^o$ direction the projectile and target molecule have an interaction time less than the rotational period of the system then reaction is classified as direct. Obviously, other criteria can be used for direct and complex formation mechanisms. However, the above is a convenient experimental approach.

In beam experiments the amount of available energy that has gone into internal energy of the ionic product is simply determined from the difference of the kinetic energy of the products and reactants. The distribution of internal energy between the rotational and vibrational degrees of freedom of the

© 0-8412-0399-7/78/47-066-222$05.00/0

ionic molecular product is not easy to determine due to limitations of instrumental resolution. However, the current state of the art permits a surmounting of this obstacle so that in the near future one can expect to be able to see vibrational structure in the energy spectra of the ionic products. At present one can infer the information from line widths in the energy spectra or from optical emission in those cases where such emission can be observed.

In this presentation we will report on the total reactive cross section measurements of the reaction of F^+ with molecular hydrogen and its isotopic variants, on some charge transfer processes and on the angular and energy variation of the FD^+ products from the reaction $F^+(D_2,D)FD^+$. Some preliminary cross section measurements for some reactions of F^+ with CH_4 will also be mentioned.

Experimental

The cross section measurements were made with an apparatus described previously (1). The instrument consists of a primary mass spectrometer (PMS) in tandem with a secondary mass spectrometer (SMS). The PMS is a 1 cm radius of curvature 2500 G permanent magnet mass analyzer. The SMS is a 60° magnetic sector instrument with an 8 in. radius of curvature. The detection of product ions was made by counting ions with a 17 stage electron multiplier. The gas pressure in the reaction cell was measured with an MKS Baratron differential pressure gauge. For reactions proceeding by direct mechanisms the instrument was calibrated with the reaction $Ar^+(D_2,D)ArD^+$ (2).

The spectrometer used for reactive scattering measurements has also been described elsewhere (3). Briefly, the instrument consists of two quadrupole mass filters and two hemispherical analyzers in tandem. F^+ ions were formed by electron bombardment of CF_4 in a source chamber. The ions were then extracted and focussed into the primary quadrupole mass filter. The ion beam is then decelerated into the first electrostatic analyzer. From the analyzer, the ions are accelerated and focussed into the reaction cell which contains the target gas. The cell consists of two concentric cylinders, with the inner cylinder being capable of rotation through an angle of ± 45°. The reaction cell is the center of the axis for the angular movement of the detecting system. The detection system moves in a planar direction, pivoting the inner cylinder as the system is moved by a mechanical feed through. The ions leaving the reaction cell are energy and mass analyzed in the detecting system by another 180° electrostatic analyzer and by the secondary quadrupole mass filter respectively. After this the ions proceed to the electron multiplier where they are counted. Pulse counting techniques are used to collect and store the data gathered from the angular and energy distributions for both the reactant and product ions.

Results and Discussion

Electronic State of Projectile Ion.

The F^+ ions were produced by electron bombardment of CF_4. In general, under such conditions one expects the resulting beam to be composed of ions in more than one electronic state (4).

In order to determine the beam composition the attenuation of the beam was studied in a suitable gas. If a single electronic state is present the beam attenuation can be represented by a simple exponential expression $I = I_o\exp(-n\sigma l)$, where I is the beam intensity when the number density is n and I_o is the corresponding intensity when n = 0, σ is the total cross section for all processes leading to attenuation, and l is the reaction path length. A semilogarithmic plot of I/I_o vs. pressure gives a straight line. If two states are present the attenuation is given by the sum of two exponentials $I = I_o\exp(-n\sigma_1 l) + I_o(1-f)\exp(-n\sigma_2 l)$, where f is the fractional abundance of the excited state. A semilogarithmic plot now gives a curved line which can be resolved into its two simple exponential components and the abundances of the two states can be determined. Figure 1(a) gives typical results obtained for electron bombardment of CF_4 with electrons of various energies. It is clear that the fractional abundance of the upper state can be varied by varying the electron energy. However, in lowering the electron energy there is a concomitant decrease in the yield of the F^+ ion, a situation which is frequently not tolerable. In some instances (5) one of the states could be eliminated by simply raising the gas pressure in the primary ion source. This approach was not successful here presumably because the ground state of F^+ is a ^3P and the first excited state ^1D and to deactivate the excited state to the ground state would involve a singlet triplet transition which is forbidden. In such instances (6) the addition of a paramagnetic impurity accomplished the desired result. In Fig. 1(b) one has the attenuation results when 30% NO was added to the CF_4. To within experimental error the slope is a straight line indicating the presence of only one electronic state.

By studying the variation in abundance of the excited state with electron energy and then plotting the abundance vs. electron energy one can identify the excited state of F^+ as the ^1D state from the appearance potential. The electron energy scale was calibrated using neon.

Charge Transfer Reaction $F^+(Ne,F)Ne^+$.

Probably the simplest reaction of a gaseous positive ion is charge transfer. Fig. 2 gives the cross section measurement of

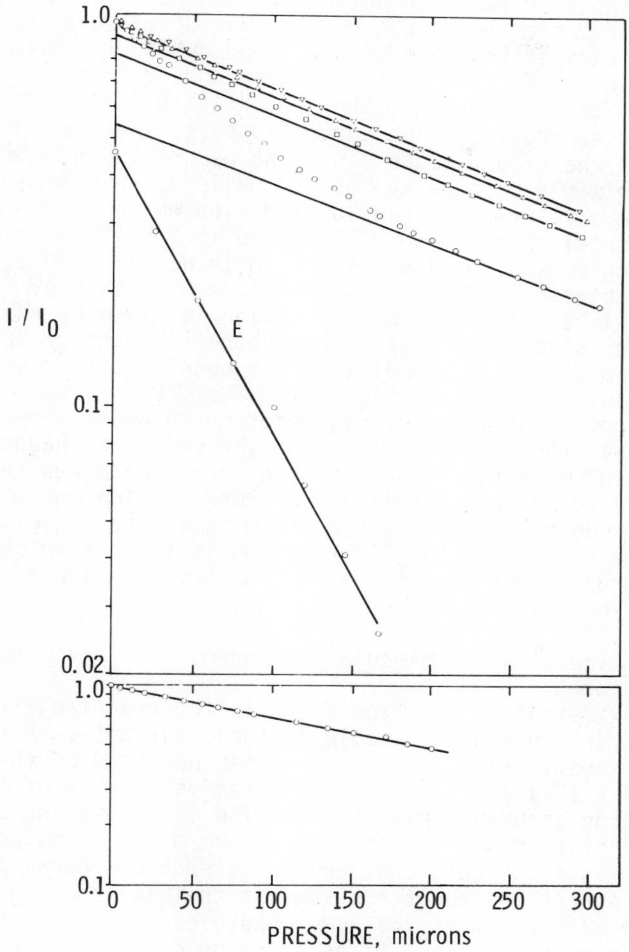

Figure 1. Upper figure: attenuation curves for F^+ produced from CF_4 by electron bombardment. Neon was the attenuating gas. The ion source pressure was 20 μ and the F^+ ion energy was 15 eV. The electron energy was 100 eV (⊙), 80 eV (□), 60 eV (△), and 50 eV (▽). E represents the attenuation of ions in the excited state. Lower figure: atenuation of F^+ in neon when the F^+ ions are produced by electron bombardment of a 70%–30% CF_4–NO mixture. All other parameters were the same as for the upper figure.

the charge transfer reaction between F^+ and Ne (7). The curve in Figure 2 represented by the circles was obtained by measuring the charge transfer reaction with F^+ produced by electron bombardment of a CF_4 + NO mixture which judging from the attenuation curves gives only the ground state (^3P) of F^+. This charge transfer cross section curve supports the fact that the F^+ ion is in the ground state since the use of the ground state reactant ion would be expected on the basis of thermodynamic data to give a reaction threshold of 8.4 eV which agrees with the experimental observation. The curve in Fig. 2 represented by the triangles was obtained with F^+ obtained from 70 eV electron bombardment of CF_4. In this case a mixture of ^3P and ^1D states exists but the beam composition is given by the corresponding attenuation data. By subtracting the contribution of the ground state and normalizing one obtains the cross section behavior of the reaction $F^+(^1D)(Ne,F)Ne^+$. The appearance potential measurements referred to above showed that this state was the ^1D state of F^+ and the charge exchange results are consistent with this. Both of the charge exchange reactions are expected to be endothermic. The cross section curves consequently go through a broad maximum as a function of energy; however, the excited state (^1D) reaction is the one which has a smaller energy defect and its maximum occurs in a lower energy region as expected.

Reaction of the F^+ With Molecular Hydrogen.

The next simplest reaction of F^+ is the reaction with molecular hydrogen and its stable isotopic variants. A typical result is given in Fig. 3 where the cross sections of the reaction of $F^+(^3P)$ and $F^+(^1D)$ are plotted vs. center of mass kinetic energy assuming that the reaction is proceeding by a direct mechanism involving one atom of the diatomic molecule. The measurements indicate that the cross sections corresponding to the two states are about the same at the higher energies but diverge measurably at the lower energies. The fact that a divergence between the two states is to be expected is borne out by statistical phase space calculations the results of which are illustrated by the dashed lines in the figure. The calculated cross sections are too high but the shapes of the curves are consistent with experimental observations as is the fact that the two cross sections diverge at lower energies. The following considerations illustrate the factors that contribute to this behavior. Use of Shuler's (8) adiabatic correlation rules indicates that the following reactions are permitted in our energy domain:

$$F^+(^3P_g) + H_2(^1\Sigma^+) \rightarrow FH^+(^2\Pi) + H(^2S_g) \quad \Delta H = -2.78 \text{ eV} \qquad (1)$$

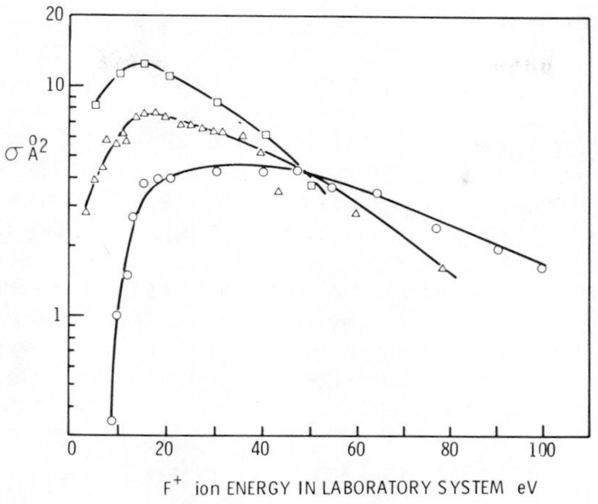

Figure 2. Charge exchange cross sections for the reaction $F^+(Ne,F)Ne^+$. ^3P state, ⊙; $^3P + {}^1D$ states, △; ^1D state, ⊡.

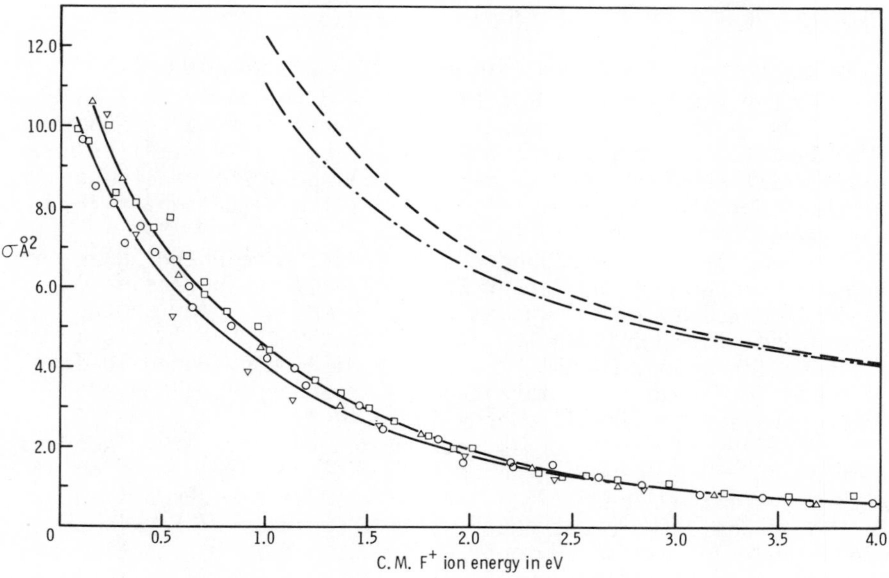

Figure 3. Cross sections for the reactions of $F^+(^3P) + H_2$ (●) and $F^+(^3P) + D_2$ (▼) and $F^+(^1D)$ with H_2 (■) and D_2 (▲). The dashed curves give the corresponding results assuming the phase space theory of ion–molecule reactions.

$$F^+(^3P_g) + H_2(^1\Sigma^+) \rightarrow FH^+(^2\Sigma^+) + H(^2S_g) \quad \Delta H = +0.35 \text{ eV} \quad (2)$$

$$F^+(^1D_g) + H_2(^1\Sigma^+) \rightarrow FH^+(^2\Pi) + H(^2S_g) \quad \Delta H = -5.37 \text{ eV} \quad (3)$$

$$F^+(^1D_g) + H_2(^1\Sigma^+) \rightarrow FH^+(^2\Sigma^+) + H(^2S_g) \quad \Delta H = -2.24 \text{ eV} \quad (4)$$

The dissociation energy (9) of $FH^+(^2\Pi)$ is 3.59 eV and of $FH^+(^2\Sigma)^+$ 0.45 eV. The heats of reaction are obtained from known thermochemical data (10).

For reactions involving the excited state $F^+(^1D_g)$ statistical phase space theory predicts a higher cross section than for the ground state $F^+(^3P_g)$ reaction because of the lower exothermicity of the latter reaction. From the conservation of energy we can write down (11)

$$E_t = E_f + E(n_f) + [M_f(M_f + 1)\hbar^2]/2I_f + Q_{fi},$$

where E_t is the total energy, E_f the relative energy of the final products, M_f the rotational quantum number of the final diatomic product, I_f the two body moment of inertia of the final product, and Q_{fi} is the exothermicity of the reaction.

Furthermore, the requirement that the strong coupling complex be able to dissociate over the final state orbital angular momentum barrier restricts L_f by

$$L_f(L_f + 1)\hbar^2 \leq 4\mu_f C_f^{1/2}[1/3E_f]^{1/2},$$

where L_f is the orbital angular momentum quantum number.

So for the same diatomic products and the same total energy, if Q_{fi} is increased, the conservation of energy requires that E_f will tend to become smaller; and, hence, L_f will be smaller and the orbital angular momentum barrier will be lower and the cross section correspondingly higher. This is the situation at the low energy.

As the translational energy increases, more energy goes into internal energy and the factor En_f increases so increasing Q_{fi} does not cause as much lowering of E_f as in the lower energy case. Therefore, this factor as well as the probability that the dissociation energy of the $^2\Sigma^+$ product will be exceeded leading to dissociation tends to make the two cross sections converge as the projectile kinetic energy increases.

Finally, as far as cross section measurements are concerned, mention should be made of the isotope effect in the reaction of F^+ with molecular hydrogen (7). Typical results are summarized in Fig. 4 where the cross sections as a function of F^+ kinetic energies are given for FH^+ and FD^+ products for the reaction of F^+ with HD. It is clear that at low energies (below 8 eV) the production of FD^+ is favored and above this energy FH^+ is favored. It will also be noted that the isotope ratio becomes rather large at the higher energies. Yuan

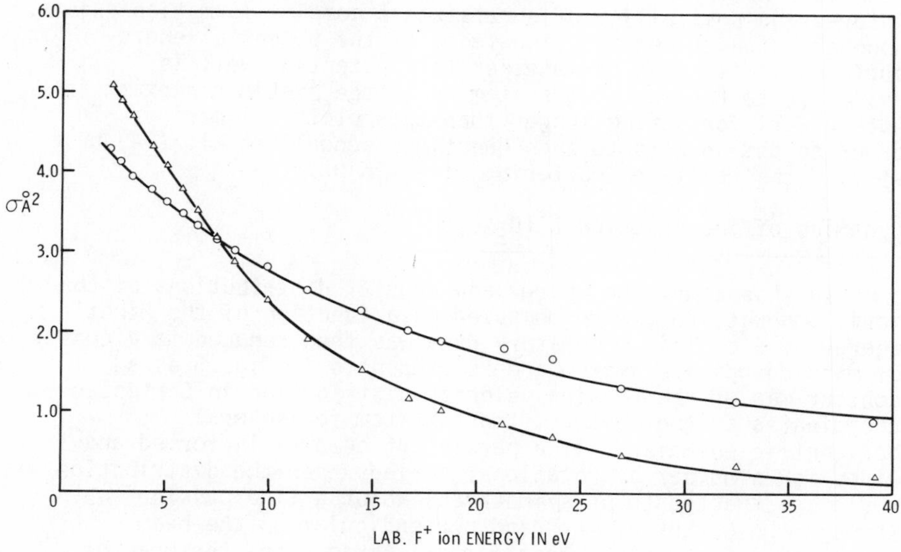

Figure 4. Cross sections of F^+ ion kinetic energy for the reaction $F^+(^3P)$ + HD: FH(○) and FD^+ (△).

and Micha (12) have used such isotope effects with considerable success to elucidate the collision dynamics of stripping reactions such as $Ar^+(H_2,H)ArH^+$. Corresponding studies of the reaction $F^+(H_2,H)FH^+$ are in progress.

The latter reaction is an interesting ion-molecule reaction from a mechanistic point of view since the H_2F^+ ion has been observed (13) as a stable ion arising from the reaction $HF^+(HF,H)H_2F^+$. The existence of the ion, H_2F^+, shows that there is a deep potential well for the F^+ - H_2 system; hence, the interesting possibility of a persistent complex formation exists. However, without detailed knowledge of the potential energy surfaces it is not clear whether this potential well is available to the reacting system or if the system can exist in such a well for a time longer than the rotational period. In order to get answers to such questions Wendell et al. (14) have studied the reactive scattering of F^+ in D_2.

Dynamics of the Reaction $F^+(D_2,D)FD^+$.

In this study the energy and angular distributions of the ionic product, FD^+, were measured as a function of the kinetic energy of F^+. This laboratory data was then reduced to a center of mass coordinate system and is presented in Fig. 5 as a contour map of the FD^+ ion velocity distribution in Cartesian coordinates in the center of mass system for several barycentric energies. If a persistent complex is formed and lives for a number of rotational periods then the distribution would be expected to be symmetrical about a line passing through the center of mass and perpendicular to the beam direction. It is clear that for all three cases the peak of the velocity distribution is forward of the center of mass. In the case of the lowest energy (0.21 eV) there is a high degree of symmetry to the distribution but the peak is slightly forward of the center of mass. This indicates that there is a strong interaction between all three particles involved in the collision. This interaction does not last much more than $\sim 10^{-12}$ sec. As the F^+ kinetic energy increases the center of the distribution occurs progressively further from the center of mass and the reaction is proceeding by a direct process which is approximated by a stripping mechanism. In this model of the reaction mechanism as the F^+ approaches, the D_2 is strongly polarized and as the distance further decreases there is an electron jump to the F^+ followed by reaction to form FD^+ (12). The process proceeds in such a manner that the neutral D atom experiences very little momentum change and hence acts as a "spectator."

Another type of information that can be obtained from these measurements is Q the translational exoergicity (which is the difference between the translational energy of the products and

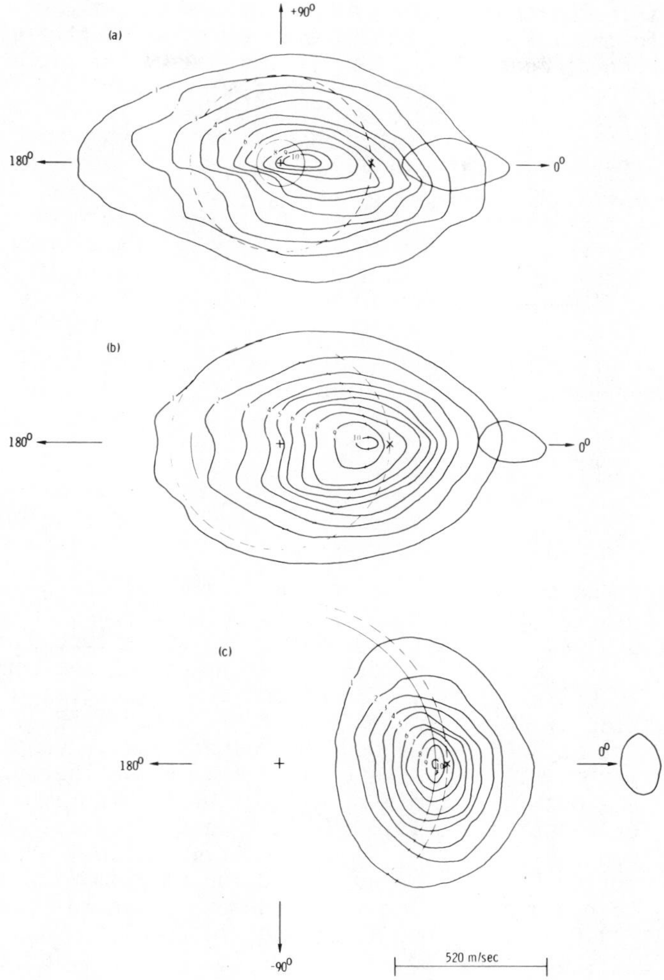

Figure 5. FD^+ ion velocity distributions shown in Cartesian coordinates in the center of mass system for barycentric energies (a) 0.21 eV, (b) 0.39 eV, and (c) 0.91 eV. (+), system's velocity of the center of mass; (×), most probable velocity of the ideal spectator stripping product at 0°. (●), most probable values of Q = −0.20, −0.26, −0.55 for (a), (b), and (c), repsectively; (⊙), the spectator stripping values of Q = −0.12, −0.21, −0.50 for (a), (b), and (c), respectively. The single contour lines to the right of each product distribution correspond to the 50% intensity profile of the primary ion beam.

reactants). From the definition of Q and the application of the principles of conservation of energy and momentum one can relate Q to the masses and the kinetic energies of the projectile (E_1) and product ions (E_3). For the case where the product ions are proceeding in the projectile ion beam direction the expression for Q for the $F^+(D_2,D)FD^+$ reaction is

$$Q + \frac{m_1 + 2m}{m} E_3 + \frac{m_1(m_1 + m) - 2m^2}{(m_1 + 2m)m} E_1 -$$

$$2\left[\frac{m_1(m_1 + m)\ E_1E_3}{m}\right]^{1/2}$$

where m_1 is the projectile ion mass and m is the mass of the D atom. Conservation of energy also permits the following relation for Q

$$T_R + U_R - \Delta H = T_p + U_p$$

$$Q = U_R - U_p - \Delta H$$

T and U represent the translational and internal energies of the reactants (R) and products (P), respectively and ΔH is the heat of reaction. The last expression indicates that Q reaches a limiting value when the internal energy reaches the bond dissociation energy of FD^+. The dissociation energies of the $^2\Pi$ and $^2\Sigma^+$ states of FD^+ have been accurately determined by Berkowitz (9) as 3.46 eV and 0.46 eV respectively. Coupling this information with ΔH as determined from known thermodynamic data (10) one obtains a minimum value of Q of -0.68 eV and -0.81 eV for reactions to form FD^+ in $^2\Pi$ and $^2\Sigma^+$ electronic states respectively. Fig. 6 gives the translational exoergicity for the reaction $F^+(D_2,D)FD^+$ as a function of primary ion energy. Q appears to reach a limiting value corresponding to approximately 0.8 eV and one is tempted to conclude that the product FD^+ is being formed in $^2\Sigma^+$ state. However, information such as the width of the FD^+ line in the energy spectrum and careful cross section measurements indicate that the $^2\Sigma^+$ state is not formed to a significant degree. This apparent conflict is currently being investigated in the author's laboratory and the preliminary indications are that the observed lowering of the apparent value stems from the presence of a dynamic rotational barrier to dissociation of FD^+ which results from the rotational excitation of the FD^+ ion product during the course of the reaction.

There is another interesting feature to the plot in Fig. 6. Q has been measured for a number of three atom systems and it

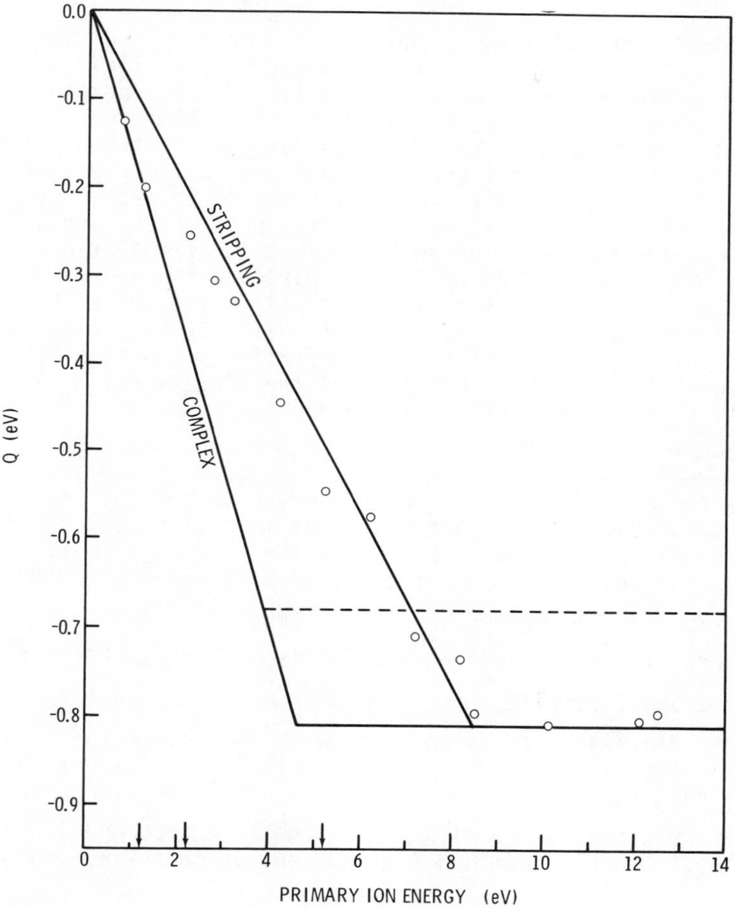

Figure 6. Translational exoergicity for the reaction $F^+(D_2,D)FD^+$ as a function of primary ion energy in the laboratory system. Arrows indicate energies at which contour maps were made.

approximately obeys the relation (15)

$$Q = Q_o - BT$$

where T is the projectile ion kinetic energy and Q_o is the value of Q at T = 0. For an ideal stripping reaction $Q_o = 0$. For many ion-molecule reactions involving hydrogen molecules Q_o has been reported to have positive values of about 0.15 eV (16). It is clear from Fig. 6 that Q remains negative down to energies <0.7 eV in the laboratory system and that its limiting value at T = 0 appears to be zero. This suggests that at low primary ion energy, assuming that the reactants are in their ground states, that the bulk of the heat of reaction to produce $FD^+(^2\Pi)$ is converted into internal energy of the product ion and a negligible amount is converted in product ion translational energy. Since $\Delta H = -2.78$ eV and the dissociation energy is 3.46 eV and the product ion line width is only 0.285 eV at a projectile energy of 1.30 eV and width of 0.067 eV (FWHM) it appears that ΔH is deposited into a relatively few high lying vibrational levels of the FD^+ ion and the vibrational population in the product ion is highly inverted.

Reaction of F^+ with CH_4.

As far as the author is aware there have been no reports of F^+ ion beam studies with complex target molecules with the exception of some preliminary work on the $F^+ + CH_4$ reaction. Using ground state F^+ Jones et al. (17) have made some observations on the production of CH_3^+ and HF^+ from these reactions. The cross section for CH_3^+ formation was 0.39 $Å^2$ at 2.5 eV and fell with increasing F^+ energy and about 20 eV the curve flattened off to about 0.09 $Å^2$ remaining flat to 60 eV which was the highest energy studied. It is of interest to compare this reaction to the corresponding C^+ reaction where (18) at 2.5 eV CH_3^+ is formed with a 7 $Å^2$ cross section and it falls with kinetic energy and then flattens out at cross section of about 3 $Å^2$. Both of these similar reactions probably proceed by hydride ion abstraction at low energy and dissociative charge transfer at the higher energies. The differences in the cross sections and the position of the plateau are probably due to the greater exothermicity of the F^+ reaction. This results in a greater internal energy deposition in the intermediate with corresponding greater fragmentation into simpler products.

The reaction involving the formation of HF^+ is very similar to the corresponding C^+ reaction. The F^+ reaction cross section is essentially constant from 9 eV to 60 eV at a value of 0.07 $Å^2$. The C^+ reaction also gave approximately a constant cross section of about 0.3 $Å^2$ up to 200 eV. These reactions very likely arise from dissociative charge transfer processes but other mechanisms are possible.

Acknowledgement

The work of the author as reported in this article was supported by the U.S. Energy Research and Development Administration.

Abstract

F^+ ion beam studies using simple atomic or molecular targets are summarized. The charge transfer between F^+ and Ne, the reaction F^+ with molecular hydrogen and its stable isotopic variants, dynamics of the $F^+(D_2,D)FD^+$ reaction and some reactions of F^+ with methane as a function of projectile ion kinetic energy are covered and examined for the information that they give on reaction mechanism and internal energy of the product ion.

Literature Cited

1. Weiner, E. R., Hertel, G. H. and Koski, W. S., J. Am. Chem. Soc. (1964), 86, 788.
2. Hyatt, D. and Lacmann, K., Z. Naturforsch (1968), A23, 2080.
3. Ausloos, P., (Ed.), "Interaction Between Ions and Molecules," 6, 673, Plenum Press, New York, 1975.
4. Turner, B. R., Rutherford, J. A. and Compton, D. M. J., J. Chem. Phys. (1968), 48, 1602.
5. Wilson, P. S., Rozett, R. W. and Koski, W. S., J. Chem. Phys. (1970), 52, 5321; Lindemann, E., Rozett, R. W. and Koski, W. S., J. Chem. Phys. (1972), 56, 5490.
6. Cotter, R. J. and Koski, W. S., J. Chem. Phys. (1973), 59, 784.
7. Lin, K. C., Cotter, R. J. and Koski, W. S., J. Chem. Phys. (1974), 61, 905.
8. Shuler, K. E., J. Chem. Phys. (1953), 21, 624.
9. Berkowitz, J., Chem. Phys. Letts. (1971), 11, 21.
10. Franklin, J. L., Dillard, J. G., Rosenstock, H. M., Herron, J. T., Draxl, K. and Field, F. H., Natl. Stand. Ref. Data Ser. (1969), 26.
11. Pechukas, P., Light, J. C. and Rankin, C., J. Chem. Phys. (1966), 44, 794.
12. Yuan, J. M. and Micha, D. A., J. Chem. Phys. (1976), 64, 1032.
13. Dibeler, V. H., Walker, J. A. and McCulloh, K. E., J. Chem. Phys., (1969), 51, 4230.
14. Wendell, K., Jones, C. A., Kaufman, Joyce, J. and Koski, W. S. J. Chem. Phys. (1975) 63, 750.
15. Herschbach, D. R., Appl. Opt. Suppl. (1965), 2, 128; Herschbach, D. R., Adv. Chem. Phys. (1966), 10, 319.
16. Connally, C. M. and Gislason, E. A., Chem. Phys. Letts. (1972), 14, 103.

17. Jones, C. A., Semo, N. and Koski, W. S., unpublished work.
18. Wilson, P. S., Rozett, R. W. and Koski, W. S., J. Chem. Phys. (1970), 52, 5321.

9

Energy Disposal in Reactions of Fluorine Atoms with Polyatomic Hydride Molecules as Studied by Infrared Chemiluminescence

DENIS BOGAN*

Chemical Dynamics Branch, 6180, Naval Research Laboratory, Washington, DC 20375

D. W. SETSER

Department of Chemistry, Kansas State University, Manhattan, KS 66506

Understanding reactive collisions at the microscopic level is a long standing goal of chemical kinetics. The potential energy is the cause of chemical change and is responsible for rearrangements of chemical bonds and for the energy states of the product molecules. Through study of the energy disposal, i.e., the partitioning of energy to the various possible quantum states of the products, properties of the potential energy surface can be inferred. The principal experimental methods for measuring energy disposal have been crossed molecular beams (1,2), laser emission or probing (3,4), and infrared chemiluminescence. The basis for infrared chemiluminescence experiments is the analysis of the infrared emission emanating from vibrationally excited molecules. The results and interpretations of infrared chemiluminescence studies of the reactions of F atoms with polyatomic hydride molecules are reviewed and discussed herein.

The infrared chemiluminescence technique has been pioneered by J. C. Polanyi (5-21). The University of Toronto group has concentrated on three-body reactions and more recently on reactions with vibrationally and translationally excited reagents (14-21). At Kansas State University attention has been directed to reactions of F atoms with polyatomic hydrides (22-33); a large body of data has been obtained and information theory has been used for interpretation (31,32). Other groups working in the HF infrared chemiluminescence field are those of Jonathan (40,41), (University of Southhampton), McDonald (34-39), (University of Illinois) and Y. K. Vasiljev (42), (Academy of Sciences, USSR). Interesting work with CO infrared chemiluminescence has been done by Smith (43,44) and others (38,45).

---

*Postdoctoral fellow, Kansas State University, 1972-1974; National Research Council Associate, 1974-1976; Staff Scientist from 1976 at Naval Research Laboratory.

HF is particularly amenable to infrared chemiluminescence studies for the following reasons:

a) it has the largest transition probabilities of any diatomic molecule,

b) the large vibrational and rotational energy spacings facilitate the assignment of vibrational and rotational distributions,

c) the activation energies for F atom reactions are low (46), 0-2 kcal/mole; hence, the reactions are fast giving strong IR emission,

d) fluorine atom reactions are highly exothermic and give strongly non-Boltzmann HF product distributions.

There also are practical needs for understanding the dynamics of reactions producing HF, since this knowledge may facilitate selection of better fuels and optimization of operating conditions for HF lasers.

McDonald and coworkers (34,36,37) have studied infrared chemiluminescence from polyatomic product molecules formed by addition-displacement reactions of the type

$$F + CH_2{=}CHX \rightarrow CH_2{=}CHF + X \quad (X = H,\ CH_3,\ Cl,\ Br), \tag{1}$$

as well as reactions with benzene and substituted benzene molecules. Valuable insight into the energy distribution in polyatomic product molecules from unimolecular reactions has been obtained from this work (47).

Basics of the Infrared Chemiluminescence Technique

The experiments involve mixing fluorine atoms with a hydride reagent with observation of the emission from the resulting HF product.

$$F + HR \rightarrow HF^{\dagger}_{vJ} + R \tag{2}$$

Fluorine atoms usually are produced by a microwave discharge through CF_4 or SF_6. Figure 1 shows a schematic diagram of a vessel with liquid nitrogen cooled inner walls that has been used for "arrested relaxation" experiments. Such a vessel arrests vibrational relaxation and partially arrests rotational relaxation. The maximum fraction of the surface should be cooled for arrested relaxation studies with HF because collisions of $HF^{\dagger}_{vJ}$ with warm surfaces result in some relaxation with subsequent reentry of those molecules into the cone of sight of observation. Mirrors usually are placed inside the vessel to more efficiently collect the infrared radiation.

Infrared radiation is observed under steady state conditions, and the following equations can be written for each vibrational (as shown) or rotational level (6).

$$0 = \frac{d[HF_v^\dagger]}{dt} = k_v[RH][F] + \sum_{v'} A_{v'-v}N_{v'} + \sum_{v'} k_{z,v'-v}N_{v'} - \sum_{v''} A_{v-v''}N_v$$

$$- \sum_{v''} k_{z,v-v''}N_v - k_pN_v \qquad (3)$$

$N_v \equiv [HF_v^\dagger]$ = concentration of HF in state v

$k_v[RH][F]$ = rate of formation of $HF_v^\dagger$ from the chemical reaction

$A_{v'-v}N_{v'}$ = radiative rate into state v from v'

$k_{z,v'-v}N_{v'}$ = collisional rate into state v from v'

$A_{v-v''}N_v$ = radiative rate out of v into v''

$k_{z,v-v''}N_v$ = collisional rate out of v into v''

k_pN_v = rate of removal by pumping of $HF_v^\dagger$

The steady state expression for $[HF_v^\dagger]$ is

$$[HF_v^\dagger] = \frac{k_v[RH][RF] + \sum_{v'} A_{v'-v}N_{v'} + \sum_{v'} k_{z,v'-v}N_{v'}}{k_p + \sum_{v''} A_{v-v''} + \sum_{v''} k_{z,v-v''}} \qquad (4)$$

If the rate of removal by pumping is significantly greater than any of the radiative or collisional terms then the steady state becomes

$$[HF_v^\dagger] = \frac{k_v[RH][F]}{k_p} \; . \qquad (5)$$

With proper design of the reaction vessel and with adjustment of flow rates, it is possible in practice to cause $k_v[RH][F]$ and k_p to dominate the steady state expression (eq. 5). Since $[HF_v^\dagger] \alpha I_v$, the measured emission intensity from level v, then $k_v \alpha I_v$ for such experimental conditions. A similar analysis can be derived for the rotational populations of a given v level. Unfortunately, some rotational relaxation is always present, even in the best experiments. For the arrested relaxation apparatus, with molecules other than H_2, CH_4 or SiH_4, the principal pumping, under typical conditions, is cryopumping by the liquid nitrogen cooled walls.

Since the $HF^\dagger$ emission intensity is weak under arrested relaxation conditions, good light collection optics and a low f/number spectrometer must be used. For work at Kansas State University, the apparatus consisted of a Welsh cell light collection system (48,49) within the cold wall vessel, a 0.25 meter f/4.3 grating monochromator, HR-8 lock-in amplifier and cooled (77°K) PbS photo-conductive detector. The alignment procedure for the Welsh cell has been described (50). In our work the optimization of alignment was done using a flame simulator constructed of nichrome wire wrapped on a glass cane framework and placed in the reagent mixing region. The absorption due to atmospheric water was minimized by enclosing the fore-optics and spectrometer in a plexiglass box which was continuously swept with pneumatically dried air.

Some work also has been done using the "measured relaxation" technique (18,33,40,41) which can give initial vibrational distributions and relative total rate constants for different RH molecules. This technique usually involves a flowing afterglow apparatus with bulk flow velocity in the 100 m sec^{-1} regime (33,40,41). In these experiments the reactants are mixed at various distances, which can be converted to times, upstream of an observation window. The observed rotational populations are relaxed to a Boltzmann distribution; however, the vibrational relaxation can be followed as a function of time and can be extrapolated to obtain the initial populations at zero time.

The $HF^\dagger$ intensity from a given v',J' level is given by (51).

$$I_{v'',J''}^{v',J'} = [HF^\dagger_{v',J'}]A_{v'',J''}^{v',J'}; \tag{6}$$

A = Einstein coefficient for spontaneous emission, the primed and double primed quantum numbers refer to the upper and lower states of the radiative transition, respectively. The photon flux (intensity) associated with each rotational level is obtained from the observed intensity after correction for the variation of the spectral slit function and the overall spectrometer response with wavelength. The spectral slit function is often given by the spectrometer manufacturer and can be determined experimentally (along with wavelength calibration) by scanning mercury lines in higher orders. The spectral slit function is triangular, i.e., a discrete line is observed as a triangle with width at half-height of S_{eff}, which is the effective spectral slit width. The spectrometer response function is determined from a calibrated black body source; for an ideal black body the energy incident upon an aperture of 2π steradians is (52):

$$R^{\circ}_{w}dw = 2\pi hc^2w^3 \frac{dw}{[\exp(\frac{hcw}{kT}) -1]} \quad (7)$$

and the photon radiancy is

$$R^{N}_{W} = \frac{R^{\circ}_{w}}{hcw} = \frac{2\pi cw^2 dw}{[\exp(\frac{hcw}{kT}) -1]} ; \quad (8)$$

h = Planck's constant, c = speed of light and w = frequency in cm^{-1}. Since determination of relative populations is equivalent to relative number of photons, R^N_W is used. For a spectrometer with a triangular slit function a continuous source, such as a black body, gives a signal at a fixed w, that is proportional to the square of the slit width. For a line spectrum, such as a vibration-rotation spectrum, the intensity at a fixed w has a first power dependence on the slit width. The above experimental observations can be understood from a detailed analysis of the effects of spectral slit width on radiation of line and continuum characteristics (52). If the spectral slit width of the instrument changes with w, these considerations must be included in acquiring the response curve.

The Einstein coefficients, A, of equation 6 were calculated from (9).

$$A^{v',J'}_{v'',J''} = \frac{64\pi^4 w^3}{3h(2J'+1)} |M^{v',J'}_{v'',J''}|^2 \, |m| \quad (9)$$

$M^{v',J'}_{v'',J''}$ is the electric dipole matrix element for the vibration-rotation transition and can be separated into two terms;

$$|M^{v',J'}_{v'',J''}|^2 = |R^{v'}_{v''}|^2 \, F^{v',J'}_{v'',J''} \quad (10)$$

where $R^{v'}_{v''} = < v'|\mu|v'' >$ is the vibrational transition matrix element, $|m| = J'' + 1$ for R branch ($\Delta J = -1$) transitions and J'' for P branch ($\Delta J = +1$) transitions, $F^{v',J'}_{v'',J''}$ = the vibration-rotation interaction factor. We used the tabulated values of Herbelin and Emmanuel (53) for $R^{v'}_{v''}$, and calculated the $F^{v'}_{v''}$ values from Herman, Rothery and Rubin (54). For room temperature experiments with Boltzmann rotational populations, the simple two term $F^{v',J'}_{v'',J''}$ expression of Herman and Wallis is adequate (55). The Herman and Wallis expression begins to deviate from the Herman, Rothery and Rubin expression at J $\cong$ 12 and the difference becomes increasingly serious with increasing J. Because of the w^3 dependence of

A, accurate line positions are important and were calculated from the Dunham expression (56). Appendix I contains a summary of the Einstein coefficients and line positions for the $\Delta v=1$ and $\Delta v=2$ HF transitions originating from levels up to $v'=5$, $J'=18$.

For the analysis of low resolution HF spectra, the above information is used to obtain $HF^{\dagger}_{vJ}$ relative populations via iterative computer simulation of the observed spectrum. The computer program accepts estimated vibrational-rotational distributions, and then

i) applies the A coefficients to obtain line strengths,

ii) performs the convolution of spectral slit and spectrometer response functions,

iii) computes the resultant spectrum as a sum of individual, triangular peaks, and

iv) plots the spectrum.

The computed spectrum is compared to the observed spectrum and adjustments in the relative populations are made, and this process is repeated until a satisfactory fit is obtained. Many vibrational-rotational bands are overlapped, and the use of the effective spectral slit width in the calculation leads to a fitting of peak areas. The final set of HF relative populations is the result of fitting fundamental and first overtone spectra and averaging of several spectra. A least squares criterion, on peak height, was used to determine when the iterative simulation process had converged. An example of the agreement of experimental and simulated spectra is shown as Figure 2. For higher resolution spectra in which the rotational lines are resolved, the systematic conversion of the area of each rotational line (corrected for spectrometer response) to relative populations is tedious but straightforward.

Vibrational Energy Disposal

In order to specify the energy disposal the mean energy available,

$$\langle E \rangle = -\Delta H^\circ_0 + E_a + \langle E_{th} \rangle, \tag{11}$$

in a reactive collision is required; ΔH°_0 is the standard enthalpy change at 0°K, E_a is the activation energy and $\langle E_{th} \rangle$ is the mean thermal energy of reactive collisions (57). The vibrational energy in the polyatomic molecule is assumed to be unavailable for disposal to $HF^{\dagger}$, thus $\langle E_{th} \rangle = 5/2\ RT$ for a diatomic or linear triatomic reagent and 3 RT for a polyatomic reagent. The activation energy was set at 1.0 kcal/mole for olefinic and aromatic C-H bonds and 0.5 kcal/mole for other hydrocarbons and reactive

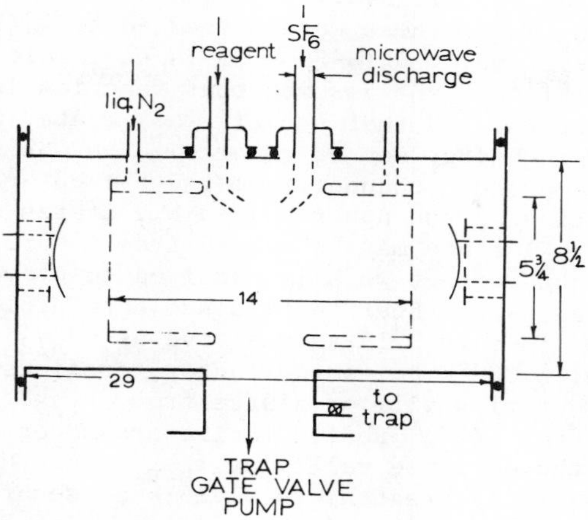

Figure 1. Schematic of an arrested relaxation reaction vessel used at Kansas State University, taken from (24). **See** *(11) for a drawing of the reaction vessel used by the Toronto group.*

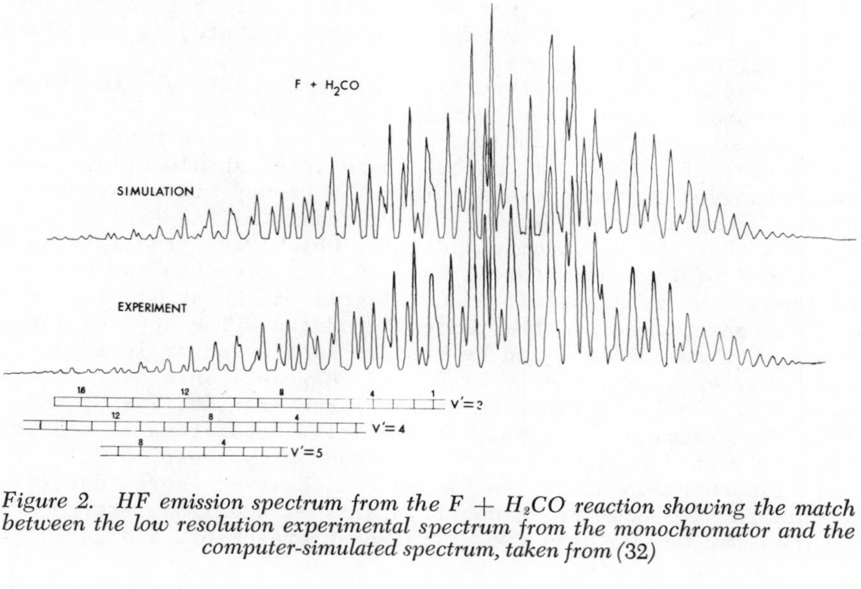

Figure 2. HF emission spectrum from the F + H_2CO reaction showing the match between the low resolution experimental spectrum from the monochromator and the computer-simulated spectrum, taken from (32)

hydrides. The values selected by Perry and Polanyi (20) were used for H_2, HD and D_2. If available, standard bond energies were used for D°_{298}(R-H); these were corrected to D°_0(R-H) by subtraction of 1.4 kcal/mole, D°_0(H-F) = 135.4 kcal/mole (58).

This calculation of <E> assumes that reaction is from the $F(^2P_{3/2})$ ground state, and that the $F(^2P_{1/2})$ state, which is 1.15 kcal/mole higher in energy can be ignored. The observation (59) of $Br(^2P_{1/2})$ as a product channel from the F + HBr reaction suggests that formation of the upper spin-orbit states can be important. The $Br(^2P_{1/2})$ state may be formed from the reaction of the $F(^2P_{1/2})$ state, but further work is required to prove this. Two recent theoretical papers bear on this subject. Tully (73) predicted that $^2P_{1/2}$ is only 0.14 as reactive as $^2P_{3/2}$ at 300°K; and this, combined with 300°K Boltzmann factor, predicts that only 2% of reactive trajectories will originate from $^2P_{1/2}$. Both Tully (73) and Jaffe et al. (74) predict facile crossovers between the two surfaces in the entrance valley at $r_{H-F} > 2Å$. Uncertainties in bond energies are as great as the 1.15 kcal separation between the F atom states and ascertaining the presence of $F(^2P_{1/2})$ from the highest observed $HF^\dagger$ energy would be difficult.

A summary of the HF vibrational distributions, <E>, $<f_V(HF)>$ and the highest observed HF energy from F + HR reactions is given in Appendix II. The column labelled thermochemical available energy was calculated from (11). The column labelled observed cutoff is the energy of the highest observed HF vib-rotational state. Data from different laboratories and from deuterated molecules are included in order to facilitate quantitative comparison. No attempt has been made to assess the quality of the data except to omit work known to be suspect or to add comments as footnotes. The v=0 populations, given for some entries, were obtained from surprisal plots (62) or from rule-of-thumb estimate, as described in the original publications.

In general, the $HF^\dagger$ vibrational distributions have the form of a bell shaped curve with a maximum at $f_V \approx 0.5$; i.e., v=2 for ~100 kcal/mole R-H bonds and v=3 or 4 for weaker R-H bonds; $f_V = E_V/\langle E \rangle$. In almost all cases, there is an absolute inversion between the v=1 and v=2 levels. The $<f_V>$ values are remarkably constant, ranging from 0.45-0.70 in virtually all cases. This indicates that the polyatomic radicals, which have greatly varying numbers of internal states, have little effect on the HF vibrational energy disposal; hence, little of the available energy is partitioned to the radical products. The obvious inference is that the radical can be treated as a structureless particle for vibrational energy disposal. However, on closer inspection, subtle differences appear. For example, F + H_2 and F + CH_4 have similar <E>; but, the relative population of v=3 is markedly lower for CH_4 and $<f_V>$ is consequently lower, 0.70 vs 0.60. The thermochemical energy and the observed energy cutoff coincide for H_2; however, E(cutoff) falls ~2 kcal/mole short of E(thermo) for CH_4. These two differences presumably indicate (24) that some energy is retained in the CH_3 group as it relaxes from the tetrahedral geometry of methane to the planar geometry of methyl. The effect is a conformational stabilization of CH_3 occurring late in the exit channel. Since the H atom is rapidly transferred from the C to F, such an effect is maximized. For some reagents, much larger effects arising from this feature have been observed. The best examples are F + toluene (30,31) and F + propene (35). In these cases, the effects are much more pronounced because of the much larger resonance stabilization energy of the benzyl and allyl radicals.

Dynamical complications can obscure the simplicity of the above interpretation. These include complex encounters, where the separation of $HF^{\dagger}_{vJ}$ from R is slow enough to permit exchange of vibrational energy between them and delayed secondary encounters where $HF^{\dagger}_{vJ}$ collides with R following a direct reaction (63,64). Complex and secondary encounters seem not too serious for vibrational energy disposal for simple polyatomics, such as methane, substituted methane, and formaldehyde; however, they appear to become increasingly serious with increasing bulkiness of the polyatomic radical, for example in the series CH_4 (slight), CH_3CH_3 (more serious), $C(CH_3)_4$ (severe)(31).

Models for Vibrational Energy Disposal

At this point it is appropriate to ask: what models exist for understanding the dynamics of hydrogen abstraction by fluorine atoms (or the similar case, Cl + HI (65)). The dominant feature of the three-body dynamics is the rapid transfer of the H atom to the attacking F (64), which results in simultaneous change in the R_{F-H} and R_{H-C} coordinates and a high degree of mixed energy release giving a high fraction of HF vibrational energy. Since the light H atom cannot transfer much momentum, the momenta of both heavy atoms, one of which is now incorporated in the newly formed HF, remain nearly the same as their values prior to the H transfer (66). After the transfer of the H to F the $HF^{\dagger}$ continues to approach the donor atom; the subsequent scattering resembles hard sphere elastic scattering (65). During the scattering, considerable interaction between the $HF^{\dagger}$ and the polyatomic fragment can occur. To the extent that this model applies, the products will have relatively small amounts of translational energy, for thermal reactants, and the reaction exoergicity must be partitioned primarily to vibration and, to a lesser extent, rotation of the product HF. Some discretion must be exercised in applying this model to F + polyatomic reactions because internal degrees of freedom of R can take up energy; nevertheless, it is clear from the observation of similar $\langle f_V \rangle$ and the inverted vibrational distributions that three-body behavior frequently dominates the energy disposal of the F + polyatomic reactions. The F + HCl and HBr reactions provide "standard" three-body systems of bond energies of ∿103 and 88 kcal/mole, respectively, for comparison to the

Appendix II. Vibrational Energy Disposal for F + RH Reactions

Reactant	Ref.	Distribution[a] P(v) v=0[a]	1	2	3	4	5	6	Available energy[b] Thermo.[c]	Obs. cutoff	$\langle f_V \rangle$[d]
Three Body Cases											
H_2	24		17	55	28				35.0	34.7	0.66
H_2	20	2	15	53	30				35.0	34.7	0.66
H_2	41		14	49	37				35.0	34.7	0.70
H_2	42		19	55	26				35.0	33.6	0.65
H_2	35		16	57	27				35[e]	34	0.66
HD (HF)	20	4	20	67	9				34.3[f]	34.1	0.59
HD (DF)	20	3	8	23	41	25			36.1[f]	34.3	0.63
D_2	20	2	7	23	43	25			35.2	34.2	0.64
HCl	17(31)	11	24	52	13				35.6	35	0.53
HCl	40		30	60	10				35.6	g	
HBr	40(31)	8	18	29	26	19			51	g	0.53
HBr	59	6	11	23	30	30			51	g	0.58
HI	40(31)	5	12	14	14	17	24	14	67	g	0.56
HI	59	4	10	11	13	19	26	17	67	g	0.60
Primary C-H											
CH_4	24(31)	2[h]	21	64	13				35	33	0.59
CH_4	41		21	64	15				35	33.6	0.61
CH_4	42		21	68	10				35	33.6	0.59

Appendix II Continued (page 2)

Reactant	Ref.	Distribution[a] P(v) v=0[a]	1	2	3	4	5	Available Energy[b] Thermo.[c]	Obs. cutoff	$<f_V>$[d]
CH_4 [g]	35		16	72	12			35	-	0.62
CD_4	26		5	29	56	10		35	33	0.62
C_2H_6	31	1[h]	12	51	36			41	37	0.59
C_2H_6	44		16	49	36			41	35	0.65
CH_3F	24	1[h]	17	61	21			36	38	0.62
CH_3Cl[i]	24	2[h]	24	37	37			39	37	0.59
CH_3Br[i]	24	2[h]	23	41	34			38	36	0.60
CH_3I	24	2[h]	26	45	27			37	39	0.59
CH_3CN[j]	31	13	34	29	24			50	37	0.36
CH_3CN[j]	44		67	33				50	-	0.30
CH_3OD (HF)	33		27	43	30			43	37	0.52
$(CH_3)SiCl_3$	106		25	53	19	2		41[k]	37	0.53
$(CH_3)_2SiCl_2$	106		17	69	13			41[k]	35	0.52
CH_3HgCH_3	106		43	47	10			41[k]	37	0.45
$Si(OCH_3)_4$	33		25	50	25			41[k]	38	0.54
$Si(CH_3)_4$	24		30	59	11			41[k]	34	0.50
$C(CH_3)_4$	31	1[h]	16	70	13			40	36	0.54
$(CH_3)_2S$[l]	25		44[l]	37[l]	19			41[k]	41	0.48
$(CH_3)_3N$	25		35	50	15			41[k]	36	0.50

Appendix II Continued (page 3)

Reactant	Ref.	Distribution[a] P(v) v=0[a]	1	2	3	4	5	Available Energy[b] Thermo.[c]	Obs Cutoff	$<f_V>$[d]
$CD_3CH_2CD_3$ (DF)	26			37	41	23		38		
CH_3CF_3	24		20	63	17			34	34	0.66
Primary C-H giving radicals with large stabilization energies										
1,3,5-$C_6D_3(CH_3)_3$ (HF)	31	3[h]	18	36	39	4		54	44	0.45
$C_6H_5CH_3$[e]	35	21[E]	41	27	11			54 (28)[e]		
$C_6H_5CH_3$	31		25	39	29	3		54 (28)	44	
o-$C_6H_4(CH_3)_2$[e]	35	22[E]	43	26	9			54 (28)[e]		
o-$C_6H_4Cl(CH_3)$[e]	35	22[E]	43	26	9			54 (28)[e]		
CH_3OCH_3	32	4[h]	23	46	27			45	39	0.48
CH_3CHCH_2[e]	35	17[E]	33	32	17	2		54 (31)[e]	45	
Secondary C-H										
$CD_3CH_2CD_3$(HF)	26			62	38			42	39	
c-C_3H_6	27		37	62	1			38	33	0.48
c-C_5H_{10}	27(31)	1[h]	19	56	23	1		45	43	0.50
c-C_6H_{12}	26(31)	1[h]	28	55	16	1		44	43	0.47
c-C_6D_{12} (DF)	26		8	32	43	17		42	36	0.52
c-C_7H_{14}	26		25	54	20	1		45	43	0.48
c-C_8H_{16}	26		27	50	20	2		45	43	0.48
c-$C_{10}H_{20}$	26		26	54	20	1		45	43	0.49

Appendix II Continued (page 4)

Reactant	Ref.	Distribution[a] P(v) v=0[a]	1	2	3	4	5	Available Energy[b] Thermo.[c]	Obs. cutoff	$<f_V>$[d]
CH_2CH_2O (ethylene oxide)	33		34	58	8			37[m]	37	0.52
1,4-dioxane	33		24	50	26	1		46	42	0.49
trioxane	33		24	64	12			46	38	0.46
Alkenes										
C_2H_4	31	8[h]	43	48				31	31	0.45
C_2H_4	44		40	54	6			31	32	0.59
C_2H_4[e]	35	22[E]	44	32	2			31[e]	-	0.41
CH_2CHCl[e]	35	20[E]	40	36	4			31[e,n]		0.45
CH_2CHBr[e]	35	19[E]	37	38	7			31[e,n]		0.48
CH_2CCl_2	35	25[E]	49	24	3			31[e,n]		0.38
ClCHCHCl[e]	35	22[E]	43	31	4			31[e,n]		0.42
Aromatic C-H										
C_6H_6	31	10[h]	63	27				28	25	0.49
C_6H_6[e]	35	21[E]	42	30	7			28	-	0.49
C_6H_5Cl[e]	35	22[E]	43	29	7			28[o]	36	0.48
C_6H_5Br[e]	35	21[E]	41	32	6			28[o]		0.49
o-$C_6H_4Cl_2$[e]	35	21[E]	42	28	9			28[o]		0.50
m-$C_6H_4Cl_2$[e]	35	22[E]	44	28	6			28[o]		0.47
p-$C_6H_4Cl_2$[e]	35	22[E]	44	26	8			28[o]		0.48

Appendix II Continued (page 5)

Reactant	Ref.	Distribution[a] P(v) v=0[a]	1	2	3	4	5	Available Energy[b] Thermo.[c]	Obs. cutoff	$<f_V>$[d]
Aldehydes, Ketones, Phenol										
H_2CO[p]	32	9[h]	25	29	24	13		52	52	0.43
CD_3CHO (HF)	32	8[h]	24	41	23	4		52	43	0.40
C_6H_5CHO	32			68	29	3		52(28)	42	
C_6H_5OH	31		69	24	6	1		53(28)	45	
Inorganic Polyatomic Hydrides										
GeH_4	29	1[h]	10	18	29	40	2	58[m]	58	0.57
PH_3	25		19	26	27	23	5	58[m]	58	0.51
SiH_4	25	2[h]	14	26	42	16		50[m]	50	0.57
H_2O_2	32	10[h]	34	26	20	10		49	44	0.42
H_2O	25		100					19	17	
H_2S	25		28	30	26	16		47	45	0.53
NH_3	25		55	45				34	27	0.48
N_2H_4[q]	25		48	39	9	3	1	62	52	0.30

a. The vibrational distributions are normalized to $\sum_v P(v) = 100$. In some cases estimates of the P(0) population have been made by extrapolation of a linear surprisal plot. If the surprisal analysis was not done in the original work, the reference in parenthesis should be consulted. In other cases P(0) was estimated by extrapolation of the distribution to v=0; these are denoted by a superscript E. These estimates are considerably larger than those obtained from extrapolation of the surprisal plots.

b. For room temperature experiments unless noted otherwise; the units are kcal/mole. The values are rounded to the nearest kcal/mole unless the bond energies are very well known.

c. This is the <E> given by eq. 11 in the text. If more than one type of C-H bond is present, the <E> for the weaker bond is given first, followed by the <E> for the stronger bond in parentheses. Most bond energies were taken from Benson and O'Neal (82).

d. If P(0) is listed in the table, it was used to calculate $\langle f_V \rangle = \sum_{v}^{v^*} P(v) f_V$.

e. McDonald and coworkers use thermal dissociation of F_2 as the F atom source and the available energy could be slightly higher than the values quoted from other laboratories using microwave dissociation of CF_4 or SF_6 as the F atom source. Within the round-off limits, the <E> are the same as the room temperature values.

f. The activation energies and zero point energies are different and lead to unequal available energies; see (20).

g. These experiments were done in a flowing afterglow apparatus which gives a Boltzmann distribution of rotational levels and a cutoff level is not observable.

h. These values were obtained by extrapolation of the model III surprisal plots. If model II or I were used, the relative population of v=0 would be higher by factors of ∿2 and 3 respectively.

i. Recently acquired results (80) using an improved reaction vessel and interferometric observation of the emission suggest the v_3/v_2 ratio is somewhat higher than given in the table.

j. The addition of F to CH_3CN followed by HF elimination may compete with direct abstraction.

k. The bond energy was assumed to be the same as for C_2H_6.

l. The v_1/v_2 ratio probably is too high. See ref. (32) for updating of the $(CH_3)_2O$ data. Similar changes probably apply to $(CH_3)_2S$.

m. The highest observed $HF^{\dagger}_{vJ}$ energy was set equal to the thermochemical energy.

n. The bond energy was assumed to be the same as for C_2H_4.

o. The bond energy was assumed to be the same as for C_6H_6.

p. Very fast V-V relaxation without an appreciable change in J may have affected these data; thus, $\langle f_V \rangle$ is a lower limit.

q. Poor quality data due to weak emission; reaction channels other than direct abstraction may contribute to the observed emission.

polyatomic systems.

Product distributions can be obtained via classical trajectory calculations on semi-empirical, potential energy surfaces (63). "Computer experiments" are performed on the surface using selected distributions of reactant states. For comparison with experimental data, the classical product energies are "requantized" by dividing the classical continuum of HF vibration-rotation states into bins which correspond to the known v', or J' energy states. Comparison of the calculated and experimental energy distributions reveals the appropriateness of the trial surface and adjustments can be made to develop a more correct surface. Extensive trajectory calculations have been done for $F + H_2$ (67-71), F + HCl (17) and Cl + HI (65) which is similar to F + HCl. A representative (not optimized) LEPS surface for F + HC is shown in Figure 3. The results from such a surface are in moderate accord with the vibrational energy disposal for $F + CH_3X$ reactions. Although the surface is of the very repulsive type, a large fraction of the exoergicity is released as E_V because of the mixed energy release characteristic of the F-H-C mass combination. Trajectory calculations for F + HCl (17) are in good agreement with experimental vibrational and rotational distributions of $HF^{\dagger}$. For F + HCl and Cl + HI the trajectory calculations and the experimental data show considerable inverse correlation between E_V' and E_R' (vide infra).

Trajectory calculations provide detailed microscopic insight for three-body reactions. However, such calculations for larger systems are much more difficult, as well as costly. Another approach is the use of information theory, particularly surprisal analysis. These methods have been developed by Levine, Bernstein and coworkers for three-body reactions (62,72,73) and we have extended the analysis in order to deal with polyatomic radical product (31). Surprisal analysis is a quantitative framework for comparison of an observed distribution to a reference (prior) distribution. A natural and convenient reference is the statistical phase space distribution (74). Accessible states are those formed by trajectories which conserve energy and angular momentum. The limitations to the product states allowed by angular momentum is difficult to evaluate (75,76). Fortunately for fast reactions, the large impact parameters contribute a large amount of orbital angular momentum to the reactive collision and this permits the neglect of the angular momentum constraint in the computation of the prior distribution (32).

The surprisal associated with the observation of a given vibrational level is (62)

$$I(f_V) = -\ln P(f_V)/P^{\circ}(f_V); \qquad (12)$$

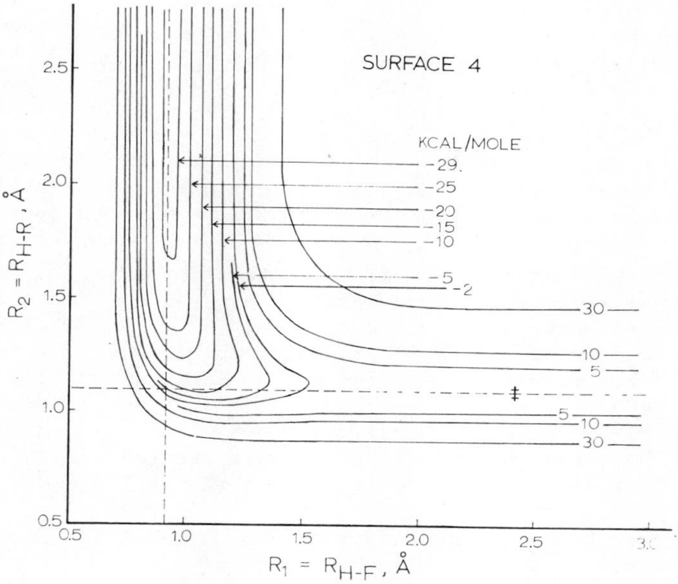

Figure 3. A LEPS potential energy surface (collinear) representing the F + HC three-body reaction, from (64). This particular surface has a low potential energy barrier which occurs at a large R_{HF} *distance.*

where P° is the prior distribution and P is the observed distribution. The surprisal is a local measure of the difference between the experimental and prior populations. The surprisal will be zero if the experimental and prior populations are in exact agreement. For an experiment with fixed available energy E, $P^\circ(f_V)$ is defined by

$$P^\circ(f_V) = \rho(v,E)/\sum_{v=0}^{v*} \rho(v,E) \tag{13}$$

The maximum accessible v is v* and ρ(v,E) is the density of product states at energy E, with v quanta fixed in HF vibration. The density of HF + R product states is (31)

$$\rho(v,E) = \sum_J (2J+1) \int_{E_I=0}^{E - E_{v,J}(HF)} \rho_I(E_I)\rho_T(E-E_{vJ}(HF) - E_I)\, dE_I. \tag{14}$$

The rotational degeneracy of HF in state J is 2J + 1, $\rho_I(E_I)$ is the internal state density of R at energy E_I and $\rho_T(E_T)$ is the translational state density for $E_T = E-E_{vJ}-E_I$. The evaluation of (14) can be done at several levels of completeness. We have chosen 3 prior models, identified by the Roman numerals below (31).

I. The three-body model, where R is treated as a structureless particle. In this model $\rho_I(E_I) = 1.0$ and $E_I = 0$.

II. The three-body model extended to allow R to acquire rotational but not vibrational energy.

III. A complete model allowing all internal degrees of freedom of R to acquire energy. For substituted methanes, it was useful to employ a reduced model III, named III-R, in which large molecules were treated as a six-body system, CH_3X, with X a structureless atom of appropriate mass.

The evaluation of (14) was done numerically for models III and III-R (31). For models I (62) and II (31), integration over J yields results which are in excellent agreement with exact summations and permits equation (14) to be reduced so as to give particularly simple expressions for the prior distributions.

$$P^\circ_I(f_V) = (1 - f_V)^{3/2}/\sum_{v=0}^{v*}(1 - f_V)^{3/2} \tag{15}$$

$$P^\circ_{II}(f_V) = (1 - f_V)^{3}/\sum_{v=0}^{v*}(1 - f_V)^{3} \tag{16}$$

In many cases, vibrational surprisals have been found to be linear functions of f_V (62). That is,

$$I(f_V) = \lambda_o + \lambda_V f_V \quad (17)$$

Thus, the experimental distribution can be expressed in terms of the parameters λ_o, λ_V and the prior distribution $P°(f_V)$

$$P(f_V) = P°(f_V) \exp [-(\lambda_o + \lambda_V f_V)] \quad (18)$$

The surprisal analysis has proven particularly useful in studying trends within a similar series of reactions. Some conclusions (31) are:

i) The surprisal plots tend to be linear for all three prior distributions unless dynamical restrictions are present. Then surprisal plots for I and II become non-linear. Extrapolation of the surprisal plots to $f_V = 0$ facilitates assignment of the relative v=0 population.

ii) The vibrational energy disposal for the CH_3X (X = F, Cl, Br, I) series are all characterized by the same λ_V. This is illustrated in Figure 4 (note the comment in the figure caption concerning new results).

iii) The CH_4, SiH_4, GeH_4 series show the same λ_V as the CH_3X series.

iv) The pairwise members of the series (CH_4 - HCl), (SiH_4 - HBr), GeH_4 - HI) have very similar vibrational energy disposal as measured by $\langle f_V \rangle$ and λ_V using model I as the prior. The bond energy of each pair is similar.

These conclusions show that three-body behavior dominates the vibrational energy disposal of the CH_3X systems, as well as SiH_4 and GeH_4. The $-\lambda_V$ values found for the above reactions are: model I, (4.7 ± 0.7); model II (10 ± 1) and model III (15 ± 1). Thus, with the aid of surprisal analysis, the product vibrational distributions for all of the above systems can be compactly expressed, within the experimental error, by the parameters λ_o and λ_V. The fact that the same vibrational disequilibrium is obtained in the above cases confirms, via information theory, that all reactions follow similar dynamics.

The population of the highest accessible level, v*, is very sensitive to slight variations in <E>; i.e., a small decrease in <E> because of an increase in D_o°(R-H) markedly lowers $P°(v^*)$. Surprisal plots (with all models) confirm this sensitivity, in that reduction of the energy causes the $I(f_{v^*})$ point to rise above the line formed by the $I(f_V)$ points, v<v* (31). The model I surprisal from the F + toluene reaction, using the full thermochemical energy, showed a very serious deviation of this sort. Since the benzyl radical is resonance stabilized, we reasoned that F + toluene might obey the three-body model if the available energy

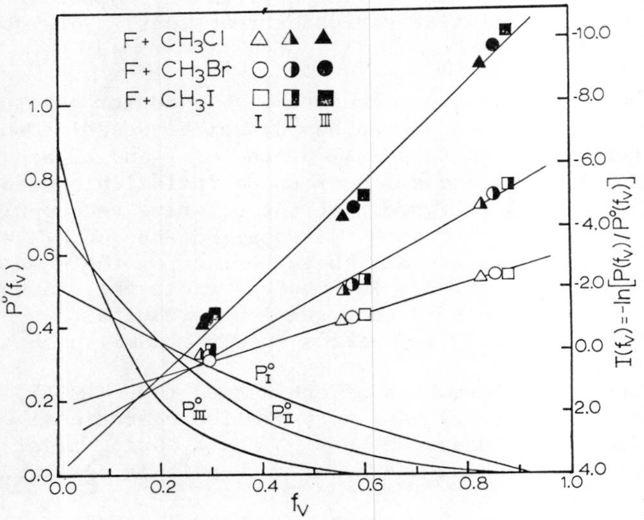

Figure 4. Vibrational surprisal plots for models I, II, and III for F + CH_3X reactions, taken from (31). Recent work suggests that the v = 3/v = 2 and v = 2/v = 1 ratios may be somewhat higher than those used in making this plot. The general conclusions (31) based upon the surprisal analyses would be unchanged; but the $-\lambda_V$ values may be increased slightly.

was reduced by the resonance stabilization energy in calculating $P°(f_V)$. This energy adjustment did linearize the model I plot giving $\lambda_V = -4.5$ in excellent agreement with plots for CH_3X, SiH_4 and GeH_4. (31) Other reactions yielding large radicals that have stabilization energies show a similar effect. (31,32)

The reactions F + ethane, neopentane, and other large hydrocarbon molecules do not give satisfactory model I surprisal plots even with adjustment of energy (31). This has been interpreted as evidence for the role of complex and delayed encounters, which are expected to be important when R is a bulky group. A further difficulty, which surprisal analysis can help to reveal, is the existence of more than one reaction channel. The rather flat vibrational distribution from F + CH_3CN has been attributed (31, 44,45) to the existence of an addition - elimination channel as well as direct abstraction. Although formaldehyde could undergo an analogous three-centered elimination process, the surprisal analysis suggests that it does not (32).

Rotational Energy Disposal

Rotational relaxation of HF is only partially stopped even in the best "arrested relaxation" experiments. In order to estimate the initial rotational distributions, some correction for relaxation is necessary. There is experimental (77,78,32) and theoretical (79) support for collisional deactivation transition probabilities of the form,

$$P_{ij} = Ag_j \exp[-(E_i - E_j)/RT]. \quad (19)$$

The probability of deactivation from level i to level j in a single collision is P_{ij}; g_j is the rotational degeneracy of level j. Detailed balancing was used to obtain the upward transition probabilities. In order to simulate experimental results, the master equation for relaxation is solved starting with an estimated initial distribution, transition probabilities from eq. 19 and a specified average number of deactivating collisions. Iterative simulation of the observed steady state distribution permits an estimation of the initial distribution. Because of the exponential dependence of the deactivation probability on the energy separation between states, the relaxation for $J \geq 10$ is inefficient. In our experience, reactions producing appreciable populations in the $J \geq 10$ levels show steady-state rotational distributions having a high J component, which is the residual of the initial distribution, a minimum at intermediate J, and a low J component having a 200–300°K Boltzmann shape.

Two examples of steady-state distributions are shown in Figures 5 and 6. The data for CH_3OCH_3 were deduced from the low resolution spectra obtained with the spectrometer (32) and simulation of the relaxation, as described above, was used to estimate the initial rotational distributions. The F + HCl reaction

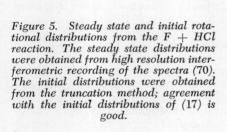

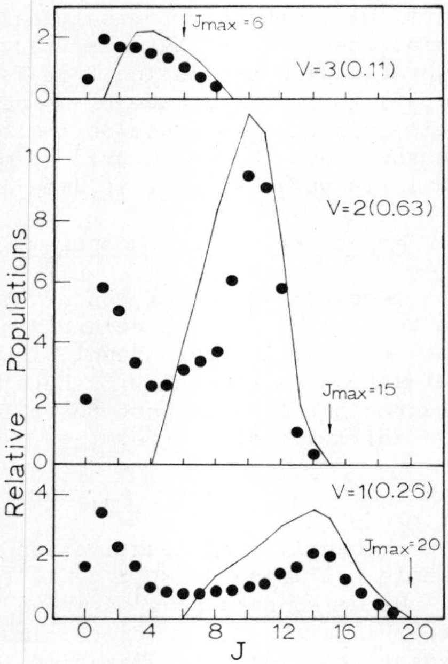

Figure 5. Steady state and initial rotational distributions from the F + HCl reaction. The steady state distributions were obtained from high resolution interferometric recording of the spectra (70). The initial distributions were obtained from the truncation method; agreement with the initial distributions of (17) is good.

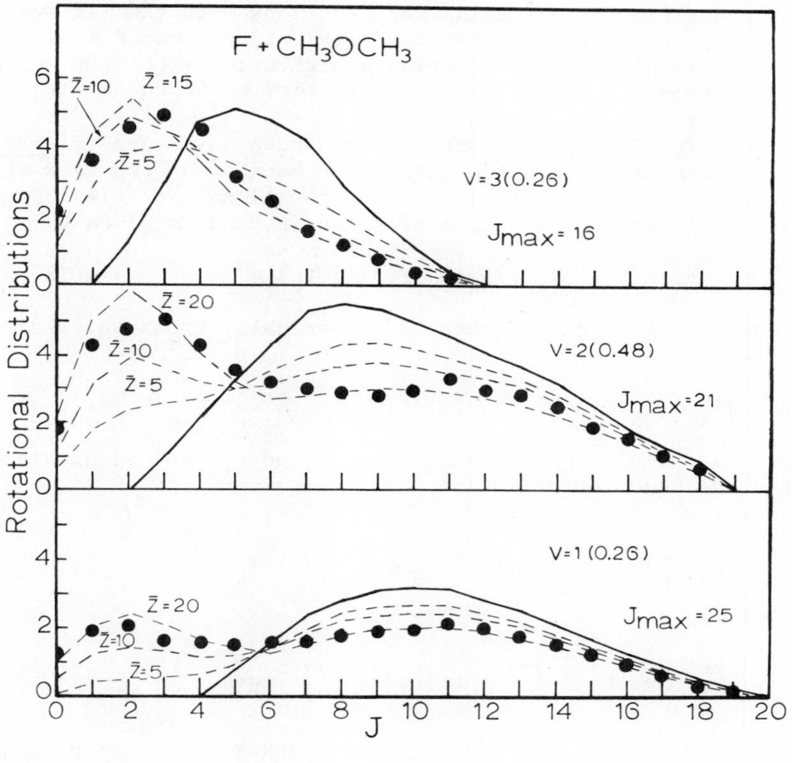

Figure 6. Steady state and initial rotational distributions from the F + $(CH_3)_2O$ reaction, from (32). The steady state values are based upon low resolution monochromator recording of the spectra. The initial distributions were estimated by simulation of the relaxation (see text). Z = the average number of collisions.

recently has been studied (80) using an interferometer to observe the emission. The ether reaction also has been reinvestigated with the interferometer and, within experimental error, the steady-state distributions of Figure 5 have been confirmed. Thus, an important point is that under essentially identical reaction conditions the steady-state rotational $HF^{\dagger}_{vJ}$ distributions can differ dramatically for reactions with similar available energies. The steady-state distributions from F + HCl are sufficiently close to the initial distributions that the truncation method (77,11) gives a satisfactory estimate of the initial rotational distributions.

Rotational surprisal analysis has been developed (81,62,72, 73,32); however, applications are rare because of the lack of reliable initial distributions. The rotational surprisal is conditional upon fixed values of f_V and E, and is given by,

$$I(f_R|f_V) = -\ln P(f_R|f_V)/P^\circ(f_R|f_V) \tag{20}$$

The value of P°_{III} can be obtained numerically from eqs. (13) and (14).

$$P^\circ_{III}(f_R|f_V) = \rho(E,v,J)/\rho(E,v) \tag{21-a}$$

As before, simple approximations to P°_I and P°_{II} can be obtained within the RRHO approximation,

$$P_I(f_R|f_V) = (2J+1)(1-f_V-f_R)^{1/2}/(1-f_V)^{3/2} \tag{21-b}$$

$$P^\circ_{II}(f_R|f_V) = (2J+1)(1-f_V-f_R)^2/(1-f_V)^3 \tag{21-c}$$

Because $I(f_R|f_V)$ is conditional upon v and E, the proper reduced variable for rotational surprisal plots is $g_R = f_R/(1-f_V)$. By convention, values of $P^\circ(f_R|f_V)$ often are expressed without the (2J+1) factor. However, since experiments do not resolve the M_J states of given J, for practical applications the prior must include (2J+1)(62,72,73). For many reactions the surprisal values are near zero and there is little rotational disequilibrium. The surprisals generally are not linear, but it is impossible to decide whether this is a problem with the estimated "initial" distributions or whether the plots are truly non-linear. Therefore, analysis in terms of the most probable rotational level, J^{mp}, has been adopted. The calculated values of g_R^{mp} for models I and II are 0.5 and 0.2, respectively. The g_R^{mp} for III is always less than 0.2 and is inversely related to the number of internal modes of the radical R.

The above analysis (hereafter the subscripts and superscripts are dropped and $\hat{g}$(Exp), $\hat{g}$(I), $\hat{g}$(II) and $\hat{g}$(III) are used) can be applied to the data (17,80,13,11,20) for the following three-body reactions:

$$F + HCl \rightarrow HF + Cl \qquad \hat{g}(Exp) = 0.50$$

$$Cl + HI \rightarrow HCl + I \qquad \hat{g}(Exp) = 0.45 \tag{22}$$

$$F + H_2 \rightarrow HF + H \qquad \hat{g}(Exp) = 0.21$$

Within experimental error, the first two reactions give $\hat{g}$(Exp) = $\hat{g}$(I), for all v levels, and there is little rotational disequilibrium. The F + H_2 reaction has different mass relationships (H + LL) and dynamics than the F + HR (H + LH) cases. For this reason, only the first two reactions are suitable three-body references for comparison with F + HR reactions.

Figures 7-8 show the triangular contour plots for the detailed relative rate constants for $HF^{\dagger}_{vJ}$ formation from F + HCl, and $(CH_3)_2O$. For the three-body case, the contour plot identifies the vibrational, rotational and tranlational energies associated with a given $HF^{\dagger}_{vJ}$ state, providing that the atom is formed in a unique electronic state. However, for the polyatomic case the residual energy, <E> - E_V(HF) - E_R(HF), can be in internal energy of the R fragment, as well as translation. The $(CH_3)_2O$ reaction is an example. The absence of emission from v=4 and the failure for any of the vibrational-rotational levels to reach to the thermochemical limit is attributed to the CH_3OCH_2 radical stabilization energy. This energy which apparently is not available to $HF^{\dagger}$, displaces the contour plot away from the diagonal line connecting f_R and f_V = 1.0. There are obvious differences between the contour plot for F + HCl and that of the polyatomic case which seem to be well beyond the uncertainties involved in extrapolating to the initial J distributions. The important differences are: a) The values of $\hat{g}$(Exp) for the polyatomic cases are ∿0.2, with the values for v=1 being somewhat less than for the other v levels. Thus, the strong inverse correlation between vibrational and rotational energy, found for the three-body F + HCl reaction, no longer seems to hold. b) For $f_V \gtrsim 0.4$ the $HF^{\dagger}_{vJ}$ distribution usually does not extend to the energy limit denoted by the diagonal line.

Plots of f_R^{mp} vs $(1-f_V)$ tend to be linear and yield a mean value of $\hat{g}$(Exp) from the slope. For the CH_3OCH_3 and CH_2O reactions (32), $\hat{g}$(Exp) was ∿0.2 which is the theoretical value of $\hat{g}$(II) (32). This suggests that the available rotational energy is partitioned statistically between $HF^{\dagger}_{vJ}$ and R. This can explain observation (a) above. A possible explanation for (b) can be obtained from consideration of the angular momentum conservation equation,

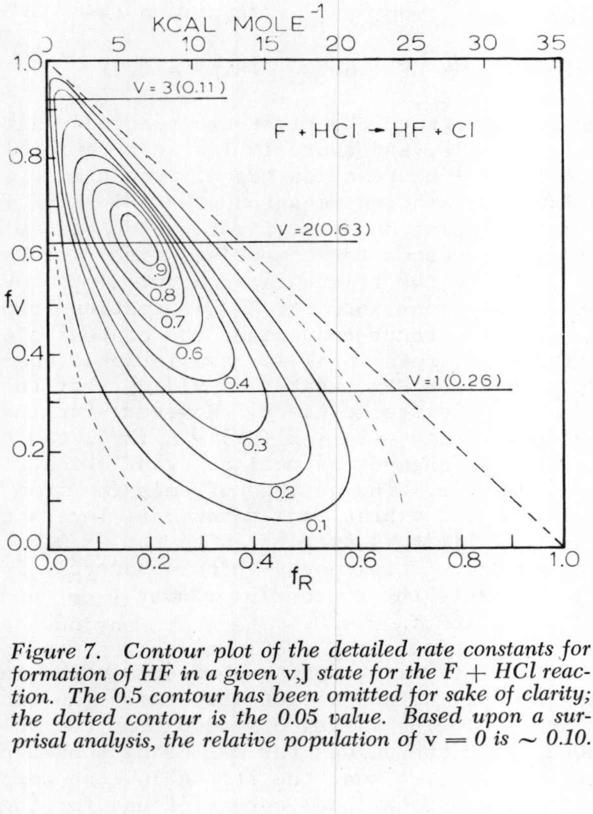

Figure 7. Contour plot of the detailed rate constants for formation of HF in a given v,J *state for the F + HCl reaction. The 0.5 contour has been omitted for sake of clarity; the dotted contour is the 0.05 value. Based upon a surprisal analysis, the relative population of* v = 0 *is ~ 0.10.*

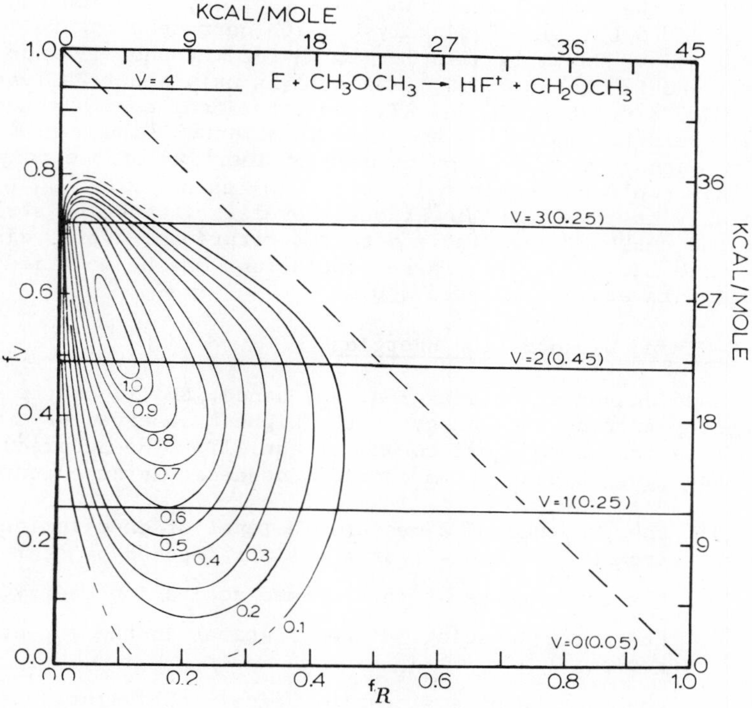

Figure 8. Contour plot of the detailed rate constants for formation of HF in given v,J states for the F + $(CH_3)_2O$ reaction, from (32). The failure of $HF^{\dagger}_{vJ}$ to acquire energy up to the thermochemical limit is attributed to the presence of the CH_2OCH_3 stabilization energy which is retained in the radical. This stabilization energy displaces the contour plot from the dotted diagonal line.

$$\vec{L} + \vec{J} = \vec{L}' + \vec{J}'_{HF} + \vec{J}'_{R}, \qquad (23)$$

$\vec{L}$ = orbital and $\vec{J}$ = rotational angular momentum; the primed quantities denote products. The measured total rate constants (46, 33) yield the thermal reaction cross section, σ, which can be converted to L (62). Typically $\vec{L} \geq 20\hbar$ corresponding to $\sigma \gtrsim 20\text{Å}^2$, however, $\vec{J}$ at 300°K is also large, $8\hbar$ for methane, $19\hbar$ for ethane and $27\hbar$ for $(CH_3)_2O$. Since J'_R is only about $2\hbar$ less than $\vec{J}$ (at 300°K for both RH and R), the rotational energy acquired by HF could be limited by the angular momentum (although $E_J(R)$ is low because of the large moments of inertia) of the polyatomic R. This could constrain $E_{vJ}(HF)$ to less than (<E >-E_V), particularly in the lower vibrational levels. It must be stressed that the currently available data for rotational energy disposal are quite limited (32). These conclusions may be modified as better data are accumulated (80).

R-H Bond Dissociation Energies

Bond dissociation energies, and hence, $\Delta H^\circ_f(R)$ can be obtained by setting the energy of the highest observed $HF^\dagger_{vJ}$ vibration-rotation state equal to <E> of eqn. 11. The accuracy of bond energies obtained from chemiluminescence is dependent upon

a) the presence of a measurable level of chemiluminescence from $HF^\dagger_{v,J}$ levels near the limit imposed by <E>,

b) the correctness of the assumed activation energy,

c) the energy spacing between J states in the cut-off region,

d) the absence of a sizeable radical stabilization energy.

Of course, normal precautions of chemical purity must be exercised. The absence of a radical stabilization energy, and also the absence of impurities (defined as two or more types of R-H bonds) and multiple reaction channels, can be inferred from a linear vibrational surprisal plot within the three-body model (I) (31,32). Requirement b, is not serious because all activation energies are nearly zero and fall in the range 0-1.5 kcal/mole (46). The rotational energy spacing increases as J(J+1) and the spacings exceed 1 kcal/mole^{-1} at J $\gtrsim$ 9. In favorable cases bond energies, accurate to 1-2 kcal/mole, can be obtained by this method. This approaches the accuracy of the best alternative methods (82). The chemiluminescence method actually gives upper limit bond energies and this is a useful check on less direct methods. Table I summarizes some of the bond energies that have been obtained from chemiluminescence data.

Table I. Bond Dissociation Energies Determined from HF Infrared Chemiluminescence.

	HF Chemiluminescence[a]			Alternate method[b,c]		
RH Bond	Highest v,J	D°_{298K}	ref.	D°_{298K}	ref.	method
benzene	2,8	≤ 113.5	31	110.8	82	kinetic (iodination)
ethylene	2,12	≤ 108.0	31	≥ 108	82	kinetic (iodination)
formaldehyde	4,14	≤ 87.4	32	87.5	82	kinetic (iodination)
ethylene oxide	3,5	≤ 105.0	33	-	-	-
cyclopropane	3,5	≤ 105.0	27	100.7	83	consensus value
PH_3	5,11	≤ 81.0	25	<81 83.9	84 88	E.A. A.P.
SiH_4	4,12	≤ 89.0	25	<92.2 80	85 86	E.A. A.P.
GeH_4	5,11	≤ 81.0	29	<92.3 87.2	85 87	E.A. A.P.

a) The chemiluminescence bond energies are upper limits.
b) Upper and lower limit values are as specified by the original references.
c) Explanation of "method" A.P. = electron impact appearance potential, E.A. = electron affinity, as measured by the electron photodetachment threshold of the radical anion.

Many values have been reported over the years for the bond energies and associated $\Delta H_f^\circ(R)$ for benzene, ethylene and formaldehyde. Our values of these quantities tend to confirm the values recommended by Benson and coworkers (82). The CH bond energy of ethylene oxide has not been reported previously. For comparison, the chemiluminescence and literature values for cyclopropane, which ought to have a similar conformational stabilization effect (if any) are shown. The reactions of F + phosphine, silane and germane are very fast and give strong IR emission with good signal to noise ratio. The values reported in Table I use the convention defined in Section C for the thermal reactive collision energy, and the bond energies should be accurate within 2 kcal $mole^{-1}$, i.e., approximately twice the J state spacing in the cutoff region. These values for PH_3, SiH_4 and GeH_4 are preferred over others currently available.

As discussed in Section C, the dynamics of F + RH reactions are very favorable for the separation of the radical stabilization energy from the remainder of <E>. Based upon this fact, we have estimated a lower limit to the benzyl radical resonance stabilization energy of 10.7 kcal $mole^{-1}$ (31). The problems associated with this estimate have been summarized (30). Nevertheless it should be noted that this is one of the few intrinsic measurements of this quantity i.e., it does not depend upon comparison of $D_0^\circ(C_6H_5CH_2\text{-}H)$ to a standard reference compound.

Energy Disposal with Excited Reagents

Experiments recently have been done using reactants with nonequilibrium energy distributions, permitting information to be obtained on selectivity of reagent energy consumption as well as specificity of product energy release (12,21). Particularly elegant experiments have been done by the fluorescence depletion technique (16,17). The technique involves steady-state formation of HCl(v) from a reaction such as Cl + HI $\rightarrow$ HCl + I and the introduction of a pulsed flow of a different atom, e.g., Br which reacts, with strong preference, for HCl(v) molecules having sufficient vibrational energy to overcome the endoergicity (16).

$$\text{Br} + \text{HCl}\ (v \geq 2) \rightarrow \text{HBr} + \text{Cl} \qquad (24)$$

Information on reagent energy consumption is obtained from the depletion spectrum of HCl in the presence of Br, and information on the product energy release is obtained from the HBr chemiluminescence spectrum. Predictions of microscopic reversibility derived from the forward (exothermic) reactions are being tested and, in general, confirmed (17-21). The general results are that vibrational energy is highly effective in promoting endothermic reactions which have the barrier in the coordinate of separation, while translational energy is less effective. For exothermic reactions, with the barrier in the coordinate of approach, the reverse is true. Rotational energy has less effect on both forward and reverse reactions.

Energy Disposal from Unimolecular Reactions

The majority of unimolecular processes yield polyatomic products. Furthermore the energy randomization question makes the internal energy distribution of the larger R fragment of interest. Analysis of the infrared chemiluminescence from vibrationally excited products ought to yield a great deal of information about the dynamics of such reactions. McDonald and coworkers have had notable success in this very difficult field (34,36,37). Polyatomic vibrational transitions have Einstein A coefficients on the order of 1 sec^{-1}, compared to approximately 200 sec^{-1} for HF, and the frequencies of these transitions lie in a region where detector sensitivities are <0.1 of the PbS detector used for HF. In addition to the above, one must work at near single collision conditions in order to insure that the observed populations will be unaffected by vibrational relaxation. These very formidable experimental problems have been overcome by an apparatus (36) having the following essential features.

a) The reagents are mixed via crossed diffuse beams in an aluminum reaction chamber where all walls are maintained at liquid nitrogen temperature.

b) A Fourier transform Michelson interferometer, having a ∿100 fold advantage in sensitivity over a grating instrument, is used to observe the emission. The entire instrument is cooled to 95°K; giving a further ca. 1000 fold improvement in signal to noise via reduction of statistical fluctuations of the background black body radiation emanating from the surfaces within the field of view of the detector.

Using this apparatus, McDonald and coworkers have studied addition-elimination reactions of halogen atoms with substituted ethylenes and substituted benzenes (34,36,37); see eq. 1. Many of these reactions also have been studied by the crossed molecular beam technique (2,89-93). The distribution of vibrational energy among the internal modes of the vinyl fluoride product from F + C_2H_4 was found, by chemiluminescence, to be non random. This was also true for F plus propene giving C_2H_3F + CH_3. Durana and McDonald have suggested that products which initially are formed with non-equilibrated internal energy can retain their non-equilibrium distributions until they are observed, provided that the initial excess energy is low, and that the density of states at that energy is $\leq 10^4$ per cm^{-1} (37). The reactions of F with ethylene and propene and Cl with C_2H_3Br differ from others such as

$$F + C_2H_3Cl \rightarrow C_2H_3F + Cl$$
$$F + C_2H_3Br \rightarrow C_2H_3F + Br \qquad (25)$$

in that the vinyl fluoride from the latter have greater internal energy and this energy was found to be equilibrated at the time of observation. The energy is thought to have randomized after the reaction. Support for this view is provided by the observation of non random distributions in C_2H_3Cl from Cl + C_2H_3Br which has a smaller amount of energy. Although the F + C_2H_4 and CH_3CHCH_2 reactions have exit channel barriers, the Cl + C_2H_3Br reaction does not.

The consensus opinion (37,47,90) is that forces, resulting from the shape of the exit channel potential, and angular momentum play an important role in determining the energy disposal in these reactions. Theory appears to give an adequate representation of the average rate of decomposition for C_2H_4F, but not the energy disposal. The exit channel effects can cause non-equilibrium product energy distributions even if the energy was equilibrated at the transition state. These studies of internal energy distributions in polyatomic molecules have opened a whole new field and many important discoveries can be expected in the future.

Infrared chemiluminescence from the HX product has been used to infer the dynamics of three- and four-centered elimination reactions (94,95). Three-centered elimination has been observed from chemically activated halomethanes (94) which were produced by recombination of an atom and a radical. The radical was the product of an abstraction process. Thus in the F + CH_2Cl_2 system, the following reactions leading to three-centered elimination of vibrationally excited HCl were observed;

$$F + CH_2Cl_2 \rightarrow HF^{\dagger}_{v,J} + CHCl_2 \qquad (26a)$$
$$F + CHCl_2 \rightarrow CHFCl_2^{*} \rightarrow HCl^{\dagger}_{v,J} + CFCl$$

where * denotes vibrational excitation arising from the exoergicity of the activating reaction. Similarly, the reactions of H atom with CCl_3Br and CF_3I allow the observation of;

$$CX_3 + H \rightarrow CX_3H^* \rightarrow HX^{\dagger}_{v,J} + CX_2 \quad (X = Cl,F). \qquad (26b)$$

In all three reaction systems, the v=1 level had the greatest relative populations (v=0 could not be observed). There is little potential energy release in these three-centered elimination reactions and the observed distributions are roughly consistent with statistical distributions. The total vibrational energy released to the carbene fragments (:CClCH$_2$Cl or :CFCH$_2$F) recently has been measured (96,97) and the magnitude of the average energy and the breadth of the distribution are consistent with statistical predictions.

The CH_3CF_3 reaction,

$$CH_3 + CF_3 \rightarrow CH_3CF_3{*} \rightarrow HF^{\dagger}_{v,J} + CH_2CF_2 \quad (27)$$

is an example of a four-centered elimination from a chemically activated molecule. In this case there is a significant potential energy release since the activation energy of the reverse reaction ($HF + CH_2 = CF_2 \rightarrow CH_3CF_3$) is ∿42 kcal/mole endoergic. The energy of the CH_3CF_3 formed by the radical recombination places the system 30 kcal/mole above the exit channel barrier. The HF product was excited up to v=4, (42.4 kcal/mole) showing that some of the potential energy, associated with the reverse reaction, was partitioned to HF vibration. Chemical laser experiments (3) have confirmed the HF vibrational distribution measured by infrared chemiluminescence. The energy contained in the olefin fragment has been assigned (98) by observation of the unimolecular rate constant, k_a, for isomerization of methylcyclobutene in the following sequence.

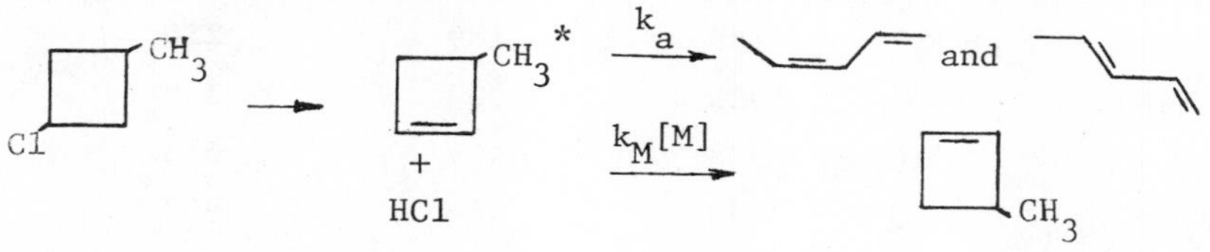

The $\langle f_V$(methylcyclobutene)$\rangle$ was 0.57 which corresponds to only 28% of the potential energy, assuming that the energy in excess of the threshold energy was partitioned statistically to methylcyclobutene. For the CF_3CH_3 reaction $\langle f_V(HF)\rangle$ was 0.13 which corresponds to only a small fraction of the potential energy. By difference a large fraction of the potential energy is released as translational energy. Recoil energy measurements (99,100) for HCl elimination from chloroalkane ions support this conclusion. Discussion of the dynamics of the HX elimination reactions is given elsewhere (101,102).

Acknowledgments

We wish to thank Dr. Keietsu Tamagake for permission to include the results from his study of the F + HCl reaction in this paper. Professor Polanyi provided preprints of recent work for which we are grateful. D.J.B. also wishes to acknowledge helpful discussions with Drs. D.S.Y. Hsu and J. W. Hudgens. The infrared chemiluminescence program at Kansas State University has been supported by the National Science Foundation (MPS75-02793).

APPENDIX I. Radiative Transition Probabilities for HF

The Einstein coefficients, eq. 9 and 10, for HF vibration-rotation transitions can be expressed as two quantities $A^{v'}_{v''}$, the pure vibrational Einstein coefficient, and $F^{v',J'}_{v'',J''}$, the vibration-rotation interaction factor, plus other terms depending upon J' and J". Values of $A^{v'}_{v''}$ from several different workers are reported in Table II.

Table II. Einstein Coefficients for HF Vibrational Transitions

v'-v"	$A^{v'}_{v''}$ (sec^{-1})			
reference	(103)	(53)	(104)	(51)
1-0	189	189	189	216
2-1	324	320	323	385
3-2	410	398	406	513
4-3	453	430	446	602
5-4	460	421	450	
2-0	23.6	23.4	22.9	
3-1	66.2	67.9	65.4	
4-2	124	131	124	
5-3	193	207	195	

In determining relative v',J' populations from iterative simulation of the spectra, relative transition probabilities are required since transition probabilities having a constant systematic error will give the same result as the correct absolute values. The published results from the group at KSU including (27) and all others prior to 1974 used $A^{v'}_{v''}$ values calculated from the method of Heaps and Herzberg (51) and $F^{v',J'}_{v'',J''}$ values from Herman and Wallis (55). Beginning with (29) and (30) all work reported from KSU, with the exception of (27), used $A^{v'}_{v''}$ from Herbelin and Emmanuel (53) and $F^{v',J'}_{v'',J''}$ from Herman, Rothery and Rubin, (54).

Table III. Einstein Coefficients for Spontaneous Emission for HF Fundamental and First Overtone Transitions

v = 1-0 R Branch

J'	J"	$A_{v''J''}^{v'J'}$ (sec^{-1})	$W_{v''J''}^{v'J'}$ (cm^{-1})
1	0	61.3	4001
2	1	71.6	4039
3	2	74.4	4075
4	3	74.6	4110
5	4	73.6	4143
6	5	71.8	4174
7	6	69.6	4203
8	7	67.1	4231
9	8	64.4	4256
10	9	61.5	4280
11	10	58.4	4302
12	11	55.3	4321
13	12	52.2	4339
14	13	49.0	4355
15	14	45.8	4368
16	15	42.6	4380
17	16	39.5	4389
18	17	36.5	4396

v = 1-0 P Branch

J'	J"	$A_{v''J''}^{v'J'}$	
0	1	193	3920
1	2	131	3878
2	3	120	3834
3	4	116	3788
4	5	114	3742
5	6	113	3694
6	7	113	3644
7	8	112	3594
8	9	112	3542
9	10	111	3490
10	11	110	3436
11	12	109	3381
12	13	108	3326
13	14	107	3270
14	15	106	3213
15	16	104	3155
16	17	102	3097
17	18	100	3038
18	19	98.2	2978

v = 2-1 R Branch

J'	J"	$A^{v'J'}_{v''J''}$	$W^{v'J'}_{v''J''}$
1	0	106	3827
2	1	125	3864
3	2	132	3899
4	3	135	3932
5	4	136	3964
6	5	135	3994
7	6	134	4022
8	7	132	4048
9	8	129	4072
10	9	126	4095
11	10	123	4115
12	11	120	4134
13	12	116	4151
14	13	113	4166
15	14	109	4178
16	15	105	4189
17	16	100	4198
18	17	96.3	4204

v = 2-0 R Branch

J'	J"	$A^{v'J'}_{v''J''}$	$W^{v'J''}_{v''J''}$
1	0	7.66	7789
2	1	9.03	7824
3	2	9.49	7856
4	3	9.64	7884
5	4	9.65	7910
6	5	9.58	7932
7	6	9.46	7951
8	7	9.31	7966
9	8	9.14	7979
10	9	8.95	7987
11	10	8.75	7993
12	11	8.54	7995
13	12	8.33	7993
14	13	8.11	7988
15	14	7.88	7980
16	15	7.66	7968
17	16	7.43	7953
18	17	7.21	7934

v = 2-1 P Branch

0	1	322	3750
1	2	216	3709
2	3	195	3666
3	4	186	3623
4	5	181	3578
5	6	178	3531
6	7	175	3484
7	8	172	3435
8	9	169	3385
9	10	167	3335
10	11	164	3283
11	12	161	3230
12	13	158	3177
13	14	155	3122
14	15	152	3067
15	16	148	3012
16	17	144	2955
17	18	140	2898
18	19	136	2841

v = 2-0 P Branch

0	1	23.8	7710
1	2	16.1	7666
2	3	14.7	7619
3	4	14.2	7569
4	5	13.9	7516
5	4	13.8	7460
6	7	13.7	7402
7	8	13.7	7341
8	9	13.7	7278
9	10	13.7	7212
10	11	13.7	7143
11	12	13.7	7072
12	13	13.7	6999
13	14	13.6	6924
14	15	13.6	6847
15	16	13.5	6767
16	17	13.4	6685
17	18	13.3	6602
18	19	13.2	6516

v = 3-2 R Branch

J'	J"	$A^{v'J'}_{v"J"}$	$W^{v'J'}_{v"J"}$
1	0	129	3659
2	1	149	3694
3	2	154	3727
4	3	154	3759
5	4	150	3789
6	5	146	3818
7	6	140	3845
8	7	134	3870
9	8	127	3893
10	9	120	3915
11	10	113	3934
12	11	105	3952
13	12	98.0	3968
14	13	90.6	3981
15	14	83.4	3993
16	15	76.3	4003
17	16	69.4	4011
18	17	62.7	4017

v = 3-1 R Branch

J'	J"	$A^{v'J'}_{v"J"}$	$W^{v'J'}_{v"J"}$
1	0	22.4	7448
2	1	26.7	7482
3	2	28.2	7512
4	3	28.9	7539
5	4	29.2	7564
6	5	29.3	7585
7	6	29.2	7603
8	7	29.0	7617
9	8	28.8	7629
10	9	28.4	7637
11	10	28.1	7641
12	11	27.7	7643
13	12	27.3	7641
14	13	26.9	7636
15	14	26.4	7627
16	15	25.9	7615
17	16	25.5	7599
18	17	25.0	7581

v = 3-2 P Branch

0	1	410	3584
1	2	280	3544
2	3	258	3504
3	4	250	3461
4	5	247	3418
5	6	246	3373
6	7	246	3328
7	8	245	3281
8	9	245	3233
9	10	244	3184
10	11	243	3134
11	12	242	3083
12	13	240	3031
13	14	238	2979
14	15	235	2925
15	16	232	2872
16	17	228	2817
17	18	224	2762
18	19	219	2706

v = 3-1 P Branch

0	1	68.5	7372
1	2	46.0	7329
2	3	41.7	7284
3	4	39.9	7236
4	5	39.0	7185
5	6	38.4	7131
6	7	38.0	7075
7	8	37.7	7016
8	9	37.4	6955
9	10	37.2	6891
10	11	36.9	6825
11	12	36.7	6756
12	13	36.4	6685
13	14	36.1	6612
14	15	35.8	6537
15	16	35.4	6460
16	17	35.0	6381
17	18	34.6	6300
18	19	34.2	6217

v = 4-3 R Branch

1	0	138	3494
2	1	158	3528
3	2	162	3560
4	3	160	3590
5	4	155	3619
6	5	148	3647
7	6	141	3672
8	7	132	3696
9	8	124	3718
10	9	115	3739
11	10	106	3757
12	11	97.4	3774
13	12	88.7	3789
14	13	80.2	3801
15	14	71.9	3812
16	15	63.9	3821
17	16	56.2	3828
18	17	49.0	3833

v = 4-3 P Branch

0	1	445	3422
1	2	307	3384
2	3	284	3345
3	4	277	3304
4	5	276	3262
5	6	276	3219
6	7	277	3175
7	8	278	3130
8	9	279	3084
9	10	279	3036
10	11	279	2988
11	12	278	2939
12	13	277	2889
13	14	275	2838
14	15	273	2787
15	16	270	2735
16	17	266	2682
17	18	262	2629
18	19	257	2575

v = 4-2 R Branch

1	0	42.1	7116
2	1	48.7	7148
3	2	50.3	7177
4	3	50.1	7204
5	4	49.2	7227
6	5	47.8	7247
7	6	46.3	7264
8	7	44.5	7277
9	8	42.7	7288
10	9	40.9	7295
11	10	39.0	7299
12	11	37.1	7300
13	12	35.2	7297
14	13	33.3	7291
15	14	31.5	7282
16	15	29.7	7270
17	16	28.0	7254
18	17	26.3	7235

v = 4-2 P Branch

0	1	135	7043
1	2	92.6	7002
2	3	85.8	6958
3	4	83.9	6912
4	5	83.6	6862
5	6	84.0	6811
6	7	84.8	6756
7	8	85.6	6699
8	9	86.5	6640
9	10	87.4	6578
10	11	88.2	6514
11	12	89.0	6448
12	13	89.6	6379
13	14	90.1	6308
14	15	90.6	6235
15	16	90.8	6161
16	17	91.0	6084
17	18	90.9	6005
18	19	90.8	5925

v = 5-4	R Branch		
1	0	134	3333
2	1	154	3365
3	2	157	3396
4	3	153	3425
5	4	148	3453
6	5	140	3479
7	6	132	3504
8	7	124	3526
9	8	115	3547
10	9	105	3566
11	10	96.2	3584
12	11	87.1	3599
13	12	78.2	3613
14	13	69.6	3625
15	14	61.3	3635
16	15	53.4	3643
17	16	46.0	3649
18	17	39.0	3653

v = 5-3	R Branch		
1	0	66.0	6792
2	1	75.7	6823
3	2	77.1	6851
4	3	76.0	6876
5	4	73.6	6898
6	5	70.6	6916
7	6	67.3	6932
8	7	63.8	6945
9	8	60.2	6955
10	9	56.6	6961
11	10	53.0	6964
12	11	49.4	6964
13	12	46.0	6961
14	13	42.6	6955
15	14	39.4	6945
16	15	36.2	6932
17	16	33.2	6916
18	17	30.4	6896

v = 5-4	P Branch		
0	1	438	3264
1	2	303	3228
2	3	282	3190
3	4	276	3151
4	5	275	3110
5	6	276	3069
6	7	278	3026
7	8	279	2983
8	9	280	2938
9	10	282	2892
10	11	282	2846
11	12	282	2798
12	13	281	2750
13	14	280	2701
14	15	277	2651
15	16	275	2601
16	17	271	2549
17	18	267	2498
18	19	262	2446

v = 5-3	P Branch		
0	1	216	6722
1	2	150	6682
2	3	140	6640
3	4	138	6595
4	5	139	6548
5	6	140	6498
6	7	143	6445
7	8	145	6390
8	9	147	6332
9	10	150	6273
10	11	152	6210
11	12	154	6146
12	13	156	6080
13	14	158	6011
14	15	159	5940
15	16	160	5867
16	17	161	5793
17	18	162	5716
18	19	162	5638

The experimental values of Sileo and Cool closely agree with the semi-empirical calculations of Meredith and Smith (104) and of Herbelin and Emanuel (53). The Einstein coefficients of Sileo and Cool, which were published during the preparation of this review, are based on a large data base and a straightforward computational procedure and their results are recommended. Sileo and Cool present an excellent discussion that places their work in perspective to previous studies.

The use of the new $A_{v''}^{v'}$ values in reanalysis of data reported prior to 1974 would not change the v=1-4 vibrational distributions beyond the experimental limit of error. However, for reactions yielding very high vibrational levels, such as $H + F_2$, the relative populations are altered significantly (53,18), by the new transition probabilities.

For $J \gtrsim 12$ it is not possible to simultaneously fit both the P- and R-branch line intensities for the same v level if the vibration-rotation interaction factors of Herman and Wallis (55) are used. The problem becomes especially severe in the region of the head formation of the R-branch. As the degree of relaxation was reduced by improvement of apparatus and technique, this problem prompted us to switch to the Herman, Rothery and Rubin (54) formulation. The resulting transition probabilities are listed in Table III. These were calculated using a computer program written by Johnson and Bogan (105). The original paper of HRR (54) contains several typographical errors in the equations; however, the graphical display of the numerical values for $F_{v'',J''}^{v',J'}$ is correct. The values of the vibration interaction factors derived from Table III, differ at high J from the values of Figure 2 in ref. 54, because we have used a more accurate value of $\theta = M_o/M_1r_e$; i.e., the ratio of the first two terms of the dipole moment expansion divided by r_e = 0.917 A. HRR (54) used θ = 1.18, we used θ = 1.32, Sileo and Cool (103) give θ = 1.287 (Table V).

Literature Cited

1. "Molecular Beam Scattering", Faraday Disc. Chem. Soc., 55, 1 (1973).
2. Farrar, J. M. and Lee, Y. T., this volume.
3. Berry, M. J., Ann. Rev. Phys. Chem., 26, 259 (1975) and this volume.
4. Hsu, D. S. Y., Umstead, M. E. and Lin, M. C., this volume.
5. Charters, P. E. and Polanyi, J. C., Disc. Faraday Soc., 33, 107 (1962).
6. Airey, J. R., Findlay, F. D. and Polanyi, J. C., Can. J. Chem., 42, 2193, 2176 (1964).

7. Anlauf, K. G., Kuntz, P. J., Maylotte, D. H., Pacey, P. D., and Polanyi, J. C., Disc. Faraday Soc., 44, 183 (1967).
8. Polanyi, J. C., J. Appl. Optics, 10, 1717 (1971).
9. Carrington, T. and Polanyi, J. C., (ed. by J. C. Polanyi) "MTP International Review of Science, Physical Chemistry", Series One, Vol. 9 (Butterworth, London, 1972).
10. Polanyi, J. C. and Sloan, J. J., J. Chem. Phys., 57, 4988 (1972).
11. Polanyi, J. C. and Woodall, K. B., J. Chem. Phys., 57, 1574 (1972).
12. Anlauf, K. G., Horne, D. S., Macdonald, R. G., Polanyi, J. C. and Woodhall, K. B., J. Chem. Phys., 59, 1561 (1972).
13. Maylotte, D. H., Polanyi, J. C. and Woodhall, K. B., J. Chem. Phys., 57, 1547 (1972).
14. Kirsch, L. J. and Polanyi, J. C. J. Chem. Phys., 57, 4498 (1972).
15. Cowley, L. T., Horne, D. S. and Polanyi, J. C., Chem. Phys. Lett., 12, 144 (1971).
16. Douglas, D. J., Polanyi, J. C. and Sloan, J. J., J. Chem. Phys., 59, 6679 (1973).
17. Ding, A. M. G., Kirsch, L. J., Perry, D. S., Polanyi, J. C. and Schreiber, L., Faraday Disc. Chem. Soc., 55, 252 (1973).
18. Polanyi, J. C., Sloan, J. J. and Wanner, J., Chem. Phys., 13, 1 (1976).
19. Douglas, D. J., Polanyi and Sloan, J. J., Chem. Phys., 13, 15 (1976).
20. Perry, D. S. and Polanyi, J. C., Chem. Phys., 12, 419 (1976).
21. Douglas, D. J. and Polanyi, J. C., Chem. Phys., 16, 1 (1976).
22. Chang, H. W., Setser, D. W. and Perona, M. J., J. Phys. Chem., 75, 2070 (1971).
23. Chang, H. W., Perona, M. J., Setser, D. W. and Johnson, R. L., Chem. Phys., Lett., 9, 587 (1971).
24. Chang, H. W. and Setser, D. W., J. Chem. Phys., 58, 2298 (1973).
25. Duewer, W. H. and Setser, D. W., J. Chem. Phys., 58, 2310 (1973).
26. Kim, K. C. and Setser, D. W., J. Phys. Chem., 77, 2493 (1973).
27. Parker, J. H., Int. J. Chem. Kinetics, 7, 433 (1975).
28. Perona, M. J., J. Chem. Phys., 54, 4024 (1971).
29. Kim, K. C., Setser, D. W. and Bogan, C. M., J. Chem. Phys., 60, 1837 (1974).
30. Bogan, D. J. and Setser, D. W., J. Am. Chem. Soc., 96, 1950 (1974).
31. Bogan, D. J. and Setser, D. W., J. Chem. Phys., 64, 586 (1976).
32. Bogan, D. J., Setser, D. W. and Sung, J. P., J. Phys. Chem., submitted (1977).
33. Smith, D. J., Setser, D. W., Kim, K. C. and Bogan, D. J., J. Phys. Chem., submitted (1977).

34. Moehlmann, J. G. and McDonald, J. C., J. Chem. Phys., 62, 3052 (1975).
35. Moehlmann, J. G. and McDonald, J. C., J. Chem. Phys., 62, 3061 (1975).
36. Moehlmann, J. G., Gleaves, J. T., Hudgens, J. W. and McDonald, J. D., J. Chem. Phys., 60, 4790 (1974).
37. Durana, J. F. and McDonald, J. C., J. Chem. Phys., 64, 2518 (1976).
38. Hudgens, J. W., Gleaves, J. T. and McDonald, J. D., J. Chem. Phys., 64, 2528 (1976).
39. Durana, J. F. and McDonald, J. D., J. Am. Chem. Soc., 98, 1289 (1976).
40. Jonathan, N., Melliar-Smith, C. M., Okuda, S. Slater, D. H., and Timlin, D., Mol. Phys., 22, 561 (1971).
41. Jonathan, N., Melliar-Smith, C. M. and Slater, D. H., Mol. Phys., 20, 93 (1971).
42. Vasiljev, Y. K., Ivanov, V. B., Makarov, E. F., Rjekenko, A. Y. and Tal'Rose, V. L., Rep. Akad. Nauk. SSR Izv. Ser. Khim, 215, 119 (1974); Izv. Akad. Nauk. SSSR Ser. Khim. No. 3, 537 (1975).
43. Airey, J. R. and Smith, I. W. M., J. Chem. Phys., 57, 1669 (1972).
44. Hancock, G., Ridley, B. A. and Smith, I. W. M., J. Chem. Soc. Faraday Trans. II, 68, 2117, 2127 (1972).
45. Arnold, S. J., Kimbell, G. H. and Snelling, D. R., Can. J. Chem., 52, 271 (1974).
46. Foon, R. and Kaufman, M., Prog. React. Kinet., 8, 81 (1975).
47. McDonald, J. D. and Marcus, R. A., J. Chem. Phys., 65, 2180 (1976).
48. Welsh, H. L., Cumming, C. and Stansbury, E. J., J. Opt. Soc. Am., 41, 712 (1951).
49. Welsh, H. L., Stansbury, E. J., Romanki, J. and Feldman, T., J. Opt. Soc. Am., 45, 338 (1955).
50. Woodall, K. B., Ph.D. Thesis, Univ. of Toronto, 1971.
51. Heaps, H. S. and Herzberg, G. Z. Physik, 133, 48 (1952).
52. Penner, S. S., "Quantitative Molecular Spectroscopy and Gas Emissivities," Addition-Wesley, Reading, MA, 1959.
53. Herbelin, J. M. and Emanuel, G., J. Chem. Phys., 60, 689 (1974).
54. Herman, R., Rothery, R. W. and Rubin, R. J., J. Mol. Spectrosc., 2, 369 (1958).
55. Herman, R. and Wallis, R. F., J. Chem. Phys., 23, 637 (1955).
56. Mann, D. E., Thrush, B. A., Lide, R. D., Ball, J. J. and Acquista, N., J. Chem. Phys., 34, 420 (1961).
57. Menzinger, M. and Wolfgang, R., Angew, Chem. Internat. Ed., 8, 438 (1969).
58. Chupka, W. A. and Berkowitz, J., J. Chem. Phys., 54, 5126 (1971).
59. Sung, J. P. and Setser, D. W., Chem. Phys. Lett., Submitted (1977).

60. Tully, J. C., J. Chem. Phys., 60, 3042 (1974).
61. Jaffe, R. L., Morokuma, K. and George, T. F., J. Chem. Phys., 63, 3417 (1975).
62. Bernstein, R. B. and Levine, R. D., (ed. R. D. Bates and B. Bederson), Adv. Atom. and Mol. Phys., 11, 216 (1975).
63. Polanyi, J. C., Accts. Chem. Research, 5, 161 (1972) and references therein.
64. Johnson, R. L., Kim, K. C. and Setser, D. W., J. Phys. Chem., 77, 2499 (1973).
65. Parr, C. A., Polanyi, J. C. and Wong, W. H., J. Chem. Phys., 58, 5 (1973).
66. Levine and R. D. and Bernstein, R. B., "Molecular Reaction Dynamics," Oxford University Press, N.Y., 1974.
67. Muckerman, J. T., J. Chem. Phys., 54, 1155 (1971); Muckerman, J. T. and Newton, M. D., 56, 3191 (1972).
68. Jaffe, R. L. and Anderson, J. B., J. Chem. Phys., 54, 2224 (1972); Jaffe, R. L., Henry, J. M. and Anderson, J. B., J. Chem. Phys., 59, 1128 (1973).
69. Blais, N. C. and Truhlar, D. G., J. Chem. Phys., 58, 1090 (1973).
70. Wilkins, R. L., J. Chem. Phys., 57, 912 (1972).
71. Polanyi, J. C. and Schreiber, J. L., Faraday Disc. Chem. Soc., 62, xxxx (1976).
72. Ben-Shaul, A., Levine, R. D. and Bernstein, R. B., J. Chem. Phys. 57, 5477 (1972).
73. Ben-Shaul, A., Chem. Phys., 1, 244 (1973).
74. Light, J. C., Faraday Disc. Chem. Soc., 44, 14 (1967).
75. Hsu, D. S. Y. and Herschbach, D. R., Faraday Disc. Chem. Soc., 55, 116 (1973); Case, D. A. and Herschbach, D. R., Mol. Phys. 30, 1537 (1976); Case, D. A. and Herschbach, D. R., J. Chem. Phys., 64, 4212 (1976).
76. Hijazi, N. H. and Polanyi, J. C., J. Chem. Phys., 63, 2249 (1975); Chem. Phys., 11, 1 (1975).
77. Polanyi, J. C. and Woodall, K. B., J. Chem. Phys., 56, 1563 (1972).
78. Ding, A. M. G and Polanyi, J. C., Chem. Phys., 10, 39 (1975).
79. Procaccia, I., Shimoni, Y. and Levine, R. D., J. Chem. Phys., 63, 3181 (1975).
80. Tamagake, K. and Setser, D. W. Unpublished results (1977).
81. Levine, R. D., Johnson, B. R. and Bernstein, R. B., Chem. Phys. Lett., 19, 1 (1973).
82. O'Neal, H. E. and Benson, S. W., "Thermochemistry of Free Radicals," in "Free Radicals," edited by J. K. Kochi, Wiley Interscience, N.Y., 1973, pp. 275-359.
83. Ferguson, K. C. and Whittle, E., Faraday Trans. Chem. Soc., 67, 2618 (1971).
84. Smyth, K. C. and Brauman, J. I., J. Chem. Phys., 56, 1132 (1972).

85. Reed, K. J. and Brauman, J. I., J. Chem. Phys., 61, 4830 (1974).
86. Darwent, B. DeB., NSRDS-NBS, 31, U.S. Dept. of Commerce (1970).
87. Saalfeld, F. E. and Svec, H. J., J. Phys. Chem., 70, 1753 (1966).
88. McAllister, T. and Lossing, F. P., J. Phys. Chem., 73, 2996 (1969).
89. Parson, J. M. and Lee, Y. T., J. Chem. Phys., 56, 4658 (1972).
90. Farrar, J. M. and Lee, Y. T., J. Chem. Phys., 65, 1414 (1976).
91. Cheung, J. T., McDonald, J. D. and Herschbach, D. R., J. Am. Chem. Soc., 95, 7889 (1973).
92. Shobatake, K., Lee, Y. T. and Rice, S. A. Rice, J. Chem. Phys., 59, 6104 (1973).
93. Shobatake, K., Lee, Y. T. and Rice, S. A., J. Chem. Phys., 59, 1435 (1973).
94. Chang, H. W., Setser, D. W. and Perona, M. J., J. Phys. Chem., 75, 2070 (1971).
95. Clough, P. N., Polanyi, J. C. and Taguchi, R. T., Can. J. Chem., 48, 2919 (1970).
96. Kim, K. C. and Setser, D. W., J. Phys. Chem., 78, 2166 (1974).
97. Holmes, B. E., Setser, D. W., and Pritchard, G. P., Int. J. Chem. Kinet., 8, 215 (1976).
98. Holmes, B. E. and Setser, D. W., J. Phys. Chem., 79, 1320 (1975).
99. Kim, K. C., Beynon, J. H. and Cooks, R. G., J. Chem. Phys. 61, 1305 (1974).
100. Kim, K. C., J. Chem. Phys., 64, 3003 (1976).
101. Holmes, B. E. and Setser, D. W., J. Phys. Chem., to be submitted (1977).
102. Holmes, B. E. and Setser, D. W., "Energy Disposal in Unimolecular and Biomolecular Reactions" in "Physical Chemistry of Fast Reactions", V-3, edited by I. W. M. Smith, Plenum Publishing Co. London (1977).
103. Sileo, R. N. and Cool, T. A., J. Chem. Phys., 65, 117 (1976).
104. Meredith, R. E. and Smith, F. G., J. Quant. Spectros. Radiat. Transfer, 13, 89 (1973).
105. Johnson, R. L. and Bogan, D. J., Fortran IV computer program for the simulation of vibration-rotation spectra of diatomic molecules.
106. Kim, K. C. and Setser, D. W., unpublished results.

Theory

10

Potential Energy Surfaces for Fluorine-Hydrogen Systems

CHARLES F. BENDER[†]*

Theoretical Atomic and Molecular Physics Group, University of California, Lawrence Livermore Laboratory, Livermore, CA 94550

HENRY F. SCHAEFER, III[†]**

Department of Chemistry and Lawrence Berkeley Laboratory, University of California, Berkeley, CA 94720

During the past decade, there has been a genuine explosion of interest in the reactions of atomic and molecular fluorine and hydrogen. As is usually the case, experimental studies have led the way, due in considerable measure to the tremendous impact of the HF laser. In addition to traditional kinetics (1), the most successful experimental techniques have been infrared chemiluminescence (2), chemical lasers (3), crossed molecular beams (4), and laser-induced fluorescence (5). This experimental research has yielded a great deal of important information concerning these elementary reactions. In addition, this work has stimulated a keen interest in the detailed understanding on the molecular level of how these simple reactions occur.

Given the above background, it is hardly surprising that there has been considerable theoretical activity directed toward fluorine hydrogen systems. From a theoretical viewpoint, the understanding of such reactions has two components: first, the potential energy surface (or surfaces) on which the pertinent reaction occurs, and secondly, the dynamics which occur given the potential surface(s). The amount of detailed experimental information now available for fluorine hydrogen reactions is sufficient to challenge the most sophisticated theorists of both the electronic structure and dynamics schools. Since dynamical considerations are taken up in a different chapter of the present volume, we will be concerned here with the potential surface half of the theoretical problem. Clearly, however, we must keep in mind the essential complementarity of these two pieces of the puzzle.

---

†This work was performed under the auspices of the U. S. Energy Research and Development Administration under contract No. W-7405-Eng-48.

*M. H. Fellow.

**J. S. Guggenheim Fellow, 1976-1977.

In an interdisciplinary volume such as the present, it does not seem appropriate to give any sort of detailed coverage to the theoretical methods currently in use (6,7). It must be noted, however, that the Hartree-Fock or Self-Consistent-Field (SCF) method remains at the core of electronic structure theory. Although SCF theory is sometimes adequate in describing potential energy surfaces, this has turned out more often not to be the case. That is, electron correlation, which incorporates the instantaneous repulsions of pairs of electrons, can have a qualitative effect on the topology of fluorine hydrogen potential surfaces.

Without further introduction, it seems appropriate to proceed to a discussion of specific systems for which *ab initio* potential surface features have been predicted. In the present paper, special emphasis will be placed on the relationship between theoretical predictions and experimental observations.

$F + H_2 \rightarrow FH + H$

Three progressively more reliable potential surfaces have been reported for this key reaction, which has become the focus of an enormous amount of scientific research. In the first (8), a double zeta (DZ) basis set was adopted. This notation implies that two basis functions are used to describe each orbital of the separated atoms. That is, for fluorine 1s, 1s', 2s, 2s', $2p_x$, $2p_x'$, $2p_y$, $2p_y'$, $2p_z$, and $2p_z'$ functions are adopted. Such a basis is clearly twice as large as the traditional minimum basis used in qualitative discussions of electronic structure.

Using the DZ basis and both SCF and configuration interaction (CI) methods, the results in Table I were obtained. It seems reasonable to conclude that the SCF results obtained with this basis are in poor agreement with experiment, while the CI results are much improved but even yet in only fair agreement. This conclusion was initially a surprise to us, since such techniques often yield rather reliable results, for example, in the prediction of equilibrium molecular structures (9).

The most obvious (in light of previous research (6,7) in this area) extension of the DZ basis is the addition of polarization functions, i.e., a set of d functions on fluorine and sets of p functions on each H atom. The basis thus obtained is labeled DZ + P and the ensuing results are summarized in Table I (10). There we see once again that the SCF approximation yields a barrier height much larger than experiment. And in fact even if one goes to a complete set of one-electron functions, the Hartree-Fock limit barrier height will be ∿25 kcal larger than the true barrier. This inherent inability of the SCF approximation to even qualitatively describe repulsive potential surfaces must be viewed as one of the most important developments of our research to date.

Table I. Some features of *ab initio* and semi-empirical potential energy surfaces for $F + H_2 \rightarrow FH + H$

Year	Surface	Saddle Point Geometry (angstroms) r (F-H)	Saddle Point Geometry (angstroms) r (H-H)	Barrier Height (kcal/mole)	Exothermicity (kcal/mole)
1971	Double Zeta (DZ) Self-Consistent-Field (SCF)[a]	1.06	0.81	34.3	-0.6
1971	DZ Configuration Interaction (CI)[a]	1.37	0.81	5.7	20.4
1972	DZ + P SCF[b]	1.18	0.836	29.3	13.2
1972	DZ + P CI[b]	1.54	0.767	1.7	34.4
1972	Muckerman V[c]	1.54	0.76	1.1	31.8
1974	Polanyi-Schreiber SE-I[d]	1.43	0.777	2.2	31.0
1976	Large Basis[e] Extended CI	1.48	0.778	3.3	31.3
	Experiment	--	--	1.6[f]	31.5 ± 0.5[g]

[a] Reference 8
[b] Reference 10
[c] Reference 11
[d] Reference 12
[e] Reference 13
[f] Experimental activation energy; see Reference 1
[g] Reference 14

Fortunately, the judicious use of CI yields a chemically reasonable DZ + P surface for F + H_2 (Figure 1). In particular, the predicted classical barrier height of 1.66 kcal is in fortuitously close agreement with the experimental activation energy. Further, the exothermicity of the BOPS surface is ∿3 kcal greater than the experimental value. Thus, the primary significance of the BOPS CI surface was that it appeared to be the first qualitatively correct ab initio surface for a chemical reaction more complicated than the prototype (15) H + H_2 system.

The earliest indication of the qualitative correctness of the BOPS surface came from the classical trajectory studies of Muckerman (11). Muckerman varied several features of the London-Eyring-Polanyi-Sato (LEPS) surface for F + H_2 to get the best agreement between predicted and experimental (2,3) FH vibrational energy distributions. Although this work was done completely independent of the BOPS ab initio study, Muckerman's "best" semi-empirical surface (his surface V, summarized in Table I) has a saddle point position essentially indistinguishable from the BOPS surface. Since the saddle point position is probably the most critical surface feature not directly accessible to experimental determination, this concurrence was especially significant.

Perhaps the next important development was the fitting of the BOPS surface to an LEPS form by Polanyi and Schreiber (12). It should be noted that this fitting was carried out against the present authors' advice, our feeling being that the BOPS surface was not sufficiently accurate to be used without adjustment in dynamical studies. Nevertheless, Polanyi and Schreiber fit the 232 collinear points to the LEPS form with an ensuing rms deviation of less than 1 kcal per point. This exercise in itself is rather striking confirmation of the LEPS form, which has only seven adjustable parameters. The most obvious weakness of the BOPS-FIT surface was a spurious 0.4 kcal attractive well in the entrance valley. In addition, of course, the BOPS-FIT surface incorporated the known 3 kcal BOPS error in the predicted exothermicity.

Collinear classical trajectory studies were then performed using the BOPS-FIT surface. By comparing with the experimental vibrational distribution, Polanyi and Schreiber (PS) concluded that although the BOPS surface was qualitatively reasonable, it does have a rather serious failing. That is, PS concluded that the BOPS surface drops too rapidly from the "shoulder" into the exit valley. It was noted, of course, that the known error of 3 kcal in the exothermicity is a major factor in this rapid drop. Our current feeling is that while this criticism of the BOPS surface may well prove to be at least partially valid, the use of classical (rather than quantum mechanical) dynamics and one (rather than three) dimension detracts from the strength of the PS conclusion.

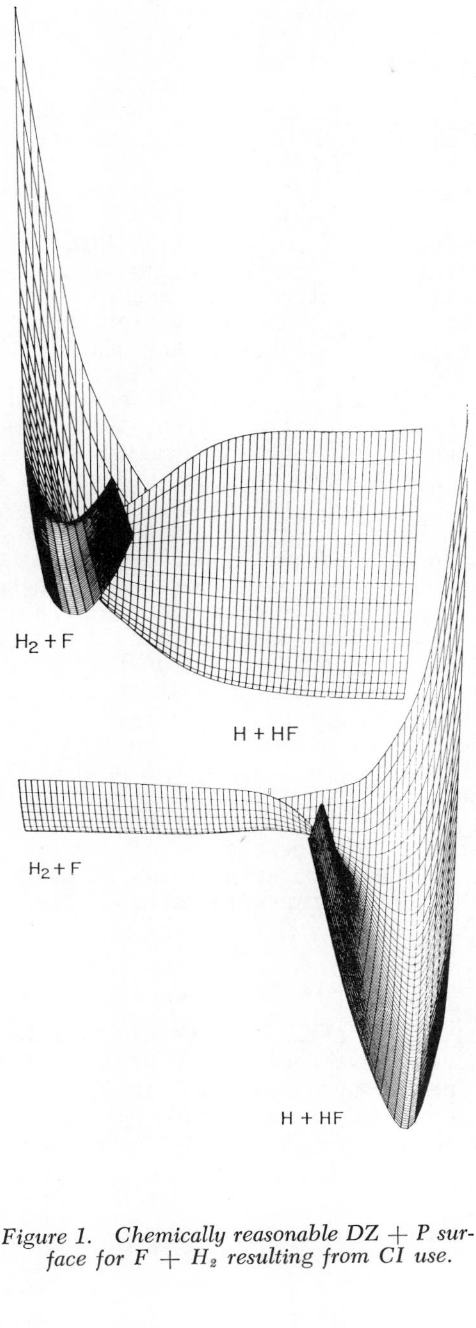

Figure 1. Chemically reasonable DZ + P surface for $F + H_2$ resulting from CI use.

Polanyi and Schreiber also suggested a "best" adjusted LEPS surface, their SE-I, and this surface is also summarized in Table I. There it is seen that the SE-I surface has its saddle point position somewhat later than either BOPS or Muckerman V. Nevertheless, it seems clear that the primary features of all three surfaces are rather similar. Both Muckerman (11) and Polanyi and Schreiber (12) have argued that the agreement between their best surfaces and the *ab initio* BOPS surface supports their use of the generalized LEPS form. We concur.

In an attempt to resolve some of the controversy arising from the BOPS surface, we decided in 1974 to attempt to determine an *ab initio* surface of sufficient reliability to be used directly or, with modest amounts of scaling, in dynamical studies. Recent developments (16) in the fully quantum mechanical treatment of A + BC reactions were another motivation for this research. However, one of us (CFB) discovered very quickly that apparent improvements in the quality of wavefunctions used led to a deterioration of the predicted FH_2 potential energy surface.

After an interval of contemplation, this research was continued in collaboration with Steven Ungemach and Bowen Liu (IBM, San Jose). Their conclusions have been presented (13) at the latest General Discussion of the Faraday Division of the Chemical Society (September 1976). Using a basis set more than twice as large as DZ + P and very large CI, USL predict a barrier of 3.3 kcal, with a suggested error of no more than one kcal. Thus, it now appears that the true barrier height is notably greater than the experimental activation energy. This idea is qualitatively supported by USL's additional prediction that there may be as much as 1.5 kcal less zero point vibrational energy at the saddle point than for separated F + H_2. Finally, it is worth noting that the USL saddle point geometry is about halfway between the Muckerman V-BOPS results and the PS SE-I prediction.

It should be clear from the above that the F + H_2 potential energy surface is likely to remain the source of much controversy for years to come. More generally, it seems apparent that the closer one looks at a particular problem, as both theoretical and experimental techniques become more sophisticated, the more fine structure comes to light.

H + FH → HF + H

As mentioned above, efforts to develop the HF chemical laser as a practical device have contributed greatly to the interest in elementary fluorine hydrogen systems. In this context, essentially every process leading to vibrational relaxation of HF must be precisely characterized. One of the simplest such processes

$$H + FH(\nu') \rightarrow HF(\nu) + H$$

has been the subject of several studies, both experimental (5) and theoretical (17). Clearly, the barrier height or activation energy for this simple exchange reaction will play a critical role in the dynamics.

Since semi-empirical potential surfaces have been so successful in describing the F + H_2 dynamics, it is not completely unreasonable to assume that the same LEPS surfaces might be suitable for H + FH. After all, the latter reaction merely corresponds to a different channel of the FH_2 surface. And, in fact, this is precisely the procedure adopted by Thompson (17) and by Wilkins in their classical trajectory studies (18) of H + FH.

Table II summarizes the predictions of a number of semi-empirical FH_2 surfaces for the H + FH barrier. The final entry indicates that rather reliable CI calculations (19), using a better than double zeta plus polarization basis set, predict a barrier of 49 kcal. BGS concluded in their paper (19) that the true collinear barrier is no less than 40 kcal. Thus, it is seen that the two "best" semi-empirical F + H_2 surfaces, Muckerman V and Polanyi-Schreiber SE-I, fail miserably for the collinear H + FH channel. This is perhaps the strongest evidence to date for the importance of *ab initio* information in potential energy surface calibration.

Two very recent (yet unpublished) papers support the qualitative conclusions of BGS. Using a very large basis, Meyer (20) has predicted a collinear barrier of 45 kcal/mole for H + FH → HF + H using the coupled electron pair approximation. Attempts to estimate the true barrier via error analysis lead to a value of ∿40 kcal. Secondly, Winter and Wadt (21) have found that the use of diffuse basis functions yields a surface which is quite "flat" with respect to the HFH bond angle. In fact, the true saddle point may occur for a bond angle less than 180° and yield a barrier as low as 35 kcal. However, this conclusion must be considered tentative at the present time.

The primary physical (or chemical, according to one's preference) conclusion is that at energies below 35 kcal the atom exchange mechanism cannot be a significant contributor to the vibrational relaxation of HF by hydrogen atoms. This conclusion seems to be given remarkable support by the recent experiments of Heidner (5). Specifically, he finds that the cross section for vibrational relaxation increases enormously between ν = 2 and ν = 3. It seems more than coincidental that the ν = 3 state of HF has enough internal energy to surmount the hypothesized barrier of 35 kcal/mole.

H + F_2 and F + HF

Polanyi and coworkers (22) have made truly impressive infrared chemiluminescence studies of this highly exothermic reaction. They find that the ratios of populations of the HF product vibra-

Table II. Barrier height and saddle point geometry for H + HF → HF + H. The saddle point is assumed to occur for a linear symmetric H-F-H geometry.

Type of Potential Energy Surface	Authors	r (H-F), Å	Barrier (kcal/mole)
Bond-energy bond-order (BEBO)	Johnston[a]	1.10	6.8
London-Eyring-Polanyi-Sato (LEPS)	Muckerman I[b]	1.04	1.0
LEPS	Jaffe and Anderson[c]	1.04	-5.2
LEPS	Muckerman II[b]	1.04	1.0
	III	1.05	1.7
	IV	1.05	2.3
LEPS	Wilkins[d]	1.04	1.4
LEPS	Thompson[e]	1.12	28.6
Semi-empirical valence bond	Blais and Truhlar[f]	1.10	14.0
Diatomics-in-molecules	Tully I[g]	1.05	14.4
	II	1.09	13.1
LEPS	Muckerman V[b]	1.04	1.2
LEPS	Polanyi and Schreiber[h]	1.05	3.5
A priori methods			
Self-consistent field	Bender, Garrison and Schaefer[i]	1.12	67.8
Configuration interaction		1.14	49.0

[a] Johnston, H. S., "Gas Phase Reaction Rate Theory" (Ronald, New York, 1966).
[b] Reference 11.
[c] Jaffe, R. L., Anderson, J. B., J. Chem. Phys. (1971), 54, (2224).
[d] Reference 18.
[e] Reference 17.
[f] Blais, N. C., Truhlar, D. G., J. Chem. Phys. (1973), 58, (1090).
[g] Tully, J. C., J. Chem. Phys. (1973), 58, (1396).
[h] Reference 12.
[i] Reference 19.

tional states (ν = 1 up to ν = 9) are 12:13:25:35:78:100:40:26: <16.

Electronic structure theory has been applied (23,24) in a manner analogous to the first four columns of Table I, and these results are summarized in Table III. A significant difference between these H + F_2 results and the earlier F + H_2 studies (8,10) is that the DZ CI barrier is lower than that obtained with the more reliable DZ + P CI. Since (for reasons discussed above) it is difficult to determine the true barrier, it must be concluded that both methods yield barriers roughly compatible with the experimental activation energy. However, the superiority of the DZ + P basis does become apparent with an inspection of the calculated exothermicities.

Both the F + H_2 and H + F_2 surfaces were found to be in harmony with the LEPS model in their angular characteristics. That is, the true saddle point does occur for a collinear geometrical arrangement. It is worth noting (8,23) that the two surfaces become repulsive at about the same rate as the angle of approach is bent away from 180°. For the F + H_2 system, e.g., it is found (8) that constraining the F-H-H angle to be 90° yields a barrier 11 kcal higher than that obtained for θ(F-H-H) = 180°.

Another part of the HF_2 surface which has been carefully explored is the collinear F + HF exchange reaction. This is another process of vital importance for an adequate understanding of the HF chemical laser. The results of OSB (25) are summarized in Table IV. There it is seen that the most reliable predicted barrier height is 24 kcal. On this basis, the true barrier was estimated to be ≥ 18 kcal. This result is much greater than the 6 kcal predicted (26) by the BEBO method. However, good agreement is found with Thompson's LEPS surface (27) calibrated for use on the H + F_2 reaction dynamics. The primary conclusion one can draw from these results is that the atom exchange mechanism is unlikely to contribute to the vibrational relaxation of the ν = 1 state of HF. However, ν = 2 relaxation could be greatly enhanced by atom exchange.

A final surface worth mentioning is that for the HF dimer. Among many theoretical studies, the most complete is that of Yarkony and coworkers (28). Although only the non-reactive part of the surface was considered, this portion is very relevant to the laser-related problems of rotational and vibrational energy transfer. The HF-HF surface has been fit to an analytical form by Alexander (29) and is being used in several dynamical studies in progress.

Concluding Remarks

It should be apparent that *ab initio* electronic structure theory is capable of genuine contributions to the understanding of simple fluorine atom reactions. This research is by no means a closed book, and an example from our current research should make

Table III. Some features of *ab initio* potential surfaces (23,24) for $H + F_2 \rightarrow HF + F$

Surface	Saddle Point Geometry (angstroms)		Barrier Height (kcal)	Exothermicity (kcal)
	r (H-F)	r (F-F)		
DZ SCF	1.56	1.49	12.2	132.4
DZ CI	2.05	1.57	1.0	88.3
DZ + P SCF	1.61	1.41	13.9	130.1
DZ + P CI	1.68	1.50	4.1	99.0
Experiment	--	--	2.4 ± 0.2[a]	102.5 ± 2.8[b]

[a]Experimental activation energy. Albright, R. G., Dodonov, A. F., Lavroskaya, G. K., Morosov, I. I., Tal'rose, V. L., J. Chem. Phys. (1969), 50, (3632).

[b]DeCorpo, J. J., Steiger, R. P., Franklin, J. L., Margrave, J. L., J. Chem. Phys. (1970), 53, (936); Chupka, W. A., Berkowitz, J., J. Chem. Phys. (1971), 54, (5126).

Table IV. Summary of theoretical potential energy surfaces (25) for the F + HF → FH + F exchange reaction.

Surface	Saddle Point Geometry r (HF), Å	Barrier Height (kcal/mole)
DZ SCF	1.087	53.8
DZ CI	1.126	21.8
DZ + P SCF	1.083	53.7
DZ + P CI	1.099	23.9

this clear. We have just begun the study of the CH_3 + F radical reaction for which several products are possible:

$$CH_3 + F \rightarrow \begin{array}{l} CH_3F \\ CH_2 + HF \\ CH_2F + H \\ CHF + H_2 \end{array}$$

This research is being done in conjunction with crossed molecular beam studies by Professor Y. T. Lee and should provide a rather detailed picture of this simple but multi-faceted reaction.

Abstract

Ab initio molecular electronic structure theory has now progressed to the point where it is capable of making genuine contributions to the understanding of simple chemical reactions. Especially noteworthy examples are the elementary fluorine hydrogen reactions pertinent to the HF chemical laser. The present paper discusses the reactions $F + H_2 \rightarrow FH + H$, $H + FH \rightarrow HF + H$, $H + F_2 \rightarrow HF + H$, and $F + HF \rightarrow FH + F$, with particular emphasis on the relationships between *ab initio* theory and experiment. Directions for future research are suggested.

Literature Cited

1. Foon, R., Kaufman, M., Progress in Reaction Kinetics (1975), 8, (81).
2. Douglas, D. J., Polanyi, J. C., Chem. Phys. (1976), 16, (1), and references therein.
3. Coombe, R. D., Pimentel, G. C., J. Chem. Phys. (1973), 59, (251, 1535).
4. Lee, Y. T., Physics of Electronic and Atomic Collisions (1972), 7, (359).
5. Heidner, R. F., Bott, J. F., J. Chem. Phys. (1975), 63, (1810).
6. Schaefer, H. F., "The Electronic Structure of Atoms and Molecules: A Survey of Rigorous Quantum Mechanical Results," (Addison-Wesley, Reading, Massachusetts, 1972).
7. Schaefer, H. F., Editor, "Modern Theoretical Chemistry," Volume 3, Methods of Electronic Structure Theory, (Plenum, New York, 1977).
8. Bender, C. F., Pearson, P. K., O'Neil, S. V., Schaefer, H. F., J. Chem. Phys. (1972), 56, (4626).
9. Lathan, W. A., Curtiss, L. A., Hehre, W. J., Lisle, J. B., Pople, J. A., Progress in Physical Organic Chemistry (1974), 11, (175).
10. Bender, C. F., O'Neil, S. V., Pearson, P. K., Schaefer, H. F., Science (1972), 176, (1412).

11. Muckerman, J. T., J. Chem. Phys. (1972), 56, (2997); Whitlock, P. A., Muckerman, J. T., J. Chem. Phys. (1975), 61, (4618).
12. Polanyi, J. C., Schreiber, J. L., Chem. Phys. Letters (1974), 29, (319).
13. Ungemach, S. R., Schaefer, H. F., Liu, B., Faraday Discussions (1976), 62, (000).
14. Johns, J. W. C., Barrow, R. F., Proc. Roy. Soc. (1959), A251, (504); Chupka, W. A., Berkowitz, J., J. Chem. Phys. (1971), 54, (5126).
15. Shavitt, I., Stevens, R. M., Minn, F. L., Karplus, M., J. Chem. Phys. (1968), 48, (2700).
16. Robinson, A. L., Science (1976), 191, (275).
17. Thompson, D. L., J. Chem. Phys. (1972), 57, (4170).
18. Wilkins, R. L., J. Chem. Phys. (1973), 58, (3038).
19. Bender, C. F., Garrison, B. J., Schaefer, H. F., J. Chem. Phys. (1975), 62, (1188).
20. Meyer, W., J. Chem. Phys., in press.
21. Winter, N. W., Wadt, W., unpublished research.
22. Polanyi, J. C., Sloan, J. J., J. Chem. Phys. (1972), 57, (4988).
23. O'Neil, S. V., Pearson, P. K., Schaefer, H. F., Bender, C. F., J. Chem. Phys. (1973), 58, (1126).
24. Bender, C. F., Bauschlicher, C. W., Schaefer, H. F., J. Chem. Phys. (1974), 60, (3707).
25. O'Neil, S. V., Schaefer, H. F., Bender, C. F., Proc. Natl. Acad. Sci. (USA) (1974), 71, (104).
26. Truhlar, D. G., Olson, P. C., Parr, C. A., J. Chem. Phys. (1972), 57, (4479).
27. Thompson, D. L., J. Chem. Phys. (1972), 57, (5164); (1974), 60, (2200).
28. Yarkony, D. R., O'Neil, S. V., Schaefer, H. F., Baskin, C. P., Bender, C. F., J. Chem. Phys. (1974), 60, (855).
29. Alexander, M. H., J. Chem. Phys. (1976), 65, 5009.

11

Thermochemistry of Fluorocarbon Radicals

ALAN S. RODGERS

Department of Chemistry, Texas A&M University, College Station, TX 77843

It may seem, at first glance, somewhat incongruous that a paper about the thermochemistry of fluorocarbon free radicals should appear in a volume nominally devoted to the kinetics of these radicals. However, for an elementary reaction such as (1)

$$CF_3 + CF_2 = CF_2 \rightleftharpoons CF_3CF_2\dot{C}F_2 \qquad (1)$$

the rate constants for the forward, k_1, and reverse, k_{-1}, reactions are related though the equilibrium constant K_1. If, as is usually the case, the rate constants are expressed in units of liters, moles and seconds, but the equilibrium constant is referenced to a standard state for gases of an ideal gas at 1 atm pressure, then

$$k_1/k_{-1} = K_1 \ (R'T)^{-\Delta n}$$

and

$$\ln \ (k_1/k_{-1}) = \Delta S^\circ_r/R - \Delta H^\circ_r/RT - \Delta n luR'T \qquad (2)$$

$R' = 0.082$ latm K^{-1} mol^{-1} while $R = 1.99$ cal K^{-1} mol^{-1} or 8.314 J K^{-1} mol^{-1} as may be appropriate. Consequently, if ΔS°_r and ΔH°_r for reaction (1) are known or can be reliably estimated, then one needs only measure one of the two rate constants and the other may be calculated. This, more often than not, can prove quite a convenience as one of the two coupled rate constants may be more easily measured than the other. Indeed, in the example shown, k_1 has been measured (1), while measurement of k_{-1} would present serious experimental difficulties. We will, in the course of this paper, estimate ΔS°_r and ΔH°_r, and thus determine a value for k_{-1}.

In addition to providing an alternate route to the determination of experimentally difficult rate constants, the analysis of the thermochemistry of free radical reactions can also provide help in selecting reaction mechanisms. In a recent example from our work (2), we had determined that the rate of reaction (3) was given by; $-d[Br_2]/dt =$

$$CH_2 = CF_2 + Br_2 \rightleftharpoons CH_2BrCF_2Br \qquad (3)$$

$k[Br_2]^{3/2}[CH_2 = CF_2]$; and that $\log(k/M^{-3/2}s^{-1}) = (7.8 \pm 0.1) - (17.8 \pm 0.3)/\Theta$. We suggested the following mechanism for this reaction:

$$Br_2 + M \rightleftharpoons 2\ Br + M \quad K_{Br_2}$$

$$Br + CH_2 = CF_2 \rightleftharpoons \dot{C}F_2CH_2Br \qquad (4)$$

$$Br_2 + \dot{C}F_2CH_2Br \rightleftharpoons CF_2BrCH_2Br + Br \qquad (5)$$

from which $k = k_5K_{Br_2}^{1/2}(k_4/k_{-4})$. From the known entropy of dissociation of Br_2 and the estimated entropy change for reaction (4), we could calculate $A_5 = 10^{9.2}\ M^{-1}s^{-1}$, which was in good agreement with similar reactions. In addition, the activation energy lead to $\Delta H^\circ_r(298) = -6.8$ kcal/mol from which we derived a value of 62.5 ± 1.5 kcal/mol for the pi bond dissociation energy in $CF_2 = CH_2$ (2). This value has recently been confirmed from independent measurements (3) so that both entropy and enthalpy calculations are consistent with the proposed mechanism. Such thermodynamic caluculations still do not assure the uniqueness of the proposed mechanisim but it does provide additional pieces of evidence that must fit the mechanistic puzzle.

Thermochemistry of Fluorocarbon Radicals

The ideal gas thermodynamic functions (S°, Cp°, $H^\circ - H^\circ_0$, etc.) for free radicals can only be determined by statistical thermodynamics (4) as the species are much too reactive for extended calorimetric measurements. We will calculate these functions in the ridgid rotor, harmonic oscillator approximation (5) and treat hindered, internal rotations by Pitzer's (6) semiclassical approximation. This approximation has been shown to give quite satisfactory agreement with exact calculations (6,7). These calculations, however, require detailed structural and mechanical data for the free radical; thus, the molecular weight, product of the principle moments of inertia, rotational symmetry, normal vibrational frequencies, reduced moments of inertia, barrier heights and symmetry for internal rotations, and ground state electronic degeneracy are all required; and this assumes that there are no low lying electronic states! Very little of these data are usually known, though difluoromethlylene ($\ddot{C}F_2$) and trifuoromethyl ($\dot{C}F_3$) are exceptions, so that much must be estimated. O'Neal and Benson (4) have suggested the "difference method" or what I would call "bond subtractivity" as one technique for the estimation of the needed data and have discussed the sensitivity of these functions to uncertainities in the estimation. They have shown that the largest sources of uncertainity in the thermo-functions have been in the estimation of the symmetry of radical (i.e., is the carbon bearing the odd electron planar or non-planar?) and in the estimation of the barrier to internal rotation about the C-$\dot{C}$ bond. Fortunately, for the radicals considered here, both the structure and barrier to internal rotation have been established by spectroscopic methods (see each radical for references) so that the major source of uncertainity in these calculations will arise from the estimates of the normal vibrational frequencies, and even here, there are experimental data for $\ddot{C}F_2$ and $\dot{C}F_3$. However, in the absence of experimental data the normal frequencies and bond lengths will be estimated by the difference method (4). In this technique, the vibrational assignments for a radical $R_f\cdot$ are estimated from the assignments of the corresponding molecule, R_fH from which the three frequencies (one stretch and two bends) associated with the C-H bond have been subtracted. The assignments for trifluoromethyl are known experimentally (see text). Table I compares these experimental assignments with those estimated from CF_3H from which the C-H stretch at 3036 cm^{-1} and two HCF bends at 1372 cm^{-1} have been subtracted.

Table I

Comparison of Estimated and Experimental Vibrational Frequencies for Trifluoromethyl.

Estimated		Experimental[a]	
cm^{-1}	degeneracy	cm^{-1}	degeneracy
1117	1	1085	1
700	1	702	1
1152	2	1251	2
507	2	510	2

a) see Table IV

It is quite clear from Table I that bond subtractivity works well for trifluoromethyl and it is particularly pleasing to note the accuracy with which the low frequencies were estimated (generally the bending modes), as these are the ones that make up the bulk of the vibrational contribution to the thermo-functions below 1500 K. We shall assume (hope?) that this estimation procedure is equally applicable to those radicals for which there is, as yet, no experimental data.

The enthalpies of formation have been obtained either from equilibrium data, as for difluoromethylene, or calculated from kinetic determinations of bond dissociation energies (generally the R_f-H bond) via eq. (6).

$$\Delta H_f^\circ(\dot{R}_f,g,298) = DH^\circ_{298}(R_f\text{-}X) + \Delta H_f^\circ(R_fX,g,298) - H_f^\circ(\dot{X},g,298) \quad (6)$$

The current, critical evaluation of the ideal gas thermodynamic data on the fluoromethanes (8) and fluoroethanes (9) will be particularly appropriate to the reduction of C-H bond dissociation energy data. Additional data on the enthalpy of formation of fluorine containing compounds will be recalculated (if necessary) to reflect the recent, direct determination of $\Delta H_f^\circ(HF\cdot nH_2O, 298)$ of Johnson, Smith and Hubbard (10). The data for the reference states, C(graphite), H_2(g) and F_2(g) are the same as those used by Rodgers et al. (8).

Difluoromethylene ($\ddot{C}F_2$): The microwave spectrum of $\ddot{C}F_2$ has been determined by Powell and Lide (11) and most recently, and in greater detail, by Kirchhoff and Lide (12). They found that $d_{C-F} = 1.303 \pm 0.0001$ A, $\angle FCF = 104.78 \pm 0.02°$, and that the ground electronic state was a singlet. This latter is in agreement with the reactivity data of Mitsch (13). The analysis of the microwave spectra was consistent with the vibrational assignment $V_1 = 1222$, $V_2 = 668$ and $V_3 = 1112\ cm^{-1}$ which is in excellent agreement with the measurements (infrared) of Lefohn and Pimentel (14) and Snelson (15).

Modica and LeGraff (16) measured the equilibrium constant for reaction (7)

$$CF_2 = CF_2(g) \rightleftharpoons 2\ \ddot{C}F_2 \quad (7)$$

from 1200 to 1800 K in a single pulse shock tube with argon as diluent. They determined the concentration of $\ddot{C}F_2$ by u.v. absorbtion spectroscopy at 2536A. The extinction co-efficient was determined at temperatures at which dissociation was essentially complete and then extrapolated to experimental temperatures. In a later investigation (17) additional values of the extinction co-efficient were obtained and the combined data resulted in, $\varepsilon_{CF_2} = (1.25 \pm 0.1) \times 10^6 cm^2 mol^{-1}$, constant over the whole temperature range. We have recalculated their earlier equilibrium results (16) using this value of ε_{CF_2} and reduced the resulting values of Keq to $\Delta H_r^\circ(298)$ using the "3rd Law" method with Gibbs energy functions for $\ddot{C}F_2$ from this work and $CF_2 = CF_2$ from the JANAF Tables (18). This gave $\Delta H^\circ_r(298) = 67.9 \pm 0.5$ kcal/mol for reaction (7) and $\Delta H_f^\circ(\ddot{C}F_2,g,298) = -44.7 \pm 0.5$ kcal/mol. Carlson (19) has also studied reaction (7) using the same experimental technique as Modica and LeGraff (16) but measured the concentration of $\ddot{C}F_2$ at 2620 A ($\varepsilon_{CF_2} = 1.60 \times 10^6 cm^2/mol$ from 1200 to 1600 K) and obtained $\Delta H_r^\circ(298) = 68.4 \pm 0.5$ kcal/mol, in excellent agreement.

Zmbov, Uy and Margrave (20) studied reaction (7) in a Knutsen cell using mass spectrometric detection for $\ddot{C}F_2$ and CF_2 =

CF_2 between 1100 and 1200 K. Their results recalculated on the same basis as those of references (16) and (19) gave $\Delta H_r^\circ(298)$ = 75.2 ± 0.5 kcal/mol for reaction (7). This corresponds to a much smaller value (about a factor of 30) for Keq for reaction (7) than found by Modica and LeGraff (16,17) or by Carlson (19); and suggests that equilibrium conditions had not been reached in their Knutsen cell.

Farber, Frish and Ko (21) have studied the reactions of fluorine with graphite in the range 2000 to 2500 K using a Knutsen cell with mass spectrometric detection of F, $\ddot{C}F_2$, $\dot{C}F_3$ and CF_4. Thus, using the thermo-functions presented here and those of Rodgers et al. (8), one can calculate $\Delta H_f^\circ(\ddot{C}F_2,g,298)$ by the "3rd Law" method from their equilibria data for reactions (8) and (9).

$$C_{(s)} + 2F_{(g)} \rightleftharpoons CF_{2(g)} \qquad (8)$$

$$CF_{4(g)} \rightleftharpoons CF_{2(g)} + 2\,F_{(g)} \qquad (9)$$

These calculations gave $\Delta H_f^\circ(\ddot{C}F_2,g,298)$ = -41.7 ± 0.5 and -44.0 ± 1 kcal/mol for reactions (8) and (9) respectively.

We have adopted a value of $\Delta H_f^\circ(\ddot{C}F_2,g,298)$ = -44.6 kcal/mol from the data of Modica and LeGraff (16,17) and of Carlson (19). This yields values of the equilibrium constant for reaction (8) with in a factor of two of those calculated from the data of Farber et al. (21), which is certainly within the accuracy of both the experiment and the limits of the rigid-rotor, harmonic oscillator approximation at 2000 to 2500 K (5). The physical and thermochemical data selected here are summarized in Table II and the ideal gas thermodynamic functions calculated to 1500 K from these data are summarized in Table III.

Table II

Selected Physical and Thermochemical Data for Difluoromethylene ($\ddot{C}F_2$).

Molecular Weight/gm mol^{-1}	50.01
Symmetry	2.0
Degeneracy, ground electronic state	1.0
Structural Parameters:	
C-F/10^{-8}cm	1.303
∠FCF/deg	104.78
Product moments of Inertia/amu^3A^6	1.071×10^4
Normal modes (degeneracy)/cm^{-1}	1222(1); 1112(1); 668(1).
Enthalpy of Formation 298K/kcal mol^{-1}	-44.6

TABLE III

Ideal Gas Thermodynamic Functions for Difluoromethylene ($\ddot{C}F_2$).

Temp K	Cp°	S°	-(G°-H°₀)/T	H°-H°₀	-ΔH°f	-ΔG°f	log K_f
	cal $K^{-1}mol^{-1}$			kcal mol^{-1}			
100.00	7.96	48.40	40.45	0.79	45.33	48.12	105.162
150.00	8.09	51.65	43.68	1.19	45.13	49.55	72.199
200.00	8.40	54.02	45.98	1.60	44.95	51.05	55.788
273.15	9.05	56.73	48.51	2.24	44.68	53.33	42.666
298.15	9.30	57.53	49.23	2.47	44.60	54.12	39.673
300.00	9.32	57.59	49.28	2.49	44.59	54.18	39.471
400.00	10.29	60.41	51.72	3.47	44.24	57.43	31.378
500.00	11.10	62.80	53.70	4.54	43.89	60.77	26.561
600.00	11.71	64.88	55.40	5.68	43.54	64.18	23.376
700.00	12.16	66.72	56.88	6.88	43.20	67.64	21.118
800.00	12.50	68.36	58.22	8.11	42.85	71.16	19.439
900.00	12.75	69.85	59.43	9.37	42.52	74.71	18.142
1000.00	12.95	71.20	60.54	10.66	42.18	78.30	17.112
1100.00	13.10	72.45	61.56	11.96	41.85	81.94	16.280
1200.00	13.21	73.59	62.52	13.28	41.52	85.59	15.588
1300.00	13.31	74.65	63.41	14.60	41.19	89.28	15.009
1400.00	13.38	75.64	64.25	15.94	40.88	92.98	14.515
1500.00	13.45	76.57	65.04	17.28	40.56	96.72	14.091

Trifluoromethyl (CF_3): The infrared spectrum of trifluomethyl has been observed in matrix isolation by Milligan and Jacox (22) and by Snelson (15). Their results were in excellent agreement and an average of their reported fundamentals have been summarized in Table I. Comparison of the fundamentals for $^{12}\dot{C}F_3$ and $^{13}\dot{C}F_3$ in valence force field calculations support a pyramidal geometry for the $\dot{C}F_3$ radical in agreement with esr studies of Fessenden and Schuler (23). We have adopted a C-F bond distance from CF_3H (8) and a $\angle FCF = 112°$; which yields an out of plane angle (angle between plane of F atoms and the C-F bond) of 17°.

The analysis of the kinetic data on the forward and reverse reactions of the type:

$$CF_3H + X \rightleftharpoons \dot{C}F_3 + HX$$

with X = Cl, Br and I (18) yields $DH°_{298}(CF_3\text{-}H) = 106.5 \pm 0.4$ kcal/mol, while Ferguson and Whittle (24) suggest that 106.7 ± 0.5 kcal/mol is the best value. We adopt this latter, which, combined with $\Delta H_f^\circ(CF_3H,g,298) = -165.7 \pm 1$ kcal/mol (8), leads to $\Delta H_f^\circ(\dot{C}F_3,g,298) = -111.1 \pm 1$ kcal/mol

These data are summarized in Table IV and used to calculate the ideal gas thermodynamic functions for trifluoromethyl which are summarized in Table V.

Table IV

Selected Physical and Thermochemical Data for Trifluoromethyl.

Molecular weight/g mol^{-1}	69.005
Symmetry	3.
Degeneracy, ground electronic state	2.
Structure	
C-F/10^{-8}cm	1.33
$\angle$FCF/deg	112.
Product moments of inertia/amu^3A^6	2.100×10^5
Enthalpy of formation 298K/kcal mol^{-1}	111.1
Normal modes (degeneracy)/cm^{-1}	1251(2), 1085(1), 702(1), 510(2).

Pentafluoroethyl ($CF_3\dot{C}F_2$): The fundamental modes of pentafluoroethyl have not been determined experimentally so that these will be estimated from the assignments of pentafluoroethane (9) using the difference method (4); thus the modes at 3008, 1393 and 1359 cm^{-1} in the assignment of CF_3CF_2H have been substracted to yield the assignment of $CF_3\dot{C}F_2$. While the bond distances and angles for pentafluoroethyl have not been measured, Meakin and Krusic (25) have studied the esr spectrum of this radical over a broad temperature range and concluded that the radical site was pyramidal, as in trifluoromethyl, and that there was a barrier to

TABLE V

Ideal Gas Thermodynamic Functions for Trifluoromethyl ($\dot{C}F_3$).

Temp / K	$C_p°$	S°	$-(G°-H_0°)/T$	$H°-H_0°$	$-\Delta H_f$	$-\Delta G_f$	log K_f
	cal $K^{-1}mol^{-1}$			kcal mol^{-1}			
100.00	8.09	52.97	45.00	0.79	110.69	109.86	240.108
150.00	8.79	56.36	48.25	1.21	110.82	109.42	159.425
200.00	9.81	59.03	50.62	1.68	110.93	108.93	119.037
273.15	11.41	62.32	53.33	2.45	111.06	108.17	86.553
298.15	11.93	63.35	54.12	2.75	111.10	107.91	79.102
300.00	11.97	63.42	54.18	2.77	111.10	107.89	78.600
400.00	13.81	67.13	56.97	4.06	111.21	106.80	58.355
500.00	15.23	70.37	59.33	5.52	111.28	105.69	46.199
600.00	16.27	73.24	61.41	7.09	111.33	104.57	38.091
700.00	17.02	75.81	63.29	8.76	111.36	103.44	32.297
800.00	17.58	78.12	65.00	10.49	111.39	102.31	27.949
900.00	17.99	80.22	66.58	12.27	111.41	101.17	24.568
1000.00	18.31	82.13	68.04	14.09	111.43	100.02	21.861
1100.00	18.55	83.89	69.40	15.93	111.45	98.90	19.650
1200.00	18.75	85.51	70.68	17.80	111.47	97.75	17.803
1300.00	18.90	87.02	71.88	19.68	111.50	96.60	16.241
1400.00	19.03	88.43	73.01	21.58	111.53	95.45	14.900
1500.00	19.13	89.74	74.08	23.49	111.57	94.30	13.740

rotation about the C-C bond of 2.85 kcal/mol. As indicated in the introduction, this is the important structural information in regard to the ideal gas thermodynamic functions. The bond distances and angles for $CF_3\dot{C}F_2$ have been taken from the structure of CF_3CF_2H (9). The enthalpy of formation of $CF_3\dot{C}F_2$ has been determined by Wu and Rodgers (26), who obtained $\Delta H_f^\circ(CF_3\dot{C}F_2,g,298) = -213.0 \pm 1$ kcal/mol

These data are summarized in Table VI and the ideal gas thermodyanimc functions for pentafluoroethyl calculated from these data are summarized in Table VII.

Table VI

Selected Physical and Thermochemical Data for Pentafluoroethyl.

Molecular weight/g mol^{-1}	119.014
Symmetry	3.0
Degeneracy, ground electronic state	2.0
Structural Parameters	
C-F (CF_3)/10^{-8}cm	1.335
C-F ($\dot{C}F_2$)/10^{-8}cm	1.345
C-C /10^{-8}cm	1.52
∠FCF (CF_3)/deg	108.14
∠FCF ($\dot{C}F_2$)/deg	109.3
∠FCC ($\dot{C}F_2$)/deg	109.6
Product moment of inertia/$amu^3 A^6$	6.926×10^6
Reduced moment of inertia/amu A^2	32.7
Barrier to internal rotation; V_3/kcal mol^{-1}	2.85
Torsional mode/cm^{-1}	67.
Enthalpy of formation/kcal mol^{-1}	-213.0
Normal modes/cm^{-1}	1309, 1218, 1111, 867, 725, 577, 523, 361, 246, 1198, 1145, 508, 413, 216.

2,2,2 Trifluoroethyl ($CF_3\dot{C}H_2$): The fundamental modes of 2,2,2 trifluoroethyl have not been determined experimentally so that these will be estimated from the assignments of 2,2,2 trifluoroethane (9) using the difference method (4); thus the modes at 3035 and 1443(2) cm^{-1} in the assignment of CF_3CH_3 have been eliminated. The esr spectrum of this radical has been studied over a broad temperature range (27,28) and its spectrum was found to be essentially invariant both with regard to line widths, as well as coupling constants. Consequently, they concluded that the radical site was planar and that there was essentially no barrier to rotation about the C-C bond. The ∠HCH has been taken as 120° and the balance of the structural parameters estimated

TABLE VII

Ideal Gas Thermodynamic Functions for Pentafluoroethyl (CF_3CF_2)·

Temp / K	C_p°	S°	$-(G^\circ-H_0^\circ)/T$	$H^\circ-H_0^\circ$	$-\Delta H_f^\circ$	$-\Delta G_f^\circ$	log K_f
	cal $mol^{-1}K^{-1}$			kcal mol^{-1}			
100.00	12.24	63.13	53.35	0.87	212.52	208.61	455.929
150.00	15.21	68.66	57.56	1.66	212.76	206.60	301.025
200.00	18.05	73.43	60.94	2.49	212.90	204.52	223.498
273.15	21.67	79.61	65.13	3.95	212.99	201.44	161.175
298.15	22.75	81.56	66.43	4.51	213.00	200.30	146.889
300.00	22.83	81.70	66.52	4.55	213.00	200.31	145.926
400.00	26.40	88.78	71.22	7.02	212.94	196.08	107.135
500.00	28.96	94.96	75.37	9.80	212.82	191.88	83.872
600.00	30.78	100.42	79.10	12.79	212.66	187.71	68.373
700.00	32.07	105.26	82.49	15.93	212.49	183.56	57.312
800.00	33.00	109.61	85.62	19.19	212.32	179.44	49.021
900.00	33.69	113.54	88.50	22.53	212.15	175.34	42.578
1000.00	34.22	117.12	91.19	25.92	211.98	171.24	37.426
1100.00	34.62	120.40	93.70	29.37	211.81	167.21	33.222
1200.00	34.94	123.43	96.50	32.85	211.66	163.14	29.712
1300.00	35.19	126.23	98.27	36.35	211.52	159.11	26.748
1400.00	35.40	128.85	100.36	39.88	211.40	155.06	24.206
1500.00	35.56	131.30	102.34	43.43	211.29	151.05	22.007

from CF_3CH_3 (9). The enthalpy of formation of 2,2,2 trifluoroethyl has been determined by Wu and Rodgers (29), who obtained $\Delta H_f^\circ(CF_3\dot{C}H_2,g,298) = -123.6 \pm 1$ kcal/mol. These data are summarized in Table VIII and the ideal gas thermodynamic functions calculated from these data are summarized in Table IX.

Table VIII

Selected Physical and Thermochemical Data for 2,2,2 Trifluoroethyl.

Molecular weight/g mol^{-1}	83.036
Symmetry	6.0
Degeneracy, ground electronic state	2.0
Structural parameters	
C-F/10^{-8}cm	1.335
C-H/10^{-8}cm	1.085
C-C/10^{-8}cm	1.53
∠FCF/deg	107.9
∠CCH/deg	120.
∠HCH/deg	120.
Product moment of inertia/amu^{3}A^6	8.122×10^5
Reduced moment of inertia/amu A^2	1.74
Barrier to internal rotation;	
V_6/kcal mol^{-1}	0.0
Enthalpy of formation/kcal mol^{-1}	-123.6
Normal modes (degeneracy)/cm^{-1}	2975, 1408, 1280, 830, 602, 3035, 1233(2), 970(2), 541(2), 365(2).

1,1-Difluoroethyl ($CH_3\dot{C}F_2$): A temperature dependent study of the esr spectrum of 1,1- difluoroethyl (28,29) has indicated that the radical site geometry is pyramidal in this radical as in trifluoromethyl (23) and pentafluoroethyl (25), and an analysis of the line widths has indicated a barrier to rotation about the C-C bond of 2.2 kcal/mol. The fundamental modes have been estimated from 1,1-difluoroethane (9) using the difference method (4). The modes at 3001, 1460 and 1360 cm^{-1} were eliminated. The bond distances and angles have also been estimated from those of 1,1-difluoroethane (9). The enthalpy of formation of 1,1-difluoroethyl has been determined by Pickard and Rodgers (30) from kinetic and thermochemical data. They obtained $\Delta H_f^\circ(CH_3\dot{C}F_2, g,298) = -72.3 \pm 2$ kcal/mol and showed that this value lead to a pi bond dissociation energy in 1,1-difluoroethene of 62.5 ± 2 kcal/mol (3) in excellent agreement with a previous, independent determination (2).

These data are summarized in Table X, and the ideal gas thermodynamic functions, calculated from these data, are given in Table XI.

TABLE IX

Ideal Gas Thermodynamic Functions for 2,2,2-Trifluoroethyl (CF_3CH_2).

Temp K	C_p°	S°	-(G°-H°₀)/T	H°-H°₀	$-\Delta H_f$°	$-\Delta G_f$°	log K_f
	cal K⁻¹mol⁻¹			kcal mol⁻¹			
100.00	9.65	57.65	48.58	0.90	122.40	119.58	261.342
150.00	11.39	61.87	52.33	1.43	122.74	118.09	172.062
200.00	13.46	65.43	55.17	2.05	123.06	116.49	127.296
273.15	16.45	70.07	58.55	3.14	123.47	114.02	91.229
298.15	17.43	71.55	59.58	3.57	123.60	113.14	82.939
300.00	17.50	71.66	59.65	3.60	123.60	113.08	82.379
400.00	20.96	77.19	63.36	5.53	124.02	109.50	59.830
500.00	23.68	82.17	66.63	7.77	124.33	105.84	46.262
600.00	25.78	86.68	69.60	10.24	124.44	102.12	37.197
700.00	27.41	90.78	72.34	12.91	124.71	98.36	30.711
800.00	28.71	94.53	74.88	15.72	124.83	94.59	25.841
900.00	29.77	97.98	77.26	18.64	124.90	90.80	22.051
1000.00	30.65	101.16	79.49	21.66	124.94	87.00	19.014
1100.00	31.38	104.12	81.60	24.77	124.95	83.24	16.538
1200.00	32.38	106.88	83.59	27.94	124.94	79.42	14.465
1300.00	32.52	109.46	85.48	31.17	124.92	75.63	12.715
1400.00	32.97	111.89	87.28	34.44	124.89	71.83	11.213
1500.00	33.36	114.18	89.00	37.76	124.85	68.04	9.914

Table X

Selected Physical and Thermochemical Data for 1,1-Difluoroethyl ($CH_3\dot{C}F_2$).

Molecular weight/g mol^{-1}	65.044
Symmetry	3.0
Degeneracy, electronic ground state	2.0
Structural parameters	
C-F/10^{-8}cm	1.345
C-H/10^{-8}cm	1.100
C-C/10^{-8}cm	1.540
∠HCH/deg	110.2
∠CCF/deg	109.4
∠FCF/deg	109.2
Product moments of inertia/$amu^3 A^6$	2.723×10^5
Reduced moment of inertia/amu A^2	3.06
Barrier to internal rotation;	
V_3/kcal mol^{-1}	2.2
Torsional mode/cm^{-1}	182.
Enthalpy of formation/kcal mol^{-1}	-72.3
Normal modes (degeneracy)/cm^{-1}	3018, 2978, 2960, 1460, 1414, 1372 1143, 1129, 868, 571, 470, 1171, 930, 383.

Discussion

While these data are limited, they can be combined with group additivity (4,31,32), to form the basis for the estimation of the thermochemistry of several classes of free radical (4,31). Some caution must be exercised, however, when this estimation procedure is applied to compounds with groups of very different polarity as here. Thus, the entropy and heat capacity of CF_3CH_3 is satisfactorily approximated by group additivity (9,32), but the enthalpy of formation is in error by nearly 8 kcal/mol (33). With this in mind then, one should expect the thermochemistry of the free radicals; $R_fCF_2\dot{C}F_2$, $R_fCF_2\dot{C}H_2$, and $R_HCH_2\dot{C}F_2$ (R_f = perfluoroalkyl and R_H = alkyl) to be satisfactorily estimated by equations 10, 11 and 12.

$$P(R_fCF_2\dot{C}F_2) = P(CF_3\dot{C}F_2) + G(R_f) + \underline{C}(C)_2(F)_2 - \underline{C}(C)(F)_3 \quad (10)$$

$$P(R_fCF_2\dot{C}H_2) = P(CF_3\dot{C}H_2) + G(R_f) + \underline{C}(C)_2(F)_2 - \underline{C}(C)(F)_3 \quad (11)$$

TABLE XI

Ideal Gas Thermodynamic Functions for 1,1-Difluoroethyl (CH_3CF_2).

Temp K	C°_p	S°	$-(G^\circ-H^\circ_0)/T$	$H^\circ-H^\circ_0$	$-\Delta H^\circ_f$	$-\Delta G^\circ_f$	log K_f
	cal K^{-1} mol^{-1}			kcal mol^{-1}			
100.00	9.62	55.76	47.28	0.84	71.06	68.86	150.504
150.00	11.24	59.96	50.83	1.36	71.37	67.69	98.637
200.00	12.90	63.42	53.56	1.97	71.69	66.42	72.582
273.15	15.25	67.80	56.80	3.00	72.15	64.42	51.547
298.15	16.03	69.17	57.78	3.39	72.30	63.70	46.697
300.00	16.09	69.27	57.85	3.42	72.31	63.65	46.369
400.00	19.04	74.31	61.35	5.18	72.85	60.67	33.152
500.00	21.55	78.84	64.40	7.21	73.31	57.58	25.168
600.00	23.61	82.95	67.15	9.48	73.68	54.39	19.814
700.00	25.32	86.73	69.68	11.92	73.99	51.15	15.971
800.00	26.74	90.20	72.03	14.53	74.23	47.87	13.079
900.00	27.95	93.43	74.23	17.27	74.41	44.57	10.823
1000.00	28.97	96.42	76.31	20.11	74.53	41.23	9.012
1100.00	29.85	99.23	78.26	23.06	74.62	37.93	7.535
1200.00	30.61	101.86	80.12	26.08	74.57	34.57	6.296
1300.00	31.26	104.34	81.89	29.18	74.70	31.23	5.250
1400.00	31.82	106.67	83.58	32.33	74.72	27.86	4.350
1500.00	32.31	108.89	85.19	35.54	74.71	24.52	3.573

$$P(R_HCH_2\dot{C}F_2) = P(CH_3\dot{C}F_2) + G(R_H) + \underline{C}(C)_2(H)_2 - \underline{C}(C)(H)_3 \quad (12)$$

In these equations, P ( ) is the particular propoerty of interest, G(R) the group contribtuions for the alkyl substitent, and $\underline{C}(C)_2(F)_2$, etc., the specific group values. One must bear in mind that intrinsic entropy (S*) is given by the group additivity (32) and that $S° = S^* - R\ln \sigma/\eta$; where σ is the rotational symmetry number and η is the number of optical isomers of the particular compound or radical.

The group contributions for perfluorocarbons have been determined from enthalpies of formation (34) recalculated using the most recent value for $\Delta H_f^\circ(HF{\bullet}nH_2O)$ (10); and from the entropies and heat capacities of CF_3CF_3 (9) and $CF_3CF_2CF_3$ (35,36). The group values thus obtained are summarized in Table XII and differ slightly from those given in reference (32).

Table XII

Group Values for the Estimation of the Thermochemical Properties of Perfluorocarbons.

Group	ΔH_f°(300K) kcal/mol	S*(300K)[a] cal/K^{-1}mol^{-1}	C_p°/calK^{-1}mol^{-1} 300	400	500
$\underline{C}(C)(F)_3$	-160.45	42.6	12.7	15.0	16.7
$\underline{C}(C)_2(F)_2$	-97.5	16.5	10.0	11.5	13.0
$\underline{C}(C)_3(F)$	-44.0	----	----	----	----

a $S^* = \text{intrinsic entropy} = S° - R\ln \sigma/\eta$

Using eq. (10), one can estimate for $CF_3CF_2\dot{C}F_2$, $\Delta H_f^\circ(300K) = -310.5$ kcal/mol; $S°(300K) = 98.5$ cal/K mol, $C_p^\circ(300K) = 32.8$ and $C_p^\circ(500K) = 37.9$ cal/K mol. The thermochemistry of reaction (1) can be

$$\dot{C}F_3 + CF_2 = CF_2 \underset{-1}{\overset{1}{\rightleftharpoons}} CF_3CF_2\dot{C}F_2 \quad (1)$$

obtained by combining these estimates with the data for $\dot{C}F_3$ given in Table V and that for $CF_2 = CF_2$ from reference (18); thus, $\Delta H_r^\circ(300K) = 42.0$ kcal/mol, $\Delta S_r^\circ(300K) = 36.7$ cal/K mol and $\overline{\Delta C_p^\circ} = 0$ cal/K mol from 300 to 500 K. Tedder and Walton (1, 37) have summarized the kinetic data on reaction (1), from which one can obtain $\log(k_1/\text{l mol}^{-1}\text{sec}^{-1}) = 7.94 - 4.6/\Theta$ at a mean temperature of 440K. When this result is combined with the above thermochemistry via eq. (2), one obtains a value for the unimolecular decomposition of the perfluoropropyl radical, reaction (-1), namely $\log(k_{-1}/sec^{-1}) = 13.9 - 45.6/\Theta$. This is the rate constant alluded to in the Introduction, and one which we have chosen to illustrate the advantages of thermodynamic data to kineticists. The estimate obtained here should be reliable to ±0.3 in the logA and ±1.5 to 2.0 kcal/mol in the activation energy.

There is a second path for the unimoleculor decomposition of the perfluoropropyl radical which is also of some interest, namely reaction (13), the decomposition to difluoromethylene.

$$CF_3CF_2\dot{C}F_2 \rightleftharpoons CF_3\dot{C}F_2 + \ddot{C}F_2 \qquad (13)$$

For this reaction, we can estimate $\Delta H^\circ_r(300K) = 52.9$ kcal/mol, $\Delta S^\circ_r(300K) = 40.8$ cal/K mol, and $\overline{\Delta C^\circ_p} = 1.0$ cal/K mol from 300 to 500 K; thus, $\log(A_{13}/A_{-13}) = 7.0$ and $E_{13}-E_{-13} = 52.1$ kcal/mol at 440 K with units of liters, moles and second for A. For this mode of decomposition, there are no experimental measurements for either rate constant, however, $\log(k_{15}/l\ mol^{-1}sec^{-1}) = 9.5$ (38, 39,40) and $\log(k_{16}/l\ mol^{-1}\ sec^{-1}) = 8.4 - 1.6/\Theta$ (41), so that a reasonable estimate for the rate constant of reaction (-13) would

$$\dot{C}F_3 + \dot{C}F_3 \longrightarrow C_2F_6 \qquad (15)$$

$$\ddot{C}F_2 + \ddot{C}F_2 \longrightarrow CF_2{=}CF_2 \qquad (16)$$

be $\log(k_{-13}/l\ mol^{-1}\ sec^{-1}) \cong 9.0 - 1/\Theta$. This and the above thermochemistry then yields; $\log(k_{13}/sec^{-1}) = 16.0 - 53/\Theta$.

Each of the rate constants for the two paths for decomposition of the perfluoropropyl radical has an uncertainty associated with it due to the estimates for this radical. Much of this uncertainty will cancel out if one compares the ratio of the rate constants for these two paths. Thus, $\log(k_{-1}/k_{13}) = -2.1 + 7.4/\Theta$. This indicates that the rate constants are equal at 770K, with reaction (13) dominating at higher temperatures. However, for larger, linear, perfluoroaklyl radicals the enthalpy of decomposition to tetrafluoroethylene is reduced to $\Delta H^\circ_r(300K) = 37.6$ kcal/mol, because $\Delta H^\circ_f(CF_3\dot{C}F_2) - \Delta H^\circ_f(\dot{C}F_3) = -102$ kcal/mol, less than the -97.5 kcal/mol for the group contribution ($\underline{C}$-$(C)_2(F)_2$). Thus for butyl and larger radicals, with other things being equal, decomposition to $CF_2{=}CF_2$ would be predicted to dominate up to 1200 K.

Acknowledgement

The author gratefully acknowledges the support of this research by the National Science Foundation under Grant No. CHE 74-22189.

Literature Cited

1. Tedder, J.M. and Walton, J.C., Accts Chem. Res. (1976), 9, 183.
2. Pickard, J.M. and Rodgers, A.S., J. Amer. Chem. Soc., (1976), 98, 6115.
3. Pickard, J.M. and Rodgers, A.S., J. Amer. Chem. Soc., in Press.
4. O'Neal, H.E. and Benson, S.W., "Free Radicals", (J. Kochi Ed.) John Wiley and Sons, New York, N.Y., (1973) II, 275.
5. Lewis, G.N., Randall, M., Pitzer, K.S., and Brewer, L. "Thermodynamics", McGraw Hill, New York, N.Y., 1961.
6. Pitzer, K.S. and Gwinn, W.D., J. Chem. Phys. (1942), 10, 428.
7. Chao, J., Rodgers, A.S., Wilhoit, R.C., and Zwolinski, B.J., J. Phys. Chem. Ref. Data (1974), 3, 141.
8. Rodgers, A.S., Chao, J., Wilhoit, R.C., and Zwolinski, B.J., J. Phys. Chem. Ref. Data (1974), 3, 117.
9. Chen, S.S., Rodgers, A.S., Chao, J., Wilhoit, R.C., and Zwolinski, B.J., J. Phys. Chem. Ref. Data (1975), 4, 441.
10. Johnson, G.K., Smith, P.N., and Hubbard, W.N., J. Chem. Thermodynamics (1973), 5, 793.
11. Powell, F.V. and Lide, D.R. Jr., J. Chem. Phys. (1966), 45, 1067.
12. Kirchhoff, W.H. and Lide, D.R Jr., J. Mol. Spectros. (1973), 47, 491.
13. Mitsch, R.A., J. Am. Chem. Soc. (1965), 87, 758.
14. Lefohn, A.S. and Pimentel, G.C., J. Chem. Phys. (1971), 55, 1213.
15. Snelson, A., High Temp. Sci. (1970), 2, 70.
16. Modica, A.P., and LeGraff, J.E., J. Chem. Phys. (1965), 43, 3383.
17. Modica, A.P. and LeGraff, J.E., J. Chem. Phys. (1966), 45, 4729.
18. "JANAF Thermochemical Tables" 2nd Ed. D.R. Stull and H. Prophet, Project Directors, Office of Standard Reference Data, National Bureau of Standards, Washington, D.C. NSRDS-NBS-37 June 1971.
19. Carlson, G.A., J. Phys. Chem. (1971), 75, 1625.
20. Zmbov, K.F., Uy, O.M., and Margrave, J.L., J. Amer. Chem. Soc. (1968), 90, 5090.
21. Farber, M., Frisch, M.A., and Ko, H.C., Trans. Faraday Soc. (1969) 65, 3202.
22. Milligan, D.E. and Jacox, M.E., J. Chem. Phys. (1968), 48, 2265.
23. Fessenden, R.W. and Schuler, R.H., J. Chem. Phys. (1965), 43, 2704.
24. Ferguson, K.E. and Whittle, E., J. Chem. Soc. Faraday Trans. I, (1972), 68, 295.

25. Meakin, P., and Krusic, P.J., J. Amer. Chem. Soc. (1973), 95, 8185.
26. Wu, E.C. and Rodgers, A.S., J. Amer. Chem. Soc., (1976), 98, 6112.
27. Edge, D.J. and Kochi, J.K., J. Amer. Chem. Soc., (1972), 94, 6485.
28. Chen, K.S., Krusic, P.J., Meakin, P. and Kochi, J.K., J. Phys. Chem., (1974), 78, 2014.
29. Chen, K.S. and Kochi, J.K., J. Amer. Chem. Soc., (1974), 96, 794.
30. Pickard, J.M. and Rodgers, A.S., J. Amer. Chem. Soc., in Press.
31. O'Neal, H.E. and Benson, S.W., Int. J. Chem. Kinetics, (1969), 1, 221.
32. Benson, S.W., Cruickshank, F.R., Golden, D.M., Haugen, G.R., O'Neal, H.E., Rodgers, A.S., Shaw, R., and Walsh, R., Chem. Rev. (1969), 69, 279.
33. Rodgers, A.S. and Ford, W.G.F., Int. J. Chem. Kinetics, (1973), 5, 965.
34. Cox, J.D., and Pilcher, G., "Thermochemistry of Organic and Organometallic Compounds", Academic Press, N.Y. (1970).
35. Plaush, A .C., Dis. Abst. (1966), B27, 440.
36. Kletskii, A.V., and Tsuranova, T.N., Kholod. Tekh. Tekhnol. (1970), 9, 42 (CA. 74:25688).
37. Tedder, J.M., and Walton, J.C. "Fluorine Containing Free Radicals", Vol. I, ACS Symposium Series, J.W. Root, Ed., 1977.
38. Ogawa, T., Carlson, G.A., and Pimentel, G.C., J. Phys. Chem. (1970), 74, 2090.
39. Basco, N., and Hawthorn, F.G.M., Chem. Phys. Letters, (1971), 8, 291.
40. Hiatt, R., and Benson, S.W., Int. J. Chem. Kinetics, (1972), 4, 479.
41. Dalby, F.W., J. Chem. Phys., (1964), 41, 2297.

12

Chemistry of High Energy Atomic Fluorine: Steady State Kinetic Theory Model Calculations for the $^{18}F + H_2$ Reaction III

EDWARD R. GRANT,† DA-FEI FENG, and JOEL KEIZER
Department of Chemistry, University of California, Davis, CA 95616

KATHLEEN D. KNIERIM,‡ and JOHN W. ROOT
Department of Chemistry and Crocker Nuclear Laboratory, University of California, Davis, CA 95616

As reviewed elsewhere in this volume (1,2) considerable recent progress has been achieved with respect to our understanding of the dynamics of atomic fluorine reactions. A central objective of chemical dynamics research involves the elucidation of coupling mechanisms by which various forms of energy affect reaction rates (3). Vibrational excitation has been shown to promote certain bimolecular processes (4). In cases that involve mode selective reagent excitation, the rate enhancement may be sufficient to provide a basis for isotope separation (5). Reactivity and energy transfer characteristics of novel electronically excited species have also been investigated (6-8).

Transient hot atom or hot radical momentum distributions arise in many nonequilibrium kinetic situations. Some examples include flash photolytic, combustion (9), explosion (10), chemical laser (11,12), radiolytic, photochemical recoil, nuclear recoil (13-15), and thick-target accelerated ion or atomic beam (16) experiments. Principally because of limitations in the available methodology, the separate consideration of translational excitation in non-ionic systems has received comparatively little attention. Molecular beam experiments have provided much of the available information pertaining to collision energy effects in reactive scattering (17). However, the beam technique can be successfully utilized only for systems that involve large reaction cross sections (1), and the energy range characteristics of available beam sources are limited (18).

Atomic recoil experiments offer an alternative procedure for investigating the chemical effects of translational excitation (19,20). Thermal (21,22) and energetic gas (23-25)

---

†Present Address: Department of Chemistry, Cornell University, Ithaca, NY 14850.
‡University of California Regents Fellow 1977-78.

and liquid phase (26,27) nuclear recoil ^{18}F studies have recently been reported. Photochemical recoil techniques (19,24) have not yet been utilized for the study of hot fluorine atom reactions, and translationally excited non-ionic polyatomic reagents have been investigated only to a limited degree (28-30).

The potential content of atomic recoil experiments has recently been shown to include useful dynamical information (13-16,19,23-27,31,32). A central interpretive difficulty, however, follows from the need for unique and accurate specifications of the reactant non-Boltzmann momentum distributions. In the present work brief reviews are given of the theoretical progress achieved thus far toward the resolution of this problem and of those aspects of the steady state hot atom kinetic theory (13) that are pertinent to modeling calculations. Detailed modeling results are presented for nuclear recoil ^{18}F atoms reacting with H_2 (15,31,33-35) in order to illustrate the level of dynamical understanding that can be obtained. Specific topics treated include the nature and significance of microscopic time dependent phenomena, the nature of the coupling between reactive cross section structure and energy ranges for hot atom reactions, and perturbations of hot reactive energy ranges by inert additives in mixture experiments.

Background Theory.

The remainder of this article is mainly concerned with the theoretical analysis of nuclear recoil hot atom chemistry experiments. Under typical laboratory conditions the recoil species are generated consecutively through irradiations having much longer duration than the characteristic hot atom mean free lifetime (23-27). It is not unusual for the individual recoil events to be isolated in real time. On this basis the early hot atom kinetic theories utilized stochastic formulations for independent recoil particle collision cascades occurring in thermally equilibrated molecular reaction systems.

In a pioneering effort (36-38) Estrup and Wolfgang (EW) adapted classical hard sphere neutron cooling theory in order to obtain a time independent stochastic kinetic treatment of hot atom reactions. The primary hot reaction yield followed from the Miller-Dodson equation (24,27) as the integral of the center-of-mass reaction probability [P(E)] over the collision density distribution function [N(E)]. Because of the explicit neglect of time dependent relaxation processes, stochastic approaches of this general type cannot yield dynamical information (14). In most EW-theory applications classical central force descriptions for elastic scattering have been adopted in order to facilitate the evaluation of N(E). In the absence of reactive perturbations, such treatments lead to N(E) distributions having the general form $(1/\alpha E)$. For classical hard spheres the logarithmic energy loss constant α follows from the hot atom and reservoir particle masses (36,37). For classical soft spheres the basic ($1/\alpha E$) form for N(E) is preserved, but α must be determined from the repulsive part of the intermolecular potential (38). For more realistic scattering descriptions and potentials and for collision energy regimes in which reactions occur, this simple classical description for N(E) requires modification.

Although Wolfgang often likened it to a nonequilibrium counterpart of the Arrhenius equation, the EW-theory does not treat temperature coupling phenomena. In fact, until rather recently (13,15,23,31) temperature insensitivity has been accepted as a general characteristic of well behaved hot atom systems (39). The principal application of EW-theory has involved the empirical analysis of absolute hot yield data measured for two component systems. Individual reactants are typically studied at progressively increasing dilution using inert gas additives under conditions of uniform sample size, total pressure, temperature and nuclear recoil production method (19,20).† The reported investigations of this type are too numerous to be catalogued here. Reactive mixtures have also occasionally been studied (37,41,42). The EW-theory data analysis involves reduced plotting procedures, which are supposed to yield information pertaining to integrated hot reaction cross sections, reaction energy range effects and reactive shadowing.‡

Refined cascade models have been incorporated in several modifications of the original EW-theory. Kostin and coworkers examined some of the basic relaxational assumptions using a mathematical model for a static collision density maintained at steady state by a constant source term (43-45).∓

†Potentially serious experimental complications that can arise at large moderator concentration include excessive recoil loss (40) and incomplete recoil ion charge exchange. Recoil loss enhances radiolytic sample degradation and reduces the available radioactivity, thereby leading to diminished experimental accuracy. The severity of these effects could be controlled through the technique of increasing the sample size in order to accommodate the addition of moderator.

‡Shadowing interpretations have often been employed in multichannel reaction systems in order to account for relative product yield variations with sample composition. Reactive additives "shadow" (deplete) yields from lower energy processes through energetic cross section components that selectively intercept the cascading hot atoms.

∓This more conventional use of the term steady state, which simply connotes time-independence for N(E), is fundamentally different from the dynamical definition that arises in the steady state hot atom kinetic theory (13,14).

Collision densities computed for more realistic molecular potentials have supported the general applicability of EW-theory with the proviso that the energy loss parameter α must be allowed to vary with collision energy. Porter's integral reaction probability (IRP) stochastic formulation utilizes generalized momentum transfer cross sections (46-49). In photodissociation recoil experiments the nonequilibrium reagent is injected directly within the hot reactive collision energy zone. A detailed IRP-theory analysis of this situation suggests that deconvoluted reaction cross sections can be derived from accurate hot yield measurements carried out over a range of initial collision energies. However, this attractive procedure for characterizing the reactive cross section energy dependence has remained controversial (50-54). Other specialized extensions of stochastic hot atom kinetic theory have also appeared (55,56).

Stochastic hot atom kinetic treatments have demonstrated varying degrees of utility with respect to the systemization of experimental results, contributing to the partial characterization of nonreactive energy loss. Although classical elastic scattering models have been generally recognized as seriously oversimplified, no more complete treatment for nonreactive collisions has yet appeared (56). A common feature of these theories is the notion that an approximate description for hot atom moderation can be incorporated into an interpretive model which then allows the characterization of reactive energy range and shadowing effects from measured product yields. However, Feng *et al*. have questioned the validity of this approach (31), and EW-type theories have been shown to be insensitive to the failure of underlying assumptions concerning nonreactive collisions (55-61). These reservations are compounded by other conceptual difficulties including the neglect of temperature coupling and microscopic time dependent phenomena. Post hot reaction unimolecular effects (19,25,62,63) constitute a potentially serious practical problem (*vide infra*) that has often been ignored in moderator experiments with polyatomic substances. Because the primary hot reaction yield (24,27) follows from the Miller-Dodson equation, all open secondary decomposition channels must be directly monitored. Other effects that can mask the significance of measured results include unimolecular collisional energy transfer (64-66) and collision induced dissociation of internally excited primary hot reaction products (67,68).†

In summary, recent theoretical calculations have supported the empirical utility of the reduced data plotting procedures proscribed by EW-theory. However, the quantitative significance

---

†Recent unpublished quasiclassical trajectory calculations suggest that collision induced product dissociation may be of minor importance in the nuclear recoil ^3H *vs*. H_2 reaction system (69).

of derived kinetic parameters is uncertain. In addition to these fundamental reservations, it also seems apparent that an adequate level of detail has generally not been provided in hot atom moderator experiments (vide infra). Reasonable agreement has recently been achieved between theoretically and experimentally derived relative integrated reaction probabilities for the nuclear recoil ^3H vs. $H_2(D_2)$ system (56). Even though a rather large number of such claims have appeared in the literature, we seriously question whether definitive a priori information pertaining to hot reactive collision energy distributions can be obtained from this type of kinetic analysis.

The Steady State Hot Atom Kinetic Theory.

In order to elucidate the dynamical features of a hot atom reaction, the microscopic time dependent collision cascade must be modeled. Provided that the required reactive and nonreactive cross section data are available, this can be accomplished using the steady state hot atom kinetic theory (13). This section contains a brief introduction to the mathematical apparatus. Later sections describe the modeling procedure and results obtained for nuclear recoil ^{18}F atoms reacting with pure (70) and inert gas moderated (71) H_2. These reaction systems have been chosen for study because of the availability of quasi-classical trajectory reaction cross sections (31) together with results from thermal (21,72) and nonthermal (23) nuclear recoil experiments.

The steady state theory begins with a fictionalized representation of the hot atom cascade in which the entire collection of atoms is assumed to be present initially. This mathematically convenient model is also rigorously applicable, provided that the recoil atoms are mutually non-interacting and that they are produced with very small total concentration and with spatial uniformity throughout the host reservoir.

The time dependent relaxation of the hot atom momentum distribution is followed with the aid of the Boltzmann equation. We begin with the formulation of the laboratory momentum distribution for hot atoms (A) undergoing relaxation in a bath of pure reservoir molecules (R) that are in thermal and mechanical equilibrium with each other and with their surroundings. For simplicity the reservoir species are assumed to be initially present in a single quantum state designated by subscript j. The generalization to systems of greater complexity is given later. The equilibrium and time (t) dependent non-equilibrium momentum distributions for species R and A are denoted $f^e_{Rj}(\underset{\sim}{P}_R)$ and $g_A(\underset{\sim}{P}_A,t)$. The time derivative of the hot atom momentum density is given by the Boltzmann equation modified for the inclusion of internal states (73).

$$\frac{\partial n_A g_A}{\partial t} = n_A(t) \sum_k \int d\Omega \int d\underset{\sim}{P}_R \, \sigma(j/k) \, (P/\mu) \, [g_A' f^{e'}_{Rk} - g_A f^e_{Rj}]$$

$$- n_A(t) \int d\Omega \int d\underset{\sim}{P}_R \, \sigma^*(j) \, (P/\mu) \, g_A \, f^e_{Rj} \qquad (1)$$

In Eq. 1 $n_A(t)$ denotes the number density of the hot atoms; P and μ the scalar center-of-mass momentum and reduced mass; $\sigma(j/k)$, the total cross section for elastic scattering and inelastic scattering to final quantum state k; $\sigma^*(j)$, the detailed state specific total reactive cross section; and the primed terms designate energy restoring collisions. The second term, which accounts for hot atom depletion due to reactions, can be integrated over $\underset{\sim}{P}_A$ to yield a time dependent chemical rate equation.

$$\frac{dn_A}{dt} = - n_A(t) \int d\Omega \int d\underset{\sim}{P}_R \int d\underset{\sim}{P}_A \, \sigma^*(j) \, (P/\mu) \, g_A \, f^e_{Rj}$$

$$= - n_A(t) \, \kappa(t) \qquad (2)$$

The rate coefficient $[\kappa(t)]$ integral in Eq. 2 is time dependent through the momentum distribution function g_A.

Equation 1 relates the changing hot atom momentum density to the distribution functions and the cross sections for elastic, inelastic and reactive scattering. We next require an explicit expression for g_A. The rate of collisional relaxation away from the initial anisotropic laboratory vector momentum distribution is extremely rapid. The spatial part of this distribution is quickly randomized, so that the average vector laboratory momentum becomes vanishingly small after only a few collisions.† In the spirit of the Chapman-Enskog treatment of nonequilibrium transport processes (74), a zeroth order approximation for g_A after a very small number of nonreactive encounters is given by a Gaussian distribution centered about the value zero.

$$g_A(\underset{\sim}{P}_A, t) = [2\pi m_A k T_A]^{-3/2} \exp[-\underset{\sim}{P}_A{}^2/2m_A k T_A] \qquad (3)$$

Here the quantities k, m_A and T_A denote Boltzmann's constant, the hot atom mass and the time dependent hot atom temperature.

---

†Depending upon the experimental method, the initial hot atom laboratory vector momentum distribution may or may not be anisotropic. The $^3He(n,p)^3H$ and ordinary ultraviolet photodissociation techniques produce isotropic distributions, whereas $^{19}F(n,2n)^{18}F$ and polarized light induced photodissociations do not.

This result is essentially equivalent to the Chapman-Enskog local equilibrium approximation, which has proven quite successful for the theoretical representation of irreversible transport processes for real gases. Reasoning by analogy, the physical basis for Eq. 3 involves the simple notion that translational relaxation occurs isotropically and much more rapidly than other relaxation modes, notably including nonthermal chemical reactions.

The validity and limitations of Eq. 3 are of central importance to the remainder of this discussion. The steady state theory represents a fundamental departure from the stochastic methodology that has been widely accepted by workers in nonequilibrium chemical kinetics (14). We therefore digress briefly to consider the nature of limitations and possible improvements for Eq. 3 along with the rationale for steady state theory modeling calculations. The appropriateness and utility of the local equilibrium approach follow from the recognition that nonthermal chemical reactions can be meaningfully described as time dependent relaxation processes that principally compete with momentum transfer. It is the detailed balancing between these effects that comprises the core of a dynamical theory.

Equation 3 represents a good approximation for situations in which momentum relaxation takes place considerably faster than nonthermal reaction. The local equilibrium model becomes increasingly inadequate as these rates approach one another, so that the present form of the steady state theory will be least accurate for systems that involve very rapid reactions. Higher order Chapman-Enskog solutions of the Boltzmann equation, which provide successive degrees of refinement, could be incorporated into the theory. Such modifications would introduce additional mathematical structure in Eq. 3, which is probably not needed except for the description of systems that closely approach true steady state behavior.† This does not occur for any of the cases of present interest (*vide infra*) or, indeed, for any known nuclear recoil reaction system. For this fundamental reason and also because of the crude level of approximation involved in our treatment of nonreactive collisions, the further refinement of Eq. 3 has not yet been considered to be worthwhile.

A paramount advantage of the above formulation for g_A is conceptual. As illustrated below, a single intuitively meaningful parameter, the hot atom temperature, provides a suitable basis for characterizing all of the salient dynamical attributes of hot atom reactions through standard nonthermal rate coefficient formalism. We feel that the principal and possibly unique value of steady state theory modeling is to provide definitive qualitative descriptions for hot atom dynamical phenomena such as reaction energy

†In the present context *steady state* signifies that the reactive and nonreactive relaxation rates exactly balance, so that both g_A and T_A cease to vary with time.

range effects, reactive shadowing, nonthermal kinetic isotope effects, internal state coupling (13,23), and other ambient temperature coupling mechanisms (15). At the present stage of theoretical development, we believe that the calculated results have semiquantitative numerical significance and that the information content of the modeling calculations has surpassed those of all but the most sophisticated hot atom experiments. A limitation of the theory is that it is not useful for direct data manipulation. Even so, the comparison of modeled *vs.* measured results can strongly influence the interpretation of nonthermal experiments.

Returning now to the development of mathematical apparatus, we next consider the important characteristics of the hot atom temperature. Equation 3 is Maxwellian, so that the average instantaneous hot atom laboratory kinetic energy ($\bar{E}_A$) is given by Eq. 4.

$$\bar{E}_A(t) = \frac{3k}{2} T_A(t) \tag{4}$$

Multiplication of this result by the total number of hot atoms [$N_A(t)$] yields an expression for the total kinetic energy (E_A), which can be solved for T_A and differentiated in order to obtain a definition for the rate of T_A relaxation.

$$\frac{dT_A}{dt} = \frac{2}{3k}\left[\left(\frac{1}{N_A}\right)\left(\frac{dE_A}{dt}\right) - \frac{3}{2}\left(\frac{kT_A}{n_A}\right)\left(\frac{dn_A}{dt}\right)\right] \tag{5}$$

The quantity (dE_A/dt) follows from averaging [$\underset{\sim}{P}_A{}^2/2m_A$] over g_A and then taking the time derivative *via* the Boltzmann equation.

$$\frac{dE_A}{dt} = N_A(t)\,[\xi_n(t) + \xi_r(t)] \tag{6}$$

Subscripts r and n in Eq. 6 denote the reactive and non-reactive terms, which follow directly from the equivalent energy expression to Eq. 1.

$$\xi_n = \sum_k \int d\Omega \int d\underset{\sim}{P}_A\,(\underset{\sim}{P}_A{}^2/2m_A) \int d\underset{\sim}{P}_R\,\sigma(j/k)\,(P/\mu)\,[g_A' f_{Rk}^{e'} - g_A f_{Rj}^{e}] \tag{7}$$

$$\xi_r = -\int d\Omega \int d\underset{\sim}{P}_A\,(\underset{\sim}{P}_A{}^2/2m_A) \int d\underset{\sim}{P}_R\,\sigma^*(j)\,(P/\mu)\,g_A f_{Rj}^{e} \tag{8}$$

The hot atom density derivative in Eq. 5 is obtained from Eq. 2.

$$\frac{dT_A}{dt} = \frac{2}{3k}\,[\xi_n(t) + \xi_r(t) + \tfrac{3}{2}k\,T_A(t)\,\kappa(t)] \tag{9}$$

The significance of the terms in Eq. 9 has been considered elsewhere (13). The quantity $\xi_n(t)$ describes the rate of cooling of the E_A distribution induced by elastic and inelastic collisions. The other terms describe the relationship between T_A and the average energy of reaction [$<\epsilon_r>$]. For $<\epsilon_r>$ values smaller than $1.5kT_A$, the combined reactive terms have a net heating effect. For systems characterized by the abrupt onset of highly efficient reaction over a narrow T_A interval, it is possible in principle for the cooling and heating rates to become exactly balanced. The derivative (dT_A/dt) then vanishes, corresponding to the establishment of a true steady state hot atom momentum distribution. As noted above, no real physical system has yet been identified that is believed to fulfill this condition. However, many nuclear recoil systems probably exhibit sufficiently enhanced reaction rates that the inhibition of cooling leads to quasi steady state behavior (13,15,23,31,32-34). The nonthermal ^{18}F vs. $H_2(D_2)$ cases are of this latter type. In general, quasi steady state (time independent) nonthermal rate coefficients can be anticipated to be useful for the analysis of data obtained for such systems.

The time dependence in Eq. 9 originates exclusively from the momentum distribution g_A, which in turn depends upon T_A.

$$\frac{dT_A}{dt} = F(T_A) \tag{10}$$

This single valued, unique relationship can be inverted in order to describe the coupling between differential time and T_A intervals.

$$dt = F^{-1}(T_A)\, dT_A \tag{11}$$

The integrated form of Eq. 11 specifies the elapsed time for a particular T_A change (75).

$$\Delta t = \int_{T_A^i}^{T_A^f} F^{-1}(T_A)\, dT_A \tag{12}$$

Here the i and f superscripts denote initial and final conditions.

For multicomponent systems the only required formalism change involves the inclusion of additional cooling and heating terms.

$$\frac{dT_A}{dt} = \frac{2}{3k} \sum_i X_i[\xi_{ni} + \xi_{ri} + \frac{3kT_A}{2}\kappa_i] \tag{13}$$

Here the index i specifies identity and X_i denotes component mole fraction. Further generalization to allow a more complex initial quantum state dependence requires the inclusion of population distribution corrections and an additional summation over index j. For reactive additives all 3 terms in Eq. 13 must be evaluated, but only the cooling term contributes to (dT_A/dt) for inert moderators.

Steady State Theory Model Calculations.

The simulation procedure first requires the detailed specification of the system to be modeled including sample composition, total pressure, ambient temperature, molecular constants, reactive cross sections and a set of T_A values.† Equation 9 is then evaluated at each T_A based upon suitable approximate representations for Eqs. 2, 7 and 8. The energy dependent elastic and inelastic cross section data required for use in Eq. 7 are not available. Accordingly, in our initial calculations Eq. 7 has been approximated by combining derived expressions for hard sphere elastic energy loss and hot atom *vs*. equilibrium Maxwellian reservoir particle collision frequency. An *ad hoc* zeroth order correction for restoring collisions, which prevents T_A from relaxing below ambient temperature (T), has been incorporated in Eq. 7 through replacement of T_A by the quantity (T_A-T).

Although restoring collisions should be relatively unimportant at energies substantially larger than kT, the above treatment of nonreactive energy loss represents our most drastic assumption. The present description of hot atom moderation is similar to those previously employed in the EW-theory (36,37) and in Monte Carlo model calculations by Koura (15). Classical trajectory calculations constitute the most probable future source for total inelastic cross sections. Available approximate treatments account for the collision energy dependence of elastic cross sections over narrow energy ranges (76). To some extent, for polyatomic substances the errors associated with our incomplete descriptions of elastic and inelastic energy loss are internally compensating. At moderate to large collision energies the elastic cross sections have been overestimated, while the inelastic values have been underestimated.

The average energy transfer [$\langle \Delta E \rangle$] between a hot atom with large average energy ($\bar{E}_o$) and a stationary reservoir particle is given by Eqs. 14 and 15.

$$\langle \Delta E \rangle = \frac{2\mu}{M} \bar{E}_o \qquad (14)$$

†The use of reservoir density reduced time units permits a simplified pressure independent representation of the primary hot atom reactions and momentum relaxation processes.

$$\bar{E}_o = \frac{3}{2}kT_A \tag{15}$$

The hot atom collision frequency (Z_{AR}) follows from integration of the hard sphere elastic cross section over the distribution functions for n_R and n_A (33,34).

$$Z_{AR} = 8\pi\mu n_R \sigma_{AR}[2\pi\mu kT_A]^{-3/2}\left[\frac{m_R}{M} + \frac{m_A T}{MT_A}\right]^{-3/2} \theta^{-2} \tag{16}$$

In Eqs. 14-16 M denotes the total mass of colliding particles and θ is given by Eq. 17.

$$\theta = \mu\left[\frac{1}{m_R kT} + \frac{1}{m_A kT_A}\right] - \frac{\mu\left[\frac{1}{kT} - \frac{1}{kT_A}\right]^2}{\left[\frac{m_R}{kT} + \frac{m_A}{kT_A}\right]} \tag{17}$$

In units of temperature the single component result for $\xi_n(t)$ follows from Eqs. 14-17:

$$\xi_n(t) = -\frac{3k\mu}{M} Z_{AR} T_A \tag{18}$$

Hard sphere elastic cross sections [σ_{AR}] have been obtained from averaged molecular force constants as determined from experimental equation of state and transport property data (76,77). The 24.2 Å^2 value for ^{18}F represents the self-collision elastic cross section for Ne. The mixed values for ^{18}F vs. H_2, Ar and Xe then follow as 25.9 Å^2, 30.1 Å^2 and 36.6 Å^2, respectively.

Reactive cross sections suitable for the simulation of energetic reactions through Eqs. 2 and 8 have been calculated via quasiclassical trajectory methods (78). In a preceding paper Feng et al. (31) demonstrated that knowledge of the potential energy surface topology (79) for the FHH collinear configuration is sufficient to determine reactive cross sections to the presently desired level of accuracy. The quantitative validity of this model for describing epithermal collision processes has been severely curtailed by the approximate formulation for $\xi_n(t)$. For this reason it has not been necessary to obtain refined reactive cross sections corresponding to thermally accessible collisions. These simulations have utilized reactive cross sections computed for Muckerman's optimized LEPS surface V (80). Analytical equations have been fitted to the trajectory cross section data through nonlinear regression analysis.

The final result for $\xi_r(t)$ is given by Eqs. 19 and 20.

$$\xi_r(t) = -\left\{8\pi\mu n_R\theta'\left[2\pi\mu kT_A\right]^{-3/2}\left[\frac{m_R}{M} + \frac{m_A T}{MT_A}\right]^{-3/2}\right.$$

$$\left. \times \int_0^\infty E^2\, \sigma^*(j)\, \exp(-\theta E)\, dE\right\}$$

$$-\left\{\frac{12\pi\mu n_R kTm_A}{M}\left[2\pi\mu kT_A\right]^{-3/2}\left[\frac{m_R}{M} + \frac{m_A T}{MT_A}\right]^{-5/2}\right.$$

$$\left. \times \int_0^\infty E\, \sigma^*(j)\, \exp(-\theta E)\, dE\right\} \qquad (19)$$

$$\theta' = \frac{d\theta}{d(1/kT_A)}$$

$$= \frac{\mu}{m_A} + \frac{2\mu(T_A - T)}{(m_R T_A + m_A T)} + \frac{\mu m_A (T_A - T)^2}{(m_R T_A + m_A T)^2} \qquad (20)$$

Similarly, $\kappa(t)$ follows from Eq. 2.

$$\kappa(t) = \left\{8\pi\mu n_R[2\pi\mu kT_A]^{-3/2}\right.$$

$$\left. \times \left[\frac{m_R}{M} + \frac{m_A T}{MT_A}\right]^{-3/2} \int_0^\infty E\, \sigma^*(j)\, \exp(-\theta E)\, dE\right\} \qquad (21)$$

Equations 16-21 permit the evaluation of Eq. 9 at each desired T_A value. The result is then inverted and inserted into Eq. 12 followed by numerical integration over the desired T_A range in order to characterize the time dependent relaxation of the hot atom momentum distribution.

The Pure H_2 Reaction System.

Calculated T_A relaxation results obtained at 300°K ambient temperature have been shown in Table I and Fig. 1. In order to remove an artificial dependence upon the reservoir pressure, density reduced time units [$(t \cdot n_R)$, molecule sec cm^{-3}] have been employed throughout the present discussion. As shown in

Table I. Reaction Rate Coefficients and T_A Relaxation Results for Pure H_2 at 300°K Ambient Temperature.

T_A [a] (°K) x 10^{-3}	$\kappa(T_A)$ [b] (cm^3 $molecule^{-1}$ sec^{-1}) x 10^9	$t \cdot n_R$ (molecule sec cm^{-3}) x 10^{-9}
10000	-	-
9000	0.077	0.0215
8000	0.090	0.0470
7000	0.107	0.0780
6000	0.131	0.116
5000	0.164	0.167
4000	0.216	0.235
3000	0.299	0.338
2000	0.452	0.517
1000	0.763	0.949
500	0.970	1.55
400	0.985	1.76
300	0.964	2.05
200	0.876	2.48
100	0.658	3.33
50	0.448	4.42
40	0.388	4.84
30	0.318	5.46
20	0.236	6.47
10	0.134	8.79
5	0.076	12.1
4	0.064	13.5
3	0.052	15.5
2	0.040	18.9
1	0.028	26.5
0.5	0.022	38.6

a. The initial value corresponding to each tabulated final T_A is given by the preceding entry.

b. *Cf*. Fig. 2.

Fig. 1 T_A exhibits an initially rapid decrease for pure H_2, levelling off at values below 10^5°K for relaxation times longer than 5 x 10^9 molecule sec cm^{-3}. The T_A relaxation results for mixed Ar/H_2 systems are considered in the following section.

The reduced time formulation for the nonthermal reaction rate is given by Eq. 22,

$$\frac{dn_A}{dt} = -\kappa(t)\, n_A(t) \qquad (22)$$

which can be integrated in order to obtain the time dependence of the hot atom density.

$$n_A(t) = n_A^\circ \exp\left[-\int_0^t \kappa(t)\, dt\right] \qquad (23)$$

Time dependent rate coefficients calculated <u>via</u> Eq. 21 (<u>cf.</u> Fig. 2) are numerically integrated using Eq. 23 in order to obtain $n_A(t)$. Reaction rates, which then follow from Eq. 22, are integrated in a similar fashion in order to produce time dependent hot yields [$Y(t^f)$].

$$Y(t^f) = \int_0^{t^f} \kappa(t)\, n_A(t)\, dt \qquad (24)$$

The time dependent hot atom densities, reaction rates and yields depicted in Figs. 3 and 4 reveal that nonthermal reaction commences almost immediately in pure H_2 because of the large magnitude of the high energy cross section. The maximum reaction rate occurs at an elapsed time of <u>ca.</u> 1.0 x 10^9 molecule sec cm^{-3}, corresponding to a real relaxation time of <u>ca.</u> 30 picoseconds at 300°K and 1000 Torr pressure. Ninety percent of the hot yield is produced at times shorter than 3 x 10^9 molecule sec cm^{-3}, and the reaction has nearly ceased following an elapsed time of 5 x 10^9 molecule sec cm^{-3}.

A 99.8% calculated total hot yield follows somewhat arbitrarily by truncating Eq. 24 at a total elapsed time (t^f) value of 2.65 x 10^{10} molecule sec cm^{-3}. From Table I this $Y(t^f)$ cutoff corresponds to an equivalent final hot atom temperature (T_A^f) of 1000°K. In agreement with the range of plausible values deduced from experiment (<u>23</u>), the thermalized ^{18}F atom fraction in pure H_2 is negligibly small. Because the present $\xi_n(t)$ model exaggerates the efficiency of cooling processes at large T_A, this calculation tends to overestimate the fraction of the initial ^{18}F atom distribution that fails to undergo nonthermal reaction.

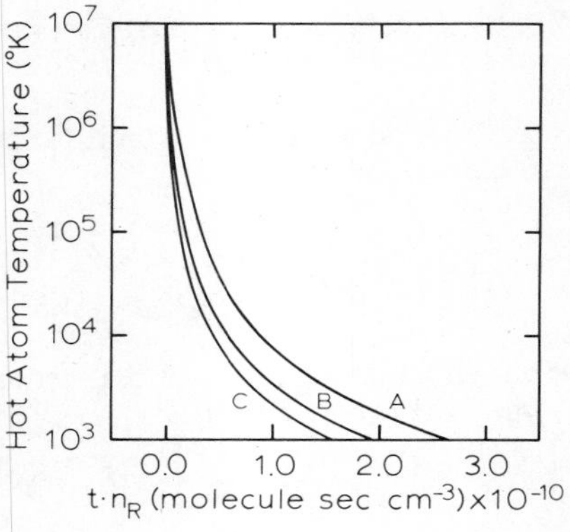

Figure 1. Relaxation time behavior of the hot atom temperature. Key to curves: A, Pure H_2; B, 50 mol %; C, 99 mol % Ar-moderated H_2 at 300°K ambient temperature.

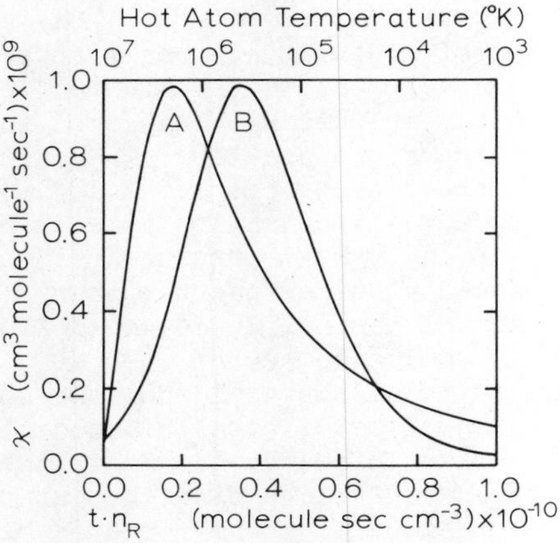

Figure 2. Time (curve A) and T_A (curve B) dependent nonthermal rate coefficients for the $^{18}F + H_2$ reaction at 300°K ambient temperature

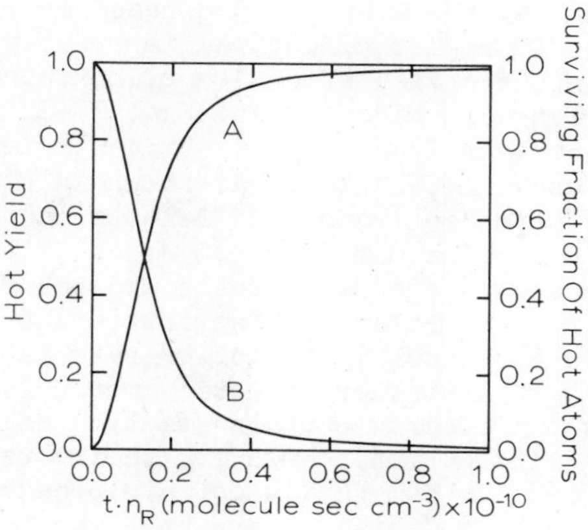

Figure 3. Time-dependent hot yields (curve A) and surviving ^{18}F atom fractions (curve B) for pure H_2 at 300°K ambient temperature

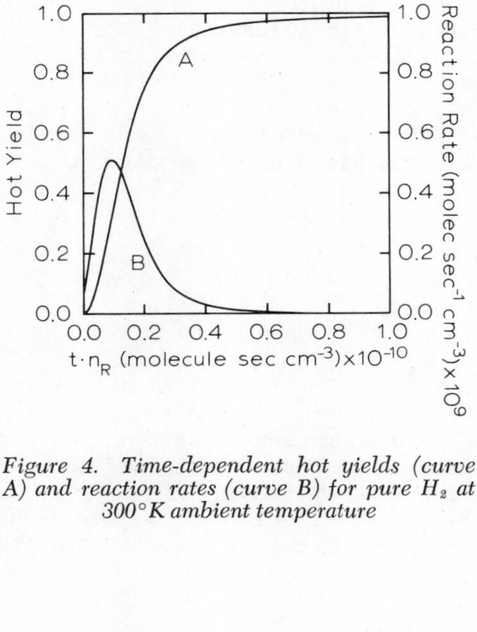

Figure 4. Time-dependent hot yields (curve A) and reaction rates (curve B) for pure H_2 at 300°K ambient temperature

From Figs. 2 and 4 the rapid decline in $n_A(t)$ at large time effectively determines the width of the reaction rate distribution (cf. Eq. 22). A problem of central importance in hot atom chemistry concerns the characterization of nonthermal reactive collision energy ranges. Figure 1 and Eqs. 10-12 show that the relationship between T_A and time, while strongly nonlinear, is single-valued and unique. Thus, as shown in Fig. 2 and Table I, the time dependent rate coefficients can also be expressed in terms of equivalent T_A dependences. In a similar fashion the temperature dependent yields and the fall off in hot atom density for the relaxing distribution have been shown in Fig. 5, and the corresponding reaction rate data in Fig. 6. Since the average ^{18}F laboratory kinetic energy is simply $1.5kT_A$, these results clearly contain reactive collision energy dependence information. However, their precise quantitative significance is somewhat uncertain because the calculated reaction rates and yields represent the behavior of relaxing Maxwellian distributions of ^{18}F atom laboratory kinetic energies. In order to specify the corresponding center-of-mass reactive collision energy distributions, the coordinate transformation and distribution unfolding problems must be solved. This has been accomplished analytically for the case of small mass hot atoms reacting with large mass reservoir species (13). Neither analytical nor numerical solutions are yet available for the ^{18}F vs. H_2 system. A slight additional ambiguity arises from the intrinsic nonlinearity of the T_A vs. relaxation time relationship, which gives rise to exaggerated relative weighting of low temperature reaction processes in T_A dependent reaction rate or yield plots.

A simple procedure for illustrating the significance of the T_A dependent results has been shown in Fig. 7. The quantity $-(dY/dLogT_A)$ follows from direct numerical differentiation of the calculated $Y(T_A^f)$ results. Integration of this distribution over any interval on the Log T_A scale gives the fraction of the hot yield corresponding to the specified T_A range.

$$Y(T_A^{min}) = \int_{T_A^{max}}^{T_A^{min}} \left(\frac{dY}{dLogT_A}\right) dLog\ T_A \qquad (25)$$

Plots of $-(dY/dLogT_A)$ vs. Log T_A thus clearly reveal the nature of the average ^{18}F laboratory kinetic energy dependence of the hot yield. This data presentation technique is especially useful for the analysis of results obtained for multicomponent and multichannel reaction systems.

We next consider how closely the nuclear recoil ^{18}F vs. H_2 system may have approached true high temperature steady state

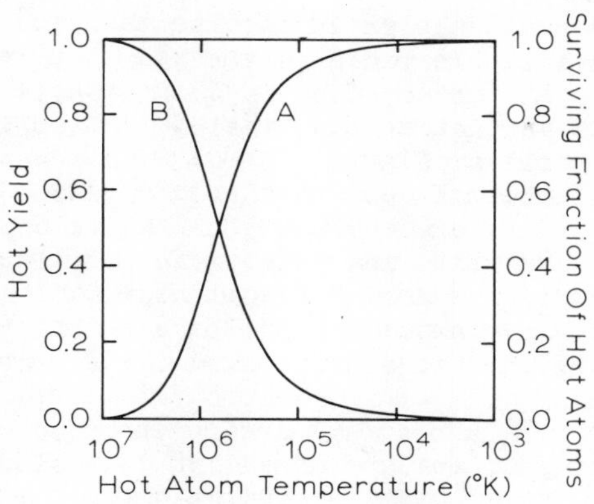

Figure 5. T_A *dependent hot yields (curve A) and surviving* ^{18}F *atom fractions (curve B) for pure* H_2 *at 300°K ambient temperature*

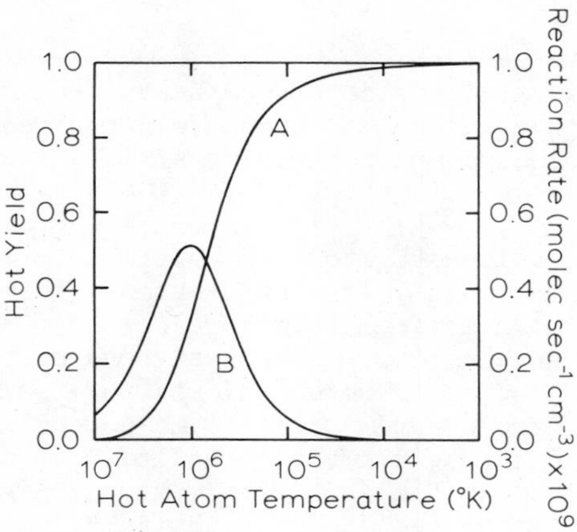

Figure 6. T_A *dependent hot yields (curve A) and reaction rates (curve B) for pure* H_2 *at 300°K ambient temperature*

conditions. An essential prerequisite to the establishment of a time invariant momentum distribution is that the reactive collision efficiency must be sufficiently large to allow dynamical balancing between heating and cooling effects. Some T_A dependent collision efficiencies have been shown in Table II. The *ca.* 25% maximum efficiency corresponds to a T_A of roughly 10^5°K and exceeds the 300°K equilibrium value by only fivefold. Detailed analysis of the cooling and heating terms in Eq. 9 provides a more critical test for the existence of a steady state. For nuclear recoil ^{18}F reacting with pure H_2 the combined heating rates never exceed 20% of the cooling rate.

Table II. Hot Atom Temperature Dependent Collision Frequencies And Reactive Collision Efficiencies.

T_A (°K) x 10^{-6}	$Z_{AR}(T_A)$ (cm³ molecule⁻¹ sec⁻¹) x 10^{10}	$\left(\frac{\kappa}{Z_{AR}}\right)$ (%)
1.00	89.0	8.6
0.50	63.0	15.4
0.100	28.5	23.1
0.050	20.4	22.0
0.0100	10.0	13.4
0.00500	7.8	9.7
0.00100	5.4	5.2
0.00030[b]	4.9	4.1
0.00030[c]	-	4.8

a. Rate coefficient values have been taken from Table I.
b. Obtained by integrating the reactive cross section over Maxwellian hot atom and reservoir particle distributions.
c. Experimental result. Ref. (21).

Koura (15) has investigated the possible formation of high temperature steady states in the nonthermal $^{18}F + H_2$ system using a Monte Carlo numerical procedure for solving the time dependent Boltzmann equation. Reactive cross section data reported from this laboratory were employed together with an energy dependent hard sphere model. Time dependent momentum relaxation, reaction rate and yield results were obtained for a variety of assumed initial hot atom momentum distributions. The momentum relaxation results are particularly revealing. Since the Monte Carlo distributions can be reasonably well represented by the local equilibrium model, these results corroborate the essential validity of the present calculational procedure. To the extent that direct comparisons can be carried out, the Monte Carlo and local equilibrium results exhibit good agreement. Both methods predict essentially 100% hot yield in pure H_2 together with the failure to establish a high temperature steady state under nuclear recoil conditions. A surprising and rather remarkable Monte Carlo result is the indication that changes in the initial hot atom momentum distribution can significantly affect the reaction dynamics. For highly energetic multichannel systems characterized by large total hot reactivity (23-27,62), it follows that different irradiation techniques may conceivably lead to different product distributions.

The present pure H_2 calculations illustrate the significance of non-Boltzmann rate coefficients for hot atom reaction systems. Since $\kappa(t)$ is a strongly varying function, Eq. 22 cannot be approximated by a first order linear differential rate expression. The mean hot atom reactive lifetime is given by Eq. 26.

$$\tau = \langle t \rangle = \frac{\int_0^{t^f} t\,\kappa(t)\,n_A(t)\,dt}{\int_0^{t^f} \kappa(t)\,n_A(t)\,dt} \qquad (26)$$

To the extent that $\kappa(t)$ is constant throughout a time interval that is comparable to τ, then the correct time dependent rate expression can be replaced by an approximate quasi steady state analog. Even in the absence of a true steady state, certain features of an average rate coefficient formulation will continue to be applicable (23,25,27,31,32). This conclusion is insensitive to the present assumptions concerning nonreactive scattering. It would be strengthened by the inclusion of a more realistic representation of the energy dependent nonreactive cross section, since a relative reactivity increase would lead to compression of the time and T_A ranges sampled by nonthermal reaction.

Moderated H_2 Reaction Systems.

Two component mixture calculations have been carried out *via* Eq. 13 for inert gas moderated nuclear recoil ^{18}F *vs.* H_2 systems (71). Care must be exercised in setting up multicomponent simulations in order to insure that the composition dependent relaxation times and reaction rates can be directly intercompared. Unless density reduced time units have been utilized, the modeled systems must correspond to conditions of constant total pressure. The T_A relaxation results shown in Fig. 1 demonstrate large cooling rate increases accompanying the addition of inert moderator. Time dependent reaction rates and yields have been shown in Figs. 8-10 for 50 mole % Ar and 99 mole % Ar and Xe moderated H_2, respectively. Total hot yields, which have been summarized in Table III, followed from numerical integration of the reaction rates to an equivalent T_A^f of 1000°K. The $Y(t^f)$ data shown in Figs. 8-10 have been plotted as fractions of the tabulated total yields obtained for 300°K ambient temperature.

The total yields exhibit pronounced negative dependences upon moderator concentration similar to those observed in actual experiments. The expected sensitivity to moderator efficiency is also apparent. The rationale for simulating the nonthermal $^{18}F + H_2$ reaction at different ambient temperatures is considered below.

Table III. Moderator Concentration Dependence of Total Hot Yields.

Moderator Concentration[a] (Mole %)	Total Hot Yield (%)			
	Ar[b]	Ar[c]	Xe[b]	He[b]
0.0	99.5	99.0	99.5	99.5
10.0	98.5	97.4	98.9	99.0
30.0	93.3	90.3	96.4	96.8
50.0	81.1	75.8	89.7	91.0
70.0	58.6	52.4	73.1	75.7
90.0	23.2	19.8	34.5	37.1
99.0	2.5	2.1	4.1	4.5

a. Hard sphere average logarithmic energy decrements (36) for H_2, He, Ar and Xe toward ^{18}F atoms are 0.208, 0.385, 0.675 and 0.251.
b. 300°K ambient temperature.
c. 10°K ambient temperature.

We next examine the hot reaction rate perturbations caused by the presence of inert moderators. For specified values of the nonthermal rate coefficient and hot atom density, the reaction rate is directly proportional to the H_2 concentration. In order to obtain information pertaining to reactive shadowing effects, it is necessary to isolate the rate changes that result from variations in the hot atom density. This can be accomplished in a qualitative fashion by scaling the calculated absolute rates to conditions of uniform H_2 density (n_{H_2}). The scaled rates shown in Figs. 8-10 represent absolute values multiplied by the correction factor ($1/X_{H_2}$). Comparison with the pure H_2 results shown in Fig. 4 reveals the occurrence of large scaled rate increases accompanying the addition of moderator. In conventional terminology these changes reflect

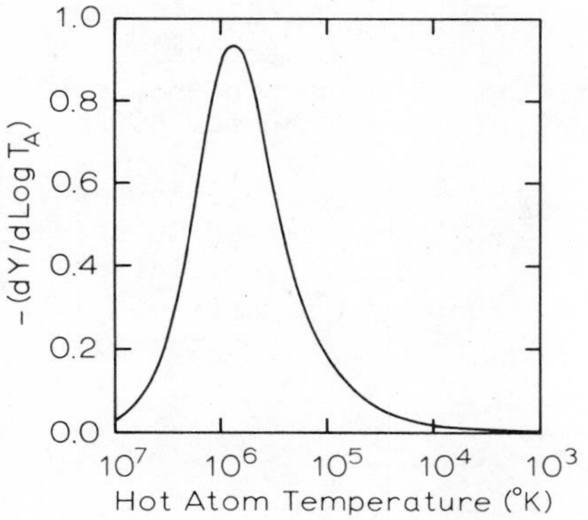

Figure 7. The −(dY/d log T_A) distribution for pure H_2 at 300°K ambient temperature

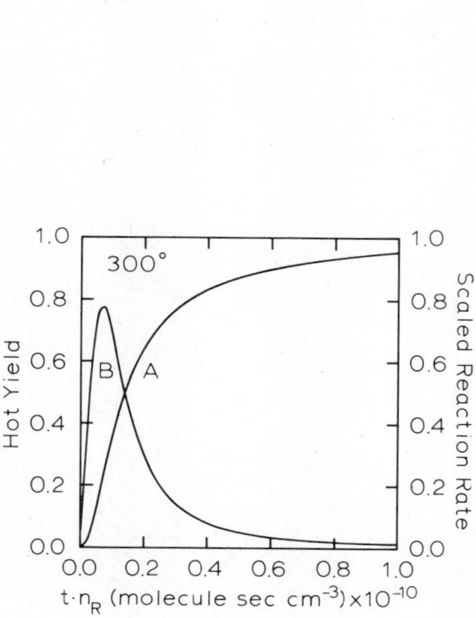

Figure 8. Time-dependent scaled hot yields (curve A) and reaction rates (curve B) for 50 mol % Ar-moderated H_2 at 300°K ambient temperature

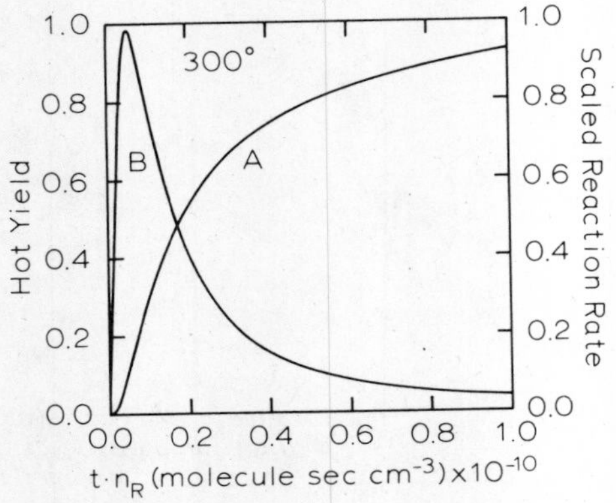

Figure 9. Time-dependent scaled hot yields (curve A) and reaction rates (curve B) for 99 mol % Ar-moderated H_2 at 300°K ambient temperature

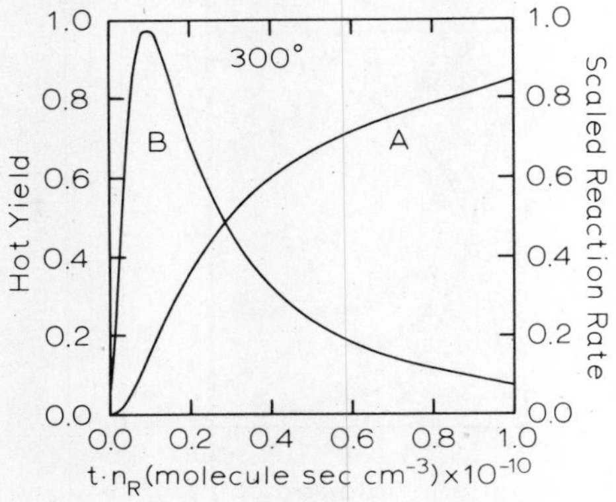

Figure 10. Time-dependent scaled hot yields (curve A) and reaction rates (curve B) for 99 mol % Xe-moderated H_2 at 300°K ambient temperature

the importance of H_2 reactive self-shadowing, which results from the rapid and efficient depletion of the hot atom density at large collision energy in the absence of moderator.

The nonthermal reaction rates exhibit significant time dependence variations associated with the presence of inert moderators. From Figs. 4, 8 and 9 the addition of 99 mole % Ar to H_2 causes the maximum reaction rate to be established roughly twofold more quickly. Presumably as a result of the rough similarity in moderating efficiencies between H_2 and Xe (cf. Table III), a similar effect does not occur in 99 mole % Xe moderated H_2. This shortening of the nonthermal reaction induction period seems primarily to reflect large total cooling rate increases.

Table IV and Figs. 4 and 8-10 show that the progressive addition of moderator also causes the reaction rate curves to become increasingly skewed to longer relaxation times. The tabulated τ values were obtained from Eq. 26 with a t^f cutoff corresponding to 1000°K T_A^f. A convenient measure of the rate distribution width is provided by $\langle t^2 \rangle$.

$$\langle t^2 \rangle = \frac{\int_0^{t^f} t^2 \kappa(t)\, n_A(t)\, dt}{\int_0^{t^f} \kappa(t)\, n_A(t)\, dt} \tag{27}$$

As revealed by these τ and $\langle t^2 \rangle$ results, the skewing of the rate distributions is strongly enhanced for Xe relative to Ar. As the total yield falls off with increasing moderator concentration, the hot atom density survives to sample the rate coefficient at longer relaxation times than is possible in pure H_2. Since rapid cooling through the epithermal region limits the effectiveness of these "slow" reactions, the skewing effect is less pronounced for Ar in comparison with Xe. We conclude that in efficiently moderated systems the length of the induction period required to establish the maximum reaction rate decreases, whereas the average reactive lifetime τ increases.

The analysis of T_A dependence results reveals the physical basis for the above noted changes in the hot atom reactive lifetime distributions, including the ambient temperature effects shown in Tables III and IV. The T_A dependent scaled reaction rates, yields and $-(dY/d\mathrm{Log}T_A)$ distributions have been plotted in Figs. 11-13. The rate data have been scaled as noted previously. Yields have been plotted as fractions of the total values calculated for the respective 300°K reaction systems. The $-(dY/d\mathrm{Log}T_A)$ distributions then follow from direct numerical differentiation of the scaled yields. The most probable $-(dY/d\mathrm{Log}T_A)$ values generally exhibit dis-

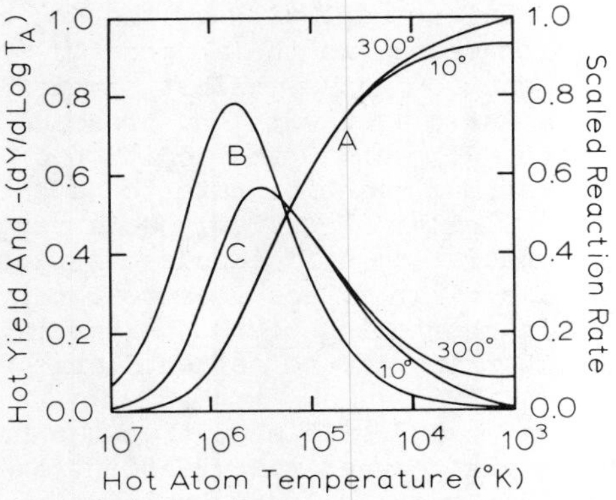

Figure 11. T_A *dependent scaled hot yields (curves A), reaction rates (curve B), and* −(dy/d *log* T_A) *distributions (curves C) for 50 mol % Ar-moderated* H_2

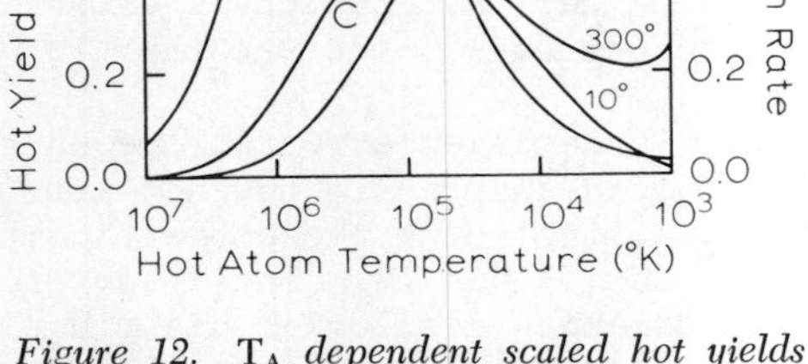

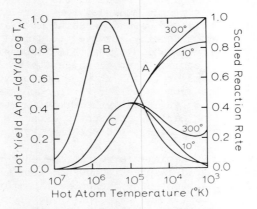

Figure 12. T_A *dependent scaled hot yields (curves A), reaction rates (curve B), and* −(dY/d *log* T_A) *distributions (curves C) for 99 mol % Ar-moderated* T_2

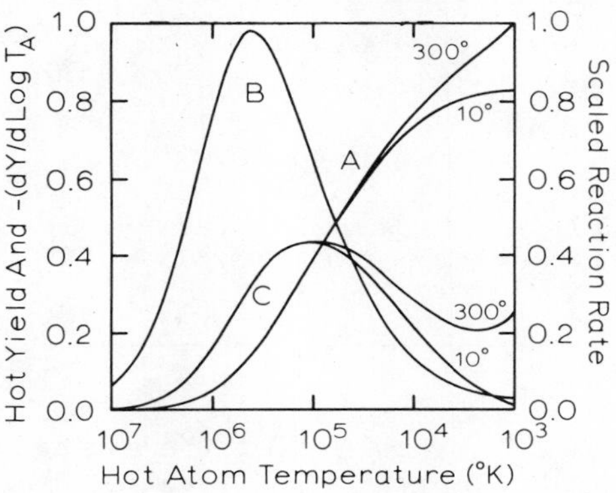

Figure 13. T_A dependent scaled hot yields (curves A), reaction rates (curve B), and —(dY/d log T_A) distributions (curves C) for 99 mol % Xe-moderated H_2

Table IV. Properties of Time Dependent Nonthermal Reaction Rate Distributions at 300°K and 10°K Ambient Temperatures.

Argon Concentration (Mole %)	Ambient Temperature 300°K		10°K	
	τ	$\langle t^2 \rangle$	τ	$\langle t^2 \rangle$
	(molecule sec cm^{-3}) x 10^{-9}			
0.0	1.68	0.59	1.64	0.51
10.0	1.80	0.78	1.70	0.60
30.0	2.15	1.23	1.85	0.76
50.0	2.53	1.65	1.99	0.86
70.0	2.85	1.97	2.07	0.89
90.0	3.10	2.17	2.10	0.88
99.0	3.19	2.22	2.10	0.87
99.0[a]	5.32	6.24	3.47	2.36

a. Xe moderator.

placement to reduced T_A, showing that the average ^{18}F laboratory kinetic energy sampled by nonthermal reaction decreases with increasing moderator concentration. At Ar concentrations below 50 mole % this effect is accompanied by simultaneous increases in the $-(dY/dLogT_A)$ distribution widths relative to the respective most probable T_A values.

As noted above we have not yet attempted to calculate average reactive collision energies in the center-of-mass coordinate system. The closest present approximation to this quantity follows from the most probable hot atom temperature [$\langle T_A \rangle$] obtained for each $-(dY/dLogT_A)$ distribution. These $\langle T_A \rangle$ results have been expressed as equivalent average ^{18}F laboratory kinetic energies [$1.5k\langle T_A \rangle$] and center-of-mass collision energies $\langle \varepsilon \rangle$ in Table V. As estimated in this fashion, the most probable reactive collision energy varies from 9.7 eV $molecule^{-1}$ in pure H_2 to 1.2 eV $molecule^{-1}$ in 99 mole % He, Ar or Xe moderated H_2. Reactive collision energy distribution widths at half maximum, which have been estimated in an analogous manner, increase by more than threefold accompanying the addition of 99 mole % He, Ar or Xe to pure H_2. The above results clearly reveal the nature of the reaction energy range changes that take place in moderated nuclear recoil hot atom chemistry experiments. In situations characterized by broad $\kappa(T_A)$ functions and large intrinsic reactivity, the addition of moderator increases the relative width of the reactive collision energy distribution and diminishes its average magnitude.

Table V. Properties of $-(dY/dLogT_A)$ Distributions at 300°K Ambient Temperature.

Moderator (Concentration, Mole %)	$\langle T_A \rangle$ (°K) x 10^{-5}	$\frac{3k}{2}\langle T_A \rangle$	$\langle \epsilon \rangle$ (eV Molecule^{-1})	ϵ Distribution Width
Ar(0.0)	7.48	96.7	9.7	22.2
(50.0)	3.27	42.3	4.3	15.2
(99.0)	0.89	11.5	1.2	9.4
Xe (99.0)	0.89	11.5	1.2	9.4
He (99.0)	0.89	11.5	1.2	9.4

Another characteristic of the $-(dY/dLogT_A)$ distributions that we believe to be general for systems of this type follows from Figs. 12 and 13. These distributions approach a pure moderator limiting form that is independent of moderator identity. The average reactive collision energy thus does not decrease indefinitely with increasing moderation. For the nonthermal ^{18}F vs. Ar/H_2 system it achieves an essentially constant value at Ar concentrations in excess of ca. 90 mole % (cf. Table IV). The form of the limiting $-(dY/dLogT_A)$ distribution is controlled by $\kappa(T_A)$, which thus governs the effective collision energy resolution that can be achieved in moderated hot atom chemistry experiments. Within the framework of the present nonreactive collision model, in the complete absence of reactive shadowing the limiting distribution should be the same for all inert moderators.

We now consider the upturn in $-(dY/dLogT_A)$ observed at small T_A values for highly moderated 300°K ambient temperature systems. The increases that occur for T_A values below 5000°K indicate the onset of contamination by epithermal reactions. Both the $\xi_n(t)$ and hot atom collision frequency (Z_{AR}) models include *ad hoc* zeroth order corrections for translational restoring collisions. The calculated low energy reaction yields can be reduced--though not eliminated--through removal of the restoring correction term from $\xi_n(t)$. However, a more effective suppression technique involves reduction of the ambient temperature to 100°K or below. From Table III and Fig. 12 the epithermal contamination approaches 20% of the total hot yield at 300°K and 99 mole % Ar concentration. In order to illustrate the quantitative importance of this effect, the 10°K $Y(T_A^f)$ and $-(dY/dLogT_A)$ results shown in Figs. 11-13 have been scaled relative to the corresponding 300°K $Y(T_A^f)$ values. Because of the limitations inherent in the present theoretical model, the numerical significance of these calculated epithermal yields is uncertain. However, the qualitative indications are valid (i) that epithermal reactions may cause interference in moderated hot atom chemistry experiments, (ii) that the epithermal yields will be strongly dependent upon the ambient temperature, and (iii) that these conclusions have general applicability except for situations that involve thermally inaccessible hot reactions. Energetic ^{3}H and ^{18}F atomic substitution processes, for example, have center-of-mass threshold energy requirements of 1.5 eV or more (24). The efficacy of competitive reaction techniques for controlling epithermal contamination will be determined in future model calculations.

An important consequence of the results shown in Table V involves the possible significance of hot yields measured in moderator experiments with polyatomic reactants. For nonthermal reactions that sample broad $\kappa(T_A)$ under conditions of large total hot reactivity, the relative decrease in the average reactive collision energy at the moderated limit may approach one order of magnitude. The secondary decomposition behavior of a polyatomic primary reaction product would then likely exhibit a marked dependence upon the moderator concentration. The present results therefore demonstrate that such experiments must include the characterization and monitoring of all open secondary decomposition channels under the full range of investigated conditions.

Finally, we briefly consider the physical significance of the T_A dependent scaled reaction rates shown in Figs. 6 and 11-13. Accompanying the addition of inert moderator, the scaled rates consistently increase at all T_A values below 10^{7}°K. This effect simply reflects the fact that the hot atom density is not significantly depleted during the relaxation process at large moderator concentration. It is particularly dramatic at small T_A values that do not contribute appreciably to the hot yield in pure H_2. However, relative to pure H_2 even the most probable scaled rate exhibits a twofold increase in the presence of 99 mole % Ar. These results provide a useful basis for elucidating the qualitative nature and role of reactive shadowing effects in single-channel nonthermal reaction systems. More complex situations can best be analyzed in terms of the dynamic interplay between the hot atom density and the available nonthermal rate coefficients.

In future research we plan to seek the development of improved models for elastic and inelastic scattering and to apply the local equilibrium steady state theory for additional simulations of nuclear and photodissociation recoil ^{3}H, ^{18}F, and ^{38}Cl reaction systems.

Acknowledgement.

The authors express appreciation to Dr. James Harrison of the Crocker Nuclear Laboratory for invaluable assistance with computer programming; to the U.C.D. Computer Center for assistance with computing costs; to Dr. James Muckerman for sponsoring the quasiclassical trajectory reaction cross section calculations; and to the U.S. Energy Research and Development Administration for financial support under contract E-(04-3)-34, agreement no. 158.

H. Literature Cited.

1. Farrar, J. M., and Lee, Y. T., this volume.
2. Bogen, D., and Setser, D. W., this volume.
3. See for example State to State Chemistry (American Chemical Society Symposium Series Monograph, Washington, D.C. 1977), Brooks, P., Editor.
4. Ding, A.M.G., Kirsch, L. J., Perry, D. S., Polanyi, J. C., and Schreiber, J. L., Disc. Faraday Soc. (1973), 55, 252, and references therein.
5. Letokhov, V. S., and Moore, C. B., Sov. J. Quant. Electron. (1976), 6, 129, 259.
6. Bergmann, K., Leone, S. R., and Moore, C. B., J. Chem. Phys. (1975), 63, 4161.
7. Houston, P. L., Chem. Phys. Letters (1977), 47, 137.
8. Stedman, D. H., and Setser, D. W., Prog. Reaction Kinetics (1971), 6, 193.
9. Riley, M. E., and Matzen, M. K., J. Chem. Phys. (1975), 63, 4787.
10. Sullivan, J. H., Invited Lecture Presented at 169th National American Chemical Society Meeting, Philadelphia, 1975.
11. Greiner, N. R., IEEE J. Quantum Electron. (1973), QE-9, 1123.
12. Gerber, R. A., Patterson, E. L., Blair, L. S., and Greiner, N. R., Appl. Phys. Letters (1974), 25, 281.
13. Keizer, J., J. Chem. Phys. (1973), 58, 4524.
14. Keizer, J., J. Chem. Phys. (1972), 56, 5958.
15. Koura, K., J. Chem. Phys. (1977), 66, 4078; ibid. (1976), 65, 3883.
16. Menzinger, M., and Wolfgang, R., J. Chem. Phys. (1969), 50, 2991.
17. See for example Proceedings of the VI International Symposium on Molecular Beams (Noordwijkerhout, Netherlands, 1977).
18. Fluendy, M.A.D., and Lawley, K. P., Chemical Applications of Molecular Beam Scattering (Chapman and Hall, London, 1973).

19. Spicer, L., and Rabinovitch, B. S., Ann. Rev. Phys. Chem. (1970), 21, 349.
20. Wolfgang, R., Prog. React. Kinetics (1965), 3, 97.
21. Mo, S. H., Grant, E. R., Little, F. E., Manning, R. G., Mathis, C. A., Werre, G. S., and Root, J. W., this volume.
22. Rowland, F. S., Rust, F., and Frank, J. P., this volume.
23. Grant, E. R., and Root, J. W., J. Chem. Phys. (1976), 64, 417.
24. Manning, R. G., Krohn, K. A., and Root, J. W., Chem. Phys. Letters (1975), 35, 544, and references therein.
25. Manning, R. G., Mo, S. H., and Root, J. W., J. Chem. Phys. (1977), 67, 636.
26. Manning, R. G., and Root, J. W., J. Chem. Phys. (1976), 64, 4926.
27. Manning, R. G., and Root, J. W., J. Phys. Chem. (1975), 79, 1478; ibid. (1977), 81, in press.
28. Riley, S. J., and Wilson, K. R., Chem. Soc. Faraday Disc. (1972), 53, 132.
29. Ting, C. T., and Weston, R. E., J. Phys. Chem. (1973), 77, 2257.
30. Chapman, S., and Bunker, D. L., J. Chem. Phys. (1975), 62, 2890.
31. Feng, D. F., Grant, E. R., and Root, J. W., J. Chem. Phys. (1976), 64, 3450.
32. Stevens, D. J., and Spicer, L. D., J. Chem. Phys. (1976), 64, 4798.
33. Grant, E. R., Ph.D. Dissertation (University of California, Davis, 1974, University Microfilms No. 75-15434).
34. Root, J. W., Nuclear Recoil Studies of Hydrogen Abstraction Reactions by Atomic Fluorine (U.S. Atomic Energy Commission Technical Report No. UCD-34P158-74-2, University of California, Davis, 1975).
35. Knierim, K. D., Grant, E. R., and Root, J. W., manuscripts in preparation.
36. Estrup, P. J., and Wolfgang, R., J. Amer. Chem. Soc. (1960), 82, 2661, 2665.
37. Wolfgang, R., J. Chem. Phys. (1963), 39, 2983.
38. Estrup, P. J., J. Chem. Phys. (1964), 41, 164.
39. Chang, H. M., and Wolfgang, R., J. Phys. Chem. (1971), 75, 3042.
40. Root, J. W., and Rowland, F. S., Radiochimica Acta (1968), 10, 104.
41. Root, J. W., and Rowland, F. S., J. Chem. Phys. (1967), 46, 4299.
42. Root, J. W., and Rowland, F. S., J. Phys. Chem. (1970), 74, 451.
43. Chapin, D. M., and Kostin, M. D., J. Chem. Phys. (1968), 48, 3067; ibid. (1967), 46, 2506.
44. Felder, R. M., and Kostin, M. D., J. Chem. Phys. (1967), 46, 3185; ibid. (1965), 43, 3082.

45. Kostin, M. D., J. Appl. Phys. (1966), 37, 791, 3801; ibid. (1965), 36, 850.
46. Porter, R. N., J. Chem. Phys. (1966), 45, 2284.
47. Porter, R. N., Disc. Faraday Soc. (1968), 44, 84.
48. Porter, R. N., and Kunt, S., J. Chem. Phys. (1970), 52, 3240.
49. Adams, J. T., and Porter, R. N., J. Chem. Phys. (1973), 60, 3354; ibid. (1973), 59, 4105.
50. Kuppermann, A., and White, J. M., J. Chem. Phys. (1966), 44, 4352.
51. Kuppermann, A., in Fast Reactions and Primary Processes in Chemical Kinetics (Interscience, New York, 1967), Claesson, S., Editor.
52. Greene, E. F., and Kuppermann, A., J. Chem. Educ. (1968), 45, 361.
53. Gann, R. G., Ollison, W. M., and Dubrin, J., J. Chem. Phys. (1971), 54, 2304.
54. Melton, L. A., and Gordon, R. G., J. Chem. Phys. (1969), 51, 5449.
55. Malerich, C. J., and Spicer, L. D., J. Chem. Phys. (1973), 59, 1577.
56. Truhlar, D. G., and Wyatt, R. E., Ann. Rev. Phys. Chem. (1976), 27, 1.
57. Urch, D. S., Radiochimica Acta (1970), 14, 10.
58. Urch, D. S., M.T.P. Internat. Rev. Sci. Ser. 1 (1972), 8, 149.
59. Carlson, R., Freedman, A., Press, G. A., and Malcolme-Lawes, D. J., Radiochimica Acta (1972), 18, 167.
60. Malcolme-Lawes, D. J., J. Chem. Phys. (1972), 57, 2476, 2481; ibid. (1972), 56, 3442.
61. Johnston, A. J., and Urch, D. S., J. Chem. Soc. Faraday I (1974), 70, 369.
62. Krohn, K. A., Parks, N. J., and Root, J. W., J. Chem. Phys. (1971), 55, 5771, 5785.
63. Urch, D. S., Radiochemistry (London) (1975), 2, 1.
64. Nogar, N. S., Dewey, J. K., and Spicer, L. D., Chem. Phys. Letters (1975), 34, 98.
65. Nogar, N. S., and Spicer, L. D., J. Phys. Chem. (1976), 80, 1736.
66. Nogar, N. S., and Spicer, L. D., J. Chem. Phys. (1977), 66, 3624.
67. Malcolme-Lawes, D. J., J. Chem. Phys. (1972), 57, 5522.
68. Malcolme-Lawes, D. J., J. Chem. Soc. Faraday II (1975), 71, 1183.
69. Valencich, T., private communication.
70. Grant, E. R., Knierim, K. D., and Root, J. W., submitted for publication.
71. Knierim, K. D., and Root, J. W., submitted for publication.
72. Grant, E. R., and Root, J. W., J. Chem. Phys. (1975), 63, 2970.

73. Snider, N. S., and Ross, J., J. Chem. Phys. (1966), 44, 1087.
74. Huang, K., Statistical Mechanics (Wiley, New York, 1963), Ch. 6.
75. Wemmer, D., and Keizer, J., unpublished results.
76. Hirschfelder, J. O., Curtiss, C. F., and Bird, R. B., Molecular Theory of Gases and Liquids (Wiley, New York, 1964).
77. Root, J. W., Ph.D. Dissertation (University of Kansas, Lawrence, 1964, University Microfilms No. 65-7004).
78. Muckerman, J. T., J. Chem. Phys. (1972), 57, 3388.
79. Bender, C. F., and Schaefer, H. F., this volume.
80. Root, J. W., and Muckerman, J. T., unpublished results.

Structure

13

Electron Spin Resonance Studies of Fluorine-Containing Radicals in Single Organic Crystals

LOWELL D. KISPERT

Chemistry Department, University of Alabama, Tuscaloosa, Alabama 35486

Electron spin resonance (esr) studies of fluorine-containing organic free radicals in X- and γ-irradiated single crystals has enabled an experimental measurement to be made of the isotropic and anisotropic hyperfine splittings and direction cosines for fluorine as well as for other nuclei containing a non-zero magnetic moment. From such data, the structure of fluorine-containing radicals has been deduced, the unpaired electron density in the 2pπ-orbitals has been calculated, the intramolecular motion exhibited by substituents of the radicals has been determined and the reaction coordinates measured for the decay and formation of radicals in oriented matrices.

Since a large body of information is known, it is the purpose of this chapter to review the important structural features of fluorine-containing radicals oriented in irradiated single crystals. It will also be demonstrated that these radicals undergo a dynamic motion over a wide range of temperatures, that oxygen can diffuse into a number of these irradiated crystals forming peroxy radicals and that a complete radical formation and decay mechanism for oriented fluorinated radicals is not known.

Historically, the first complete esr study of a fluorine-containing radical in a single crystal, that of the $\cdot CF_2CONH_2$ radical, was reported by Lontz and Gordy (1) in 1962. This was followed soon after by a study of the $\cdot CFHCONH_2$ radical by Cook, Rowlands and Whiffen (2). From such studies it is readily apparent that the fluorine hyperfine splittings are characterized by large anisotropy varying from approx. 200 to 0 Gauss. In contrast, similar studies of radicals exhibiting only proton hyperfine interactions show that the hyperfine anisotropy varies from 30 to 10 Gauss and is due to the dipolar interaction between the electron and

© 0-8412-0399-7/78/47-066-349$10.00/0

the nucleus of the attached proton (3). For an α-proton of a π-radical with C-H bond length of 1.08Å, this hyperfine anisotropy amounts to approximately 0, -10, and +10 Gauss; perpendicular to the radical plane, perpendicular to the C-H bond and parallel to the C-H bond respectively. In contrast, similar calculations in which the longer C-F bond length (~1.35Å) is substituted by the C-H bond length reduces by approximately 1/2 rather than increases the magnitude of the anisotropic fluorine-electron dipolar contribution (1). Thus the observed large anisotropy is not due to such an interaction.

A breakdown of the α-fluorine hyperfine splitting tensor into components due to different orbitals is given in Table I for the $\cdot CF_2COO^-$ radical at 300 K in irradiated $CClF_2COONa$ (4). It is apparent from Table I that the main contributor to the hyperfine anisotropy is due to approximately 10% of the unpaired electron density residing in the fluorine 2pπ orbital. The lack of cylindrical symmetry for the fluorine tensors is attributed to a competition between the dipolar interaction of the unpaired electron in the carbon 2pπ orbitals and the fluorine nucleus (item 1, Table II) with a contribution from the spin polarization of the fluorine 2pσ orbital (item 3, Table I). Since the signs of the $\cdot CF_2COO^-$ radicals are known, (5) the isotropic hyperfine splitting can be deduced by calculating the average of the principal hyperfine splittings along the X,Y,Z directions. In Table II are listed the experimental fluorine hyperfine splitting tensors for all radicals which have been reported to date, the irradiated crystal in which the radical was observed, the isotropic hyperfine splitting calculated from the observed hyperfine splitting tensor, the largest component of the anisotropic contribution 2B(F) due to the unpaired electron density in the fluorine 2pπ orbital, and the temperature at which the radical was studied. If no attempt was made to determine the signs of the hyperfine splittings, the most likely choice is given first in Table II with a possible alternate choice listed in square brackets. The value of 2B(F) was deduced by subtracting the sum of the contributions of the carbon 2pπ (item 1, Table I), the fluorine 2pσ orbital (item 3, Table I) and the fluorine 2s orbital from the observed tensor. The remainder along the direction perpendicular to the radical plane (or largest value) is given as 2B(F). Since the deviation from cylindrical symmetry is fairly consistent throughout Table II, the same values of the C 2pπ and F 2pσ contribution as given in Table I were used throughout.

TABLE I. An Estimation of the Unpaired Spin Population in Various Orbitals and Their Contributions to the Fluorine Hyperfine Splitting Tensor (in Gauss) of the $\cdot CF_2COO^-$ Radical(4) at 300 K.[(a)]

Item	Orbital	Spin Population	$A_\perp$ plane	$A_\perp$ C-F in plane	$A_{//}$ C-F
1.	C 2pπ	(+0.75)[b]	-2.4	-5.4	+7.8
2.	F 2pπ	+0.094	108.0[(c)]	-54.0	-54.0
3.	F 2pσ	(-0.011)[b]	+7.0	+7.0	-14.0
4.	F 2s	0.005	+73.5	+73.5	+73.5
	total		186.1	+21.1	+13.3
	observed		185.7	20.6	14.3

(a) The type "c" radical observed in irradiated $CClF_2COONa$ (ref. 4).

(b) $C_{2p\pi}$ = 0.75, $F_{2p\sigma}$ = -0.011, and $F_{2s} \cong$ 0.005 are estimated assuming the theoretical values of the anisotropic fluorine couplings given in Ref. 2. $F_{2p\pi}$ = 0.094 has been deduced by assuming the fluorine hyperfine splitting for an unpaired electron localized in a $2p_z$ fluorine orbital equals 1080 Gauss.

(c) This value is referred to as 2B(F) throughout this chapter.

TABLE II. Alpha - Fluorine Hyperfine Splitting Tensors (Gauss) for Fluorine Containing Radicals in Single Crystals[(a)]

Crystal	Radical	a^F isotropic	2B(F)	observed Tensor $A_\perp$ plane $A_\perp$C-F $A_\parallel$ C-F	observed Temp.	Ref.
$CF_2(COOK)_2 \cdot H_2O$	$F\text{-}\dot{C}(COOK)_2$	55.0	148.0	(208,-17,-27)[b]	300 K	6
		[or 84.0	119.4	(208,-27,-17)][b]	300 K	6
$CF_2(CONH_2)_2$	$F\text{-}\dot{C}(CONH_2)_2$	63.0	132.4	(200.0,-1,-10)	300 K	7
CFH_2CONH_2	$F\text{-}\dot{C}HCONH_2$	56.4	128.0	(189,-3.9,-16.0)	300-77 K	8
$CFCl_2CONH_2$	$F\text{-}\dot{C}ClCONH_2$	50.1	113.3	(168,-6.9,-10.7)	200-77 K	9
$CF_3CF_2CONH_2$	$F\text{-}\dot{C}(CF_3)\text{-}CONH_2$	60.3	136	(201,-8,-12)[b]	300-77 K	10
		[or 73.7	122.6	(201, +12, +8)][b]		
$NaO_2CCF_2CF_2CO_2Na$	$NaO_2CCF_2\dot{C}FCO_2Na$	71.2		(150.2,+59,+3.9)	300 K	11
$NaO_2CCF_2CF_2CO_2Na$	$NaO_2CCF_2\dot{C}FCO_2Na$ (I)[(c)]	69.3	143.0	(217,-2,-7)[b]	77 K	12
		[or 75.3	137.0	(217,+7, +2)][b]	77 K	12
''	'' (II)[(c)]	71.6	147.8	(224,-4,-5)	77 K	12
		[or 77.7	141.7	(224,+5, +4)	77 K	12
$CClF_2COONa$	$F_2\dot{C}COONa$[(d)]	73.5	108	(185.7, 20.6, 14.3)[b]	300 K	4
CF_3CONH_2	$\dot{C}F_2CONH_2$	72.7	103	(181, +23.2, +14)	300 K	13,5
CF_3CONH_2	$\dot{C}F_2CONH_2$	75.3	122.1	(202, +18, +6)[e]	77-4 K	5

CF_3CONH_2	$\dot{C}F_3$	144.6	103.8	(+ 253, +92, +88)	200-77 K	13
CF_3I	$\dot{C}F_3$	143.5	115.4	(+ 263.5, +87, +80)	4.2 K	14
CF_3COONH_4	$\dot{C}F_3$	152	97.4	(254, 113, 89)	77 K	15 (Fluorine 1)
				(260, 103, 53)	77 K	15 (Fluorine 2)
				(257, 85, 67)	77 K	15 (Fluorine 3)
$CF_2(COOK)_2 \cdot H_2O$	$\dot{C}F_2COO^-$ (I)	68.3	108.1	(181, 13, 11)[b]	300 K	16
$CF_2(COOK)_2 \cdot H_2O$	$\dot{C}F_2COO^-$ (II)	71.3	112.1	(188, 19, 7)[b]	300 K	16
CF_3COONH_4	$\dot{C}F_2COONH_4$	72.0	111.4	(+188.0, +14, +14)	300 K	15

(a) In each case the contribution from the C_{2P} and F_{2P} populations are taken to be the same as in Table I and the value of 2B(F) is choosen to give the observed value of A(F) max.

(b) signs of the principal components of the observed tensors have not been experimentally determined. However the choice of signs which best reproduce the deviations from axial symmetry was assumed in deducing A^F and 2B(F).

(c) An alternate conformation of the $NaO_2CCF_2\dot{C}FCO_2Na$ radical was initially observed before warming above 77 K.

(d) Three different crystallographic forms of the $\dot{C}F_2$-COO^- radical were observed. The data quoted is for the type "C" radical.

(e) The two fluorines are nonequivalent at 77K, however each possesses similar hyperfine splitting tensors with different direction cosines

It is important to note that the α-fluorine isotropic splitting for those radicals containing one α-fluorine generally varies between 50 and 65 Gauss; the isotropic fluorine splitting for radicals containing two fluorines varies between 72 and 76 Gauss, while the isotropic fluorine splitting for radicals containing three fluorines varies from 144 to 152 Gauss. The increase with the number of fluorine substituents has been shown (4, 13) to be due to an increase in the non-planarity of the radical at the α-carbon. This has been substantiated from measurements of the carbon-13 isotropic hyperfine splittings in solution for the radicals $\cdot CH_2F$, $\cdot CHF_2$, and $\cdot CF_3$ where $a(^{13}C)$ equals 54.8, 148.8 and 271.6 Gauss, respectively (17). Assuming that the angle, θ, between the CF bond and the normal to the axis of the π orbital for a planar radical is related to the isotropic carbon-13 splitting $a(^{13}C)$ by the following expression:

$$a(^{13}C) = a_\circ(^{13}C) + 1190(2 \tan^2 \theta) \qquad (1)$$

where $a_\circ(^{13}C) = 38.5$ Gauss, the value for a planar radical, results in the following values for θ. For $\cdot CH_2F(\theta<5°)$, $\cdot CHF_2(\theta \cong 12.7°)$; $\cdot CF_3(\theta \cong 17.8°)$: The complete carbon-13 tensor cannot be obtained with certainty from natural abundant ^{13}C satellite lines for the radicals listed in Table II. However an estimated value of the isotropic splitting can be deduced from the observed $A_{zz}(^{13}C)$, (the largest tensor value), and a reasonable value of the anisotropic part of the ^{13}C tensor (see Table III). If the anisotropic portion of the ^{13}C tensor is taken to be 2B, -B, -B where $2B \cong 45$ Gauss, then for $\cdot CFHCONH_2$ $(\theta \cong 0°)$, reference 2; $\cdot CClFCONH_2(\theta \cong 0°)$, reference 9; $\cdot CF_2COO^-(\theta \cong 8°)$, reference 4; and $\cdot CF_3(\theta \cong 18°)$, reference 13.

Table III. Carbon-13 Hyperfine Splittings for Selected Fluorine-Containing Radicals.

Radical	$A_{zz}(^{13}C)$			Temp.	Ref.
$\cdot CFHCONH_2$	85.0 Gauss	-	-	300 K	2
$\cdot CClFCONH_2$	90.6	-	-	200 K	9
$\cdot CF_2CONH_2$	147.5	-	-	77 K	13
$\cdot CF_2CONH_2$	130	67.6	66.1	300 K	13
$\cdot CF_3$	318	-	-	77 K	13

In Table IV are given the β-fluorine hyperfine tensors for the fluorine-containing radicals in irradiated crystals that have been reported so far. From the limited data one notes that the magnitude of the isotropic β-fluorine hyperfine splitting for the three equivalent fluorines of the $CF_3CFCONH_2$ radical (10) is smaller than that of the β-proton splitting usually observed for three equivalent β-protons. The reason for this has been recently discussed (20, 21). Briefly, it is a result of the hyperconjugation mechanism (by which the β-fluorine atoms obtain spin density) being less effective for β-fluorines than for β-protons. Such an effect occurs for β-fluorines of a CF_3 group because the CF_3 group orbital of π symmetry which interacts with the unpaired electron orbital is lower in energy than the corresponding protonated analogue. The overlap of the carbon p-orbital and the pseudo p-orbital of the CF_3 group is therefore less likely and the effect of hyperconjugation reduced.

Hyperconjugation is not the only mechanism which is responsible for the observed fluorine β-hyperfine splittings. The large anisotropic component [2B(F)] of 54.5 Gauss observed for one of the β-fluorines of the $CO_2\text{-}CF_2\dot{C}FCO_2^-$ radical at 77 K indicates (4) direct overlap of the p-orbital on the β-fluorine with the pπ orbital of the α-carbon. Analysis of the direction cosines for the β-fluorine tensor suggests that the p_x-orbital of the β-fluorine overlaps with the $p_z(p_\pi)$ orbital of the α-carbon at the point of maximum unpaired electron density. This introduces unpaired electron density in the β-fluorine p_x-orbital and thus large anisotropic and isotropic fluorine splittings are observed. The other β-fluorine p_x-orbital overlaps the p_z-orbital of the α-carbon along its nodal plane and thus with little unpaired density in the orbital both the anisotropic and isotropic fluorine couplings are small.

Inspection of Table IV also shows that the anisotropic components of the α- and β-fluorine tensors for some radicals vary with temperature. For instance, the 2B(F) value for $\cdot CF_2CONH_2$ at 300 K equals 103 Gauss (13, 5) while at 77 K it equals 122.1 Gauss (5). Furthermore the 2B(F) value of $CO_2^-CF_2\dot{C}FCO_2^-$ varies from 78.8 Gauss at 300 K (11) to 143.0 Gauss at 77 K (12). A temperature dependence is also observed for the anisotropic components of the β-fluorine splittings (12). Such variations with temperature will be shown in the next section to be due to one or more of the substituents of the radical undergoing an internal motion causing a portion of the hyperfine anisotropy to

TABLE IV

β - Fluorine Hyperfine Splitting (Gauss)

Crystal	Radical	a isotropic	a aniso.	tensor hfs	temp	Ref.
$CF_3CF_2CONH_2$	$CF_3\dot{C}FCONH_2$	22.3	13.7	(36,17,14)	300 K	10
$NaOOCCF_2CF_2COONa$	$CO_2^-CF_2\dot{C}FCO_2^-(\beta_1)$	34.7	28.3	(62.8,22.8,18.5)	300 K	11
$NaOOCCF_2CF_2COONa$	'' (β_2)	40.5	31.0	(71.5,26.3,22.8)	300 K	11
$NaOOCCF_2CF_2COONa$	$CO_2^-CF_2\dot{C}FCO_2^-$ (β_1)	69	54.5	(122,41,41)	77 K	4
$NaOOCCF_2CF_2COONa$	'' (β_2)	2.5	9	(11.5,-2,-2)	77 K	4
CF_3CH_2I (a)	$CF_3CH_2\cdot$	28.7	19.3	(48.0,17.7,20.4)	4.2 K	18
CF_3CONH_2	$CF_3CO\dot{N}H$	4.63	0.33	(4.96,4.51,4.43)	77 K	19

(a) In argon matrix at 4.2 K.

be averaged out at high temperature (rapid reorientation rates).

One last feature of Table II and IV is worth noting for radicals at 77 K. The 2B(F) value for the planar radicals (usually those containing one fluorine nucleus) equals ~130-140 Gauss which decreases to 120 Gauss for radicals with two fluorines (CF_2CONH_2) and to ~110 Gauss for those with three fluorines. This is a direct result of an increase in the nonplanarity of the radical with an increase in the number of fluorines.

Molecular Dynamics

The appearance of temperature dependent hyperfine splittings as found in Tables II and IV is commonly found not only for fluorine-containing radicals but for others as well (22). In fact, double resonance studies such as electron-nuclear double resonance (endor) and electron-electron double resonance (eldor) have shown (23) that a large number of radicals exhibit molecular motion over a wide range of temperatures including some that occur at 4.2 K. The presence of molecular motion is easily detected by esr studies when the radical contains α-fluorine nuclei as even a small variation of the large fluorine hyperfine anisotropy results in a measurable shift. Because this is such a common occurrence, it is important to know how to properly analyze the temperature dependent esr spectra in terms of structural parameters which characterize the radical. For this reason, a description will be given of the methods used to correctly treat the situation where the hyperfine splitting is temperature dependent, where the esr line width is dependent on M_I, and where a g value shift occurs. The first two temperature dependences will be demonstrated by the results recently reported for the $\cdot CF_2CONH_2$ and $^-O_2CCF_2\dot{C}FCO_2^-$ radicals (5, 12, 24). In these examples one (CF_2) or two groups (CF_2 and $-\dot{C}F$) of atoms respectively undergo intramolecular motion. The presence of a large g value shift will be established by analysis of the temperature dependent spectra of the $\cdot OOCF_2CONH_2$ radical (25).

The $\dot{C}F_2CONH_2$ Radical. The 35 GHz esr spectrum of the $\cdot CF_2CONH_2$ radical at 300 K consists of three allowed lines with an intensity ratio 1:2:1 (lines A, Figure 1) due to the appearance of two equivalent fluorines along the (-cos 24°, o, sin 25°) crystal direction. The 1:2:1 spectrum denoted by the letter A

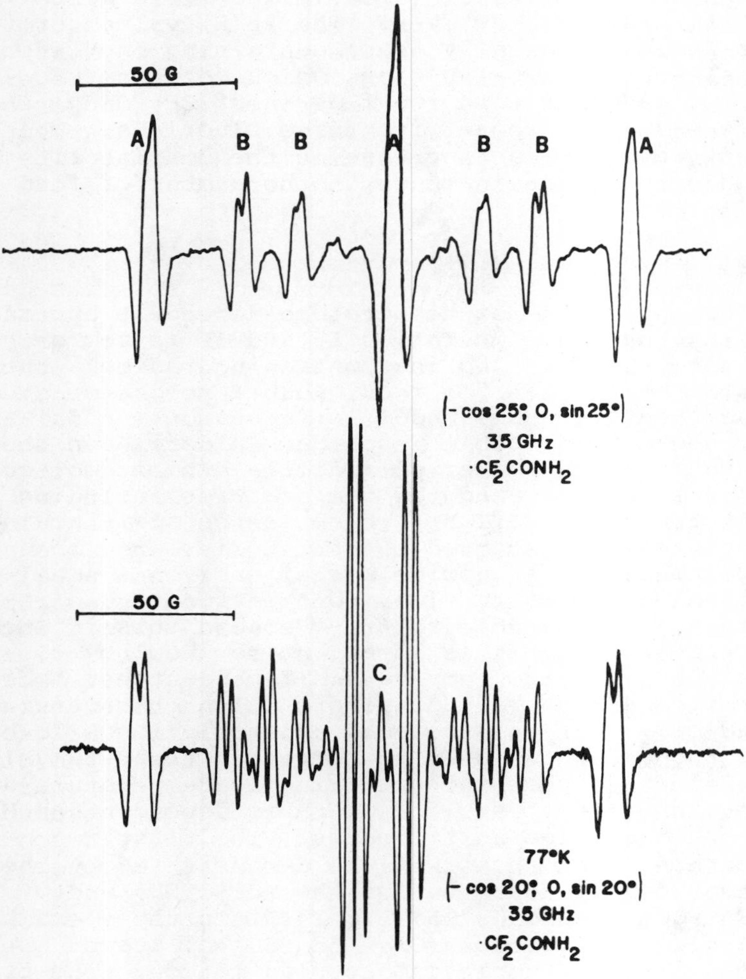

Journal of Chemical Physics

Figure 1. 35 GHz ESR spectra of $\cdot CF_2CONH_2$ *at 290°K (upper) and 77°K (lower) in irradiated crystals of* CF_3CONH_2*. The center line at 290°K separates into a resolved doublet of doublets at 77°K owing to the freezing out of the torsional oscillation of the* CF_2 *group. A, B, and C identify the* $\Delta M_I = 0$, $\Delta M_I = \pm 1$, *and* $\Delta M_I = \pm 2$ *transitions, respectively. The large difference in heights of the low and high field A lines at 77°K relative to the central A line doublets is caused by an alternating line width effect caused by a rapid in-phase motion of the* CF_2 *group above and below the CON plane. From Reference 5.*

is also observed for all other crystal directions at 300 K providing that the spectrum has been corrected for second-order effects (5). This feature alone suggests the presence of intramolecular motion as the two fluorines would not be equivalent for all crystal directions of an oriented $\cdot CF_2CONH_2$ radical. Nevertheless, the appearance of intramolecular motion for a fluorinated radical in a crystal is always assured if a change in the hyperfine splitting or spectral line width occurs as a function of temperature. Indeed as shown in Figure 1 for $\cdot CF_2CONH_2$ such things do appear as the temperature is lowered to 77 K. Notice that the central line of the 1:2:1 pattern observed at 300 K splits into a doublet of doublets at 77 K. This has been shown to be due to small resolvable N-H proton splitting (partially resolved at 300 K) split further by a larger doublet due to the presence of two non-equivalent fluorines. In addition, the line width of the high and low field esr lines is greater than that of the two central lines as evidenced by the lower height of the outer high and low field second-derivative esr lines. Furthermore as the temperature is lowered, the hyperfine splitting increases when the crystal is mounted along the a^* axis (Figure 2) and decreases along the b axis (Figure 3). It should also be pointed out that little change in hyperfine splittings occurs below 77 K. The resulting changes in the principal hyperfine splittings and direction cosines from 300 K to those at 77 K are given in Table II.

To account for this variation with temperature, one proposes that a torsional motion of the CF_2 group occurs through an angle θ since the observed principal fluorine hyperfine splittings do not equal approximately 110, 110, 20 Gauss, values typically observed for a CF_2- group rotating freely about the C-C bond in $\cdot CF_2CONH_2$. Assuming this to be the case, the torsionally averaged fluorine hyperfine splittings must be calculated. As noted in Table I, fluorine hyperfine anisotropy occurs largely due to roughly 10% of the unpaired electron density occupying the fluorine p_z-orbital. Since the fluorine p_z-orbital anisotropy in the absence of motion varies as $(1-3\cos^2\theta)$ it is only necessary to calculate the torsionally averaged value of $\cos^2\theta$ ($\langle \cos^2\theta \rangle$) as a function of temperature (5). Knowing the average value of $\cos^2\theta$ at each temperature, a hyperfine tensor based on a new average value for $(1-3 \cos^2\theta)$ can be constructed from which the hyperfine splitting at each crystal angle can be calculated. In general, a calculation of $\langle \cos^2\theta \rangle$ can be difficult if an exact quantum mechanical solution is required.

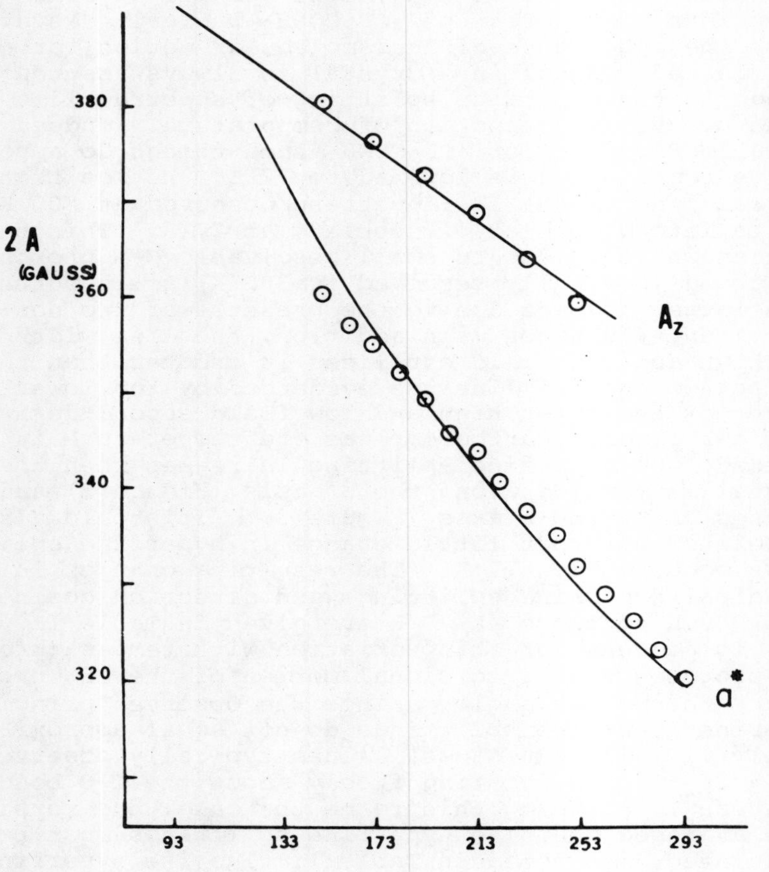

Figure 2. Plot of twice the experimental (⊙) and the calculated (——) hyperfine splitting in Gauss along the A_z principal hyperfine and the a reference axis as a function of temperature for $\cdot CF_2CONH_2$. The splittings in the A_z and a* direction decrease as a function of increasing temperature. From Reference 5.*

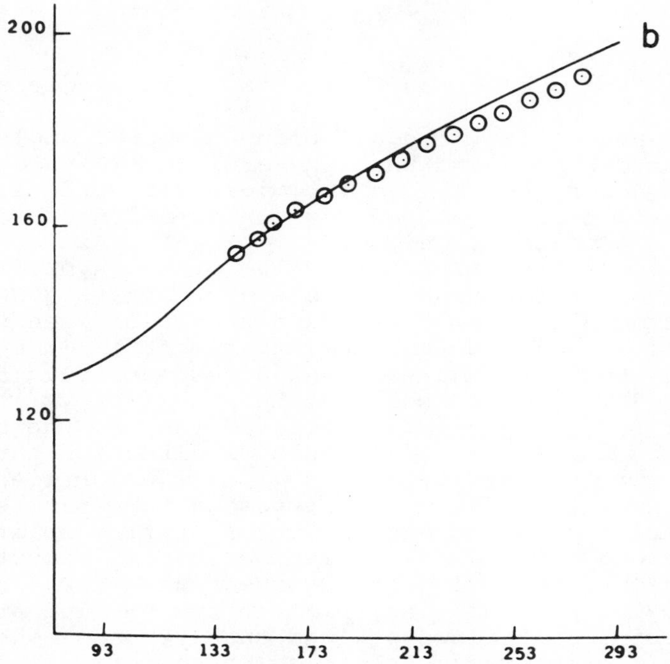

Journal of Chemical Physics

Figure 3. Plot of twice the experimental (⊙) and the calculated (——) splitting along the b crystal axis. The hyperfine splitting increases with temperature because of the onset of classical reorientation. From Reference 5.

However since a solution in this case is sought for temperatures above 120 K and a relatively high barrier to rotation is present, it has been shown (5) that a classical treatment will suffice. In such a treatment the value of $\langle \cos^2\theta \rangle$ is obtained from equation (2).

$$\int_{-\pi}^{\pi} \cos^2(\alpha+\theta_o) e^{-V(\alpha)/kT} d\alpha \Big/ \int_{-\pi}^{\pi} e^{-V(\alpha)/kT} d\alpha \qquad (2)$$

where $V(\alpha) = \frac{V_o}{2}(1 - \cos 2\alpha)$; α is in the angle of rotation and θ_o is the equilibrium dihedral angle between the C-F bond and the $C_{2p\pi}$-orbital. A value of θ_o can be obtained from an INDO molecular orbital calculation of the minimum energy conformation of $\cdot CF_2CONH_2$. Such a calculation gave $\theta_o = 76.5°$.

Based on the calculated value of the hyperfine splitting as a function of temperature, a reasonably good fit occurs along the a^* and b axis between the calculated (solid lines) and the experimental (⊙) splittings as shown in Figures 2 and 3 respectively (5). A value of 3.1 Kcal is used for the potential V_o. As expected for a torsionally averaged tensor, the splitting along the principal A_z axis (a direction parallel to the p-orbital axis) and the a^* axis which lies near the A_z axis decreases with increasing temperature. On the other hand an increase in the hyperfine splitting occurs along the b axis. This increase is expected as the torsional averaging process decreases the fluorine anisotropy. This implies that the two smaller principal fluorine splittings of less than 20 Gauss must approach the isotropic splitting which normally equals ~70 Gauss. Since one of the splittings lies near the b axis, an increase in the observed splitting with increasing temperature along that direction must follow.

Thus in general the presence of a torsional rotation for an alpha substituted CF_2 group is easily recognized as a decrease in hyperfine anisotropy along a crystal direction near to the maximum hyperfine splitting and an increase in the hyperfine splitting in an orthogonal direction.

The presence of an alternating line width at 77 K, but not at higher temperatures which are independent of magnetic field suggests an additional motion is also present. Proper analysis of the line width effect shows that it is largely dipolar in nature and requires the motion of two fluorines to be in-phase with one another (26). The particular type of motion which the radical undergoes has been suggested by an INDO molecular orbital calculation (5) which predicts that two possible potential minimum configurations of the CF_2CONH_2 radical exist; these being where the CF_2 plane makes a dihedral angle of 9° with respect to the CON plane either above or below the CON plane. This further suggests that a librational motion between these two conformations would at least be energetically favored. Additional support for an in-phase librational motion is based on the availability of hyperfine anisotropy data as a function of angle and temperature. If the temperature dependence can be largely explained by torsional motion about the C-C bond then the addition of a librational motion should not affect the calculated fit to the exprimental data.

In fact, the suggested librational motion causes little change in the calculated splitting above 77 K. This, along with the line width variation, implies that the librational motion may be very fast above 77 K and therefore difficult to detect except at low temperatures.

One word of caution should be given. A classical treatment is usually sufficient to describe the change in the esr spectrum of a radical in a single crystal as a function of temperature. However, in some instances certain features will not easily be predicted. For example, the temperature dependence of the separation between the two central esr peaks of the $\cdot CF_2CONH_2$ spectrum cannot be accounted for by using just a classical treatment. This is because the classical treatment does not take into account any time dependence of the magnetic resonance parameters. In general the observed positions of all esr lines of a radical in a single crystal will shift by a few Gauss whenever the internal molecular motion causes a time dependence of the axis of quantization of the spin system relative to the external reference axes to occur. To carry out an accurate time dependent calculation requires that the density matrix formalism be used. This is usually a reasonably complex calculation which is done only when absolutely necessary to understand all shifts in A values as a function of temperature (5).

The $^-O_2CCF_2\dot{C}FCO_2^-$ Radical. The torsional oscillation of individual radical substituents such as occurred for $\dot{C}F_2CONH_2$ is known to exist for a number of different radicals. However, if more than one substituent in a radical undergoes torsional oscillation, the motion can be coupled. Such is the case of the oriented radical $^-O_2CCF_2\dot{C}FCO_2^-$ in an irradiated crystal of $^-O_2CCF_2CF_2CO_2^-$ where the torsional oscillation motion of the CF_2 group does not occur independent of the torsional motion of the CF group (12). In addition, measurement of the anisotropic fluorine hyperfine splittings and the corresponding direction cosines makes it possible to determine the type of bonding necessary for a coupled torsional motion to occur for fluorinated radicals.

As expected for any radical undergoing torsional oscillation, the fluorine hyperfine splitting of $^{-}O_2CCF_2\dot{C}F_2CO_2^{-}$ decreases with increasing temperature along a direction near to the axis of the p-orbital containing the unpaired electron. In Figure 4, this direction corresponds to the b axis. On the other hand, an increase in splitting is observed along the a* axis, a direction near to the nodal plane of the p-orbital (Figure 4).

An additional motion is also suggested by the appearance of a variation in the line height of the four high field or the four low field second-derivative esr lines at 300 K as shown in Figure 5a. Since the second-derivative presentation has been recorded, the line height variation is a very sensitive measure of a line width change. Using this feature one should note that the largest line width occurs for line k while the narrowest occurs for line i where the four high field lines are labelled i, j, k and l respectively with increasing magnetic field. Upon cooling the crystal to 77 K (Figure 5b), an eight-line pattern also occurs, however all lines possess the same line width and are narrower than those at 300 K. Careful inspection of the data will show that the eight-line spectrum at 77 K is actually a superposition of the spectra due to two radicals denoted as I and II in Figure 5b. Furthermore, it is noted from angular rotation of the crystal that the spectra of I and II each contain three nonequivalent fluorines due to a $^{-}O_2CCF_2\dot{C}FCO_2^{-}$ radical however with tensors different from those measured at 300 K. Significantly, the α-tensor is made up of a larger 2B(F) term (Table II) than observed at 300 K while the β-tensors no longer possess nearly identical isotropic splittings of 34.7 and 40.5 Gauss but rather consists of one large isotropic value (69 Gauss) and one small isotropic value (2.5), respectively (Table III).

To explain the line width variation at 300 K, it is necessary to assume (12) that the transitions occurring for radical I and II are averaged according to the scheme given in Figure 6 for the crystal directions H // b and H // c. Notably, at 300 K lines i, j, k and l can occur at the magnetic field observed, only if

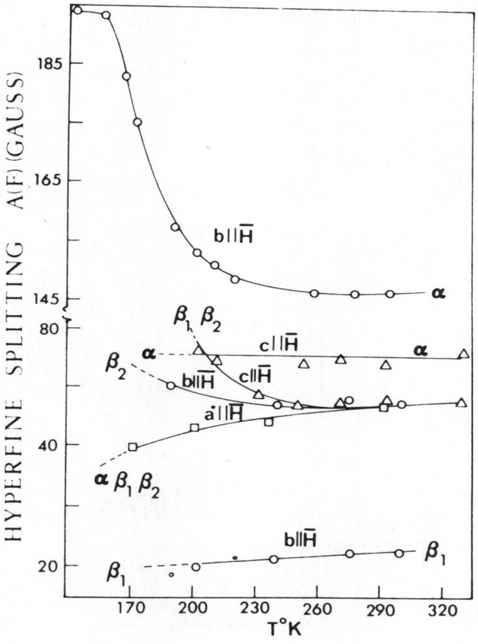

Journal of Chemical Physics

Figure 4. Temperature variation of the α, β_1, *and* β_2 *fluorine hyperfine splittings of the 300°K radical with the magnetic field parallel to a*, b, and c. The spectra become too complex to analyze where the curves are shown dashed. From Reference 12.*

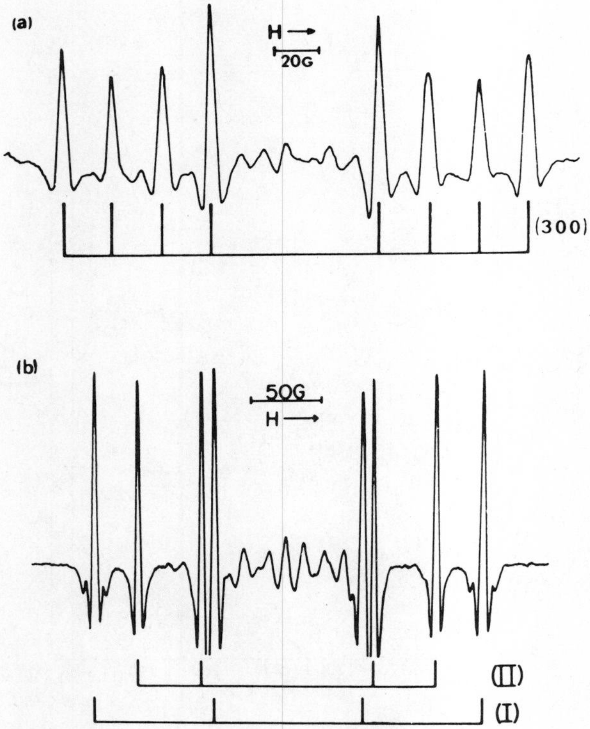

Figure 5. (a.) The second derivative ESR spectrum of irradiated sodium perfluorosuccinate at 300°K with H//b showing the eight lines caused by the average conformation of $^{-}O_2CCF_2\dot{C}FCO_2^{-}$*. (b.) The same crystal orientation as above but cooled to 77°K; the assignments of lines to conformations I and II of the radical* $^{-}O_2CCF_2CFCO_2^{-}$ *is indicated by the stick diagram. From Reference* 12.

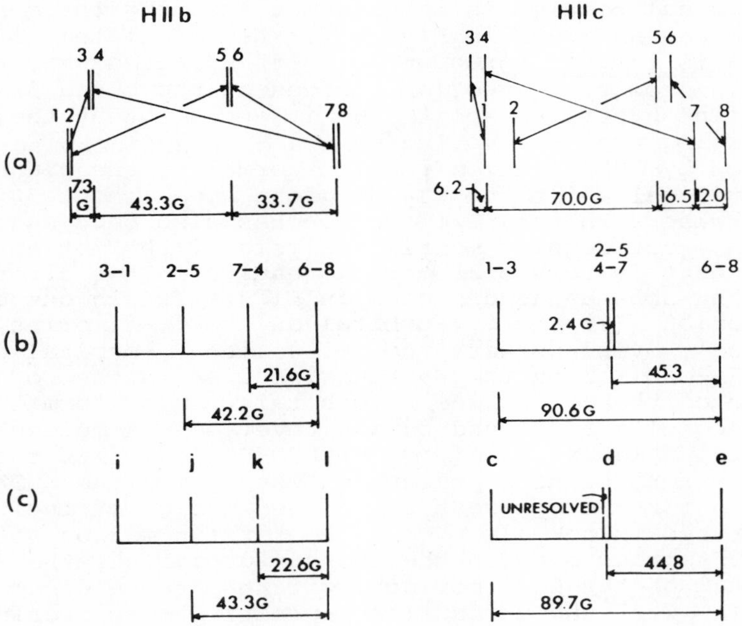

Journal of Chemical Physics

Figure 6. Diagram correlating the four transitions at high magnetic field in the 77°K ESR spectra of radical conformations I and II with those for the average radical conformation at 300°K. The small doubling of the lines for H//b is not experimentally resolved because of the small F(β_2) splitting, however this same splitting is resolved for H//c. If the transitions in (a.) are averaged according to the scheme shown, the four lines shown in (b.) are obtained. These agree with the experimental 300°K spectra shown in (c.). From Reference 12.

transitions 1 and 3; 2 and 5; 7 and 4; and 6 and 8 are averaged together. Furthermore, this specific averaging scheme also gives rise to the observed line width of lines i, j, k and l as the largest and the smallest magnetic field separation is between transitions 3 and 7; and transitions 3 and 1, respectively. If a density matrix calculation is carried out assuming the correlation diagram given in Figure 6, the calculated line width is in good agreement with the observed variation (24). From the direction cosines of the α and β-fluorine hyperfine splittings it appears that the averaging process results from a counterclockwise rotation by 70° of the direction normal to the plane of the radical while the $CF_2(\beta)$ group moves about 30° clockwise. This type of motion has also been suggested by an exact density matrix analysis of the motion (24). Besides this torsional motion results in two alternate bonding schemes as depicted in Figure 7. In one conformation (II) the $P_{x'}$-orbital on the β_1-fluorine directly overlaps with the α-fluorine P_z-orbital while the P_x-orbital on the β_2-fluorine lies in the nodal plane of the α-fluorine P_z-orbital. In conformation I, the roles of β_1 and β_2 are reversed. A molecular orbital calculation of $^-OCCF_2\dot{C}FCO_2^-$ shows that these two conformations represent an energy minimum. The reason for the apparent correlated motion of the β-fluorines is now clear. As the torsion motion rotates the CF_2 group between the two conformations, the P_x'-orbital of the β_2-fluorines switches from a direct overlap with the α-fluorine p-orbital to an overlap at the nodal position of the α-fluorine p-orbital, thus causing one very large and one very small p-fluorine isotropic splitting. The large β-fluorine anisotropy indicates that approximately 5% of the unpaired electron density occurs in the β-fluorine $P_{x'}$-orbital.

In the previous two examples, the g tensor variation with temperature and crystal orientation was not discussed since such a measurement has not been reported. Nevertheless, as expected for any radical undergoing internal motion, the g value did vary with temperature along selected orientations. Because the g anisotropy is small even for the rigid form of the radical, quite careful measurements are required if the data is to be used to identify the particular type of motion which occurs. This is usually not done as the fluorine hyperfine anisotropy variation is usually large and provides the most sensitive measure of the internal reorientation.

The $\cdot OOCF_2$ Radical. On the other hand this is not

the case when a study is carried out for peroxy radicals of the type $\cdot OOCF_2R$. The unpaired electron density in radicals of this form is localized largely on the $2p\pi$ oxygen orbital of the peroxy group and thus exhibits no hyperfine structure. Instead a large g anisotropy exists due to the spin-orbit interaction of the oxygen. This g anisotropy is sufficiently large that any internal motion exhibited by the radical over a given temperature range can be easily determined by a measurement of the g tensor at some low temperature such as 77 K where the radical exhibits no motion on an esr time scale and also at a higher temperature (300 K) where rapid internal motion exists.

The formation of peroxy radicals in irradiated perfluoro-substituted crystals is a rather common occurrence if the irradiated crystals are exposed to air. It has been observed (27) that for irradiated crystals of CF_3COONa, $CF_3CF_2CF_2COONa$, and their corresponding amides, oxygen from the air can diffuse into the irradiated crystal and react with the forerunner radical $\cdot CF_2R$, producing the $\cdot OOCF_2R$ radical over a period of time dependent largely on the diffusion rate of oxygen in the crystals. If the diffusion of O_2 is stopped before all of the $\cdot CF_2R$ has reacted, the esr spectrum will appear similar to that given in Figure 8a for irradiated CF_3CONH_2 crystals at 300 K (25). In addition to the usual three line spectrum of 1:2:1 intensity at 300 K due to $\cdot CF_2CONH_2$, a single down-field line due to $\cdot OOCF_2R$ occurs when $H \parallel a^*$. Upon cooling the crystal to 77 K (Figure 8b), a shift in g value of $\cdot OOCF_2R$ to smaller value is noted suggesting the presence of molecular motion at 300 K.

Angular rotation of the crystal at 77 K and 300 K results in the principal hyperfine splitting and the corresponding direction cosines given in Table V for $\cdot OOCF_2R$. Because only a single line occurs for the peroxide, a classical treatment of the motion is sufficient.

To deduce a model for the molecular motion exhibited by the peroxy radical several features of the data in Table V must be noted. First it is known (28) that the direction of the largest g value for a rigid peroxy radical lies parallel to the O-O bond direction. From the esr data at 77 K (supposedly the rigid form of the radical), this would be the direction of g_3. Secondly, the direction of the half-filled $2p\pi$-orbital on oxygen at 77 K is given by the direction of g_1. Thirdly, if rotation of the peroxy radical takes place about some axis at 300 K, then partial averaging results in an axially symmetric tensor. This is

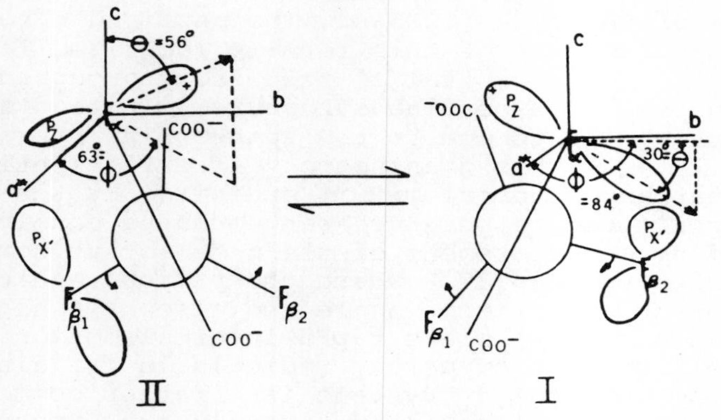

Figure 7. Newman projections along the C_α–C_β bond in $^-O_2CCF_2\dot{C}FCO_2^-$ showing conformation II and I which are stable below 130°K on an ESR time scale. Above 130°K, rapid exchange (on an ESR time scale) between conformation II and I leads to the average conformation observed at room temperature. The location of the a, b, and c axes are given relative to the α-fluorine P_z orbital. From Reference 24.*

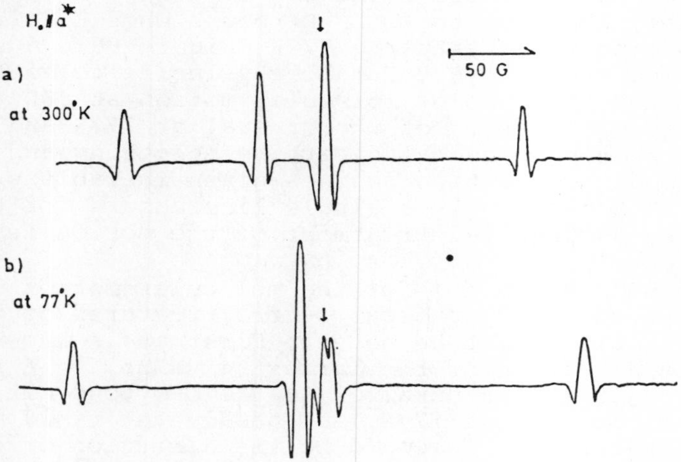

Figure 8. ESR spectra at (a.) 300°K and (b.) 77°K recorded for H//a of a single crystal of trifluoroacetamide exposed to air after γ-irradiation in vacuum. The resonance field of DPPH is indicated by arrows. Adapted from K. Toriyama and M. Iwasaki, J. Phys. Chem. *73*, 2663 (1969).

Table V. The Principal g Values and Direction Cosines for $\cdot OOCF_2CONH_2$ at 77 and 300 K.

		Principal Values	Direction Cosines with Respect to a^*, b, c		
			a^*	b	c
77 K	g_1	2.002_2	-0.290	±0.943	-0.163
	g_2	2.007_4	+0.944	±0.254	-0.208
	g_3	2.038_4 (a)	+0.155	±0.155	+0.964
	<g>	2.016_0			
300 K	g_1'	2.021_0	-0.151	±0.089	-0.985
	g_2'	2.018_2	-0.736	±0.655	+0.172
	g_3'	2.008_2	+0.660	±0.750	-0.033
	<g'>	2.015_8			

(a) g_3 = 2.0384 is located ∥ to the O-O bond direction. From Toriyama, K. and Iwasaki, M., J. Phys. Chem. (1969), 73, 2663.

approximately true at 300 K where $g_2' \cong g_1'$. Thus the direction of g_3' must be the axis of rotation. From the direction of g_3 and g_3', the angle between the O-O bond and the rotation axis is calculated to be 104°. Since this angle appears to be reasonable for the COO bond angle, the rotation axis was assigned to the C-O bond.

Assuming this to be the case, the g values at 300 K were calculated (25) from the g values obtained at 77 K for two different modes of rotation. First (1) only rotation around the C-O bond was allowed and secondly (2) an additional rotation around the O-O bond takes place simultaneously. The results given in Table VI show good agreement between calculated and observed g values for both cases (1) and (2).

Table VI. The Observed and Calculated Principal g Values for $\cdot OOCF_2CONH_2$

	Obsd.	Calcd. (1)	Calcd. (2)
$g_{\parallel}^{rot}$	2.008	2.005	2.007
$g_{\perp}^{rot}$	2.020 (a)	2.022	2.021

(a) From Toriyama, K., and Iwasaki, M., J. Phys. Chem. (1969), 73, 2663. This value is the arithmetic mean of g_1' and g_2'.

As one last note, it is also possible to deduce the complete structure of the peroxide radical as given in Figure 9 by knowing the direction cosines of the fluorine splittings for the $\cdot CF_2CONH_2$ structure. By an appropriate analysis (25), it was found that the O-O bond in the peroxy radical lies nearly parallel to the plane of the $\cdot CF_2R$ radical and that the conformation of the O-O bond to the C-C bond is trans. It appears that the oxygen molecule bonds with the half-filled $C_{2p\pi}$-orbital of $\cdot CF_2CONH_2$ without causing any change in $\cdot CF_2R$ except the change in hybridization of the orbital of the radical carbon.

Mechanism of Radical Formation

Chemical Reaction Schemes. The radiation damage scheme for protonated radicals has been extensively studied over a wide range of temperatures (22). Notably the radicals which are stable at room temperature have been found to be due to secondary products originating from some initially unstable species.

To observe thse initially unstable or primary radicals requires low temperatures (77 K) and sometimes as low as 1.5 K (29). Based on such studies, it has been possible to give a reasonably complete radiation damage mechanism for crystals consisting of C, N, O and H (22). Essentially, the main process is the ejection of an electron from a molecule to leave a positive hole (usually referred to as a molecular cation) trapped in a preferred site. The ejected electron is captured by another molecule to form a molecular anion. Both the molecular anion and cation decay at different temperatures creating a sequence of events from which other radicals or dimagnetic species are formed, some of which are stable at room temperature.

A similar study has not been carried out for fluorine-containing radicals due in part to the very great complexity of the esr spectra (31) and the difficulty of obtaining ENDOR spectra from fluorine-containing radicals. However some similarity must exist between the radiation decay mechanism of protonated radicals and that for fluorinated radicals as a number of the decay and formation processes in both instances involve the carbonyl proton in acids and the N-H proton in amides. Thus it is important to briefly review the radiation decay processes observed for the irradiated carboxylic acids and acetate crystals which do not contain fluorines.

As an example, the carbon-carbon bond breakage

observed in irradiated carboxylic acids seems to be due to a proton transfer from the cation primary followed by decarboxylation (22), as depicted in the following equation

$$RCH_2COOH \rightsquigarrow [RCH_2COOH]^+ \longrightarrow RCH_2COO\cdot + H^+ \quad RCH_2\cdot + CO_2 \qquad (3)$$

In irradiated acetate salts, where a similar proton transfer is not possible, methyl radicals have still been shown to form from the cation primary (32). Experimental evidence indicates that the cation is formed by an electron loss from $CH_3CO_2^-$ yielding $CH_3COO\cdot$ and the electron excess center $CH_3CO_2^{-2}$. However $CH_3COO\cdot$ is very unstable at any temperature above 4.2 K, and thus decomposes, yielding $\cdot CH_3$ and CO_2 with little reorientation or change in crystallographic location.

The decay of the anions (or the electron excess center) appears to proceed through several known steps. In carboxylic acids, protonation of the anion is a highly favored first step above 4.2 K. At 77 K or higher the protonated anion decays to form the acetyl radical by loss of water. This is followed by proton abstraction of a type $\cdot CH_2R$. Such a scheme is illustrated in equation (4)

$$RCH_2\dot{C}(OH)O^- \xrightarrow{H^+} RCH_2\dot{C}(OH)_2 \longrightarrow RCH_2\dot{C} = O + H_2O \qquad (4)$$

$$RCH_2\dot{C} = O \xrightarrow{RCH_2COOH} R\dot{C}HCOOH + RCH_2CHO$$

In irradiated hydrated carboxylic acid salts (32), the electron excess center $CH_3\dot{C}O_2^{-2}$ has been shown not to convert into $\cdot CH_3$ or $CH_2\dot{C}OO^-$ but rather remains stable at 77 K. In addition, the radical $\cdot CH_2COO^-$ is formed at 77 K by the proton abstraction from a neighboring CH_3COO^- molecule by $\cdot CH_3$ to form CH_4 and CH_2COO^-. On the other hand anhydrous carboxylic acid salts show (33, 34) the absence of $\cdot CH_3$ at 77 K and the presence of $\cdot CH_2COO^-$. A similar effect has also been observed if the metal salt is changed (34-38). Quite clearly, the water of crystallization and the particular salt used play a role in the radiation mechanism.

Based on these examples, the detection of some of the fluorine-containing radicals reported in Tables II and IV comes as no surprise. For instance, a number of the radicals are formed by C-C bond breakage. Thus a radical formation mechanism similar to the decay of the cation primary in irradiated alkali acetate salts would explain the appearance of $F_2\dot{C}OOK$ from $CF_2(COOK)_2$, $\cdot CF_3$ from CF_3COONH_4, $\cdot CFH_2$ and $\cdot CF_2H$ from CFH_2COONH_4 and $CF_2HCOONH_4$, respectively (39) and $^-O_2CCF_2\dot{C}F_2$ from $^-O_2CCF_2CF_2CO_2^-$ (12). Interestingly, the UV photolysis of polycrystalline lead(IV) fluoroalkanoates [(RCOO)$_4$-Pb] where R = CF_3, CH_2F, CHF_2, CF_3CF_2 and $CF_3CF_2CF_2$ shows (40) the initial photochemical processes to be the scisson of the Pb-O bond followed by rapid carboxylation to yield $\cdot R$. This alternate mechanism may be important if the samples are not irradiated in the dark. In this same study it is also found that when R = CH_2F or CHF_2, hydrogen abstraction by $R\cdot$ occurs to give $\cdot CHFCOO^-$ and $\cdot CF_2COO^-$ radicals respectively, however, no abstraction of fluorine is seen. That a perfluoro-substituted radical cannot abstract a fluorine from a neighboring molecule to yield a radical formed by C-F bond rupture has been substantiated by a study of the radiolysis of trifluoroacetic acid (41) at 77 K. The major radiolysis products of trifluoroacetic acid are CO_2, CF_3H and C_2F_6 while HF, CF_4, C_2F_4, CO and H_2 (which suggests fluorine abstraction) are either not detected or are minor products. As suggested in equation (3), it was found that CO_2 is formed by the decarboxylation of $CF_3\dot{C}OOH^+$, CF_3H is formed by proton abstraction from CF_3COOH which decomposes to yield $\cdot CF_3$ and CO_2 while some $\cdot CF_3$ recombines to give C_2F_6.

On the other hand radicals formed by C-F bond rupture could be produced by an electron dissociative attachment such as depicted in equation (5).

$$RF + e^- \longrightarrow [RF]^- \longrightarrow R\cdot + F^- \qquad (5)$$

In fact esr studies of irradiated polycrystalline samples (31) of RCF_3COOH where R = F, CF_3, CF_3CF_2, C_6H_{13}, $CF_2\dot{C}OOH$ and CF_2CF_2COOH suggest that the anion $[RCF_2\dot{C}OOH]^-$ is stable at 77 K. Subsequent thermal annealing yields $R\dot{C}FCOOH$ and the decay of the anion, thus giving support to equation (5). However, UV irradiation of samples containing $[RCF_2\dot{C}OOH]^-$ causes the anion to disappear with no identifiable radicals formed (31). Toriyama and Iwasaki (42) have shown experimentally that CF_3CONH_2 undergoes an electron dissociative attachment to form $\cdot CF_2CONH_2 + F^-$. However by irradiating a polycrystalline glass made up of a mixture of MTHF (source of trapped electrons) and perfluoro-substituted acids and esters, they found that an electron dissociative attachment did not occur due to the greater electron withdrawing tendency of the ester or carboxyl groups over that of the amide group.

Thus the loss of F^- via equation (5) may not be the reason for the loss of fluorine in perfluoro acids.

Carbon-carbon bond breakage has also been observed in irradiated acetamides. For instance, in CF_3CONH_2, $\cdot CF_3$ is formed, however, in CFH_2CONH_2 and CF_2HCONH_2, $\cdot CFH_2$ and $\cdot CF_2H$ have not been detected respectively (39, 43). In analogy to the mechanisms operative in protonated carboxylic acids $\cdot CF_3$ could be formed by the proton (H^+) transfer from the cation according to the following scheme:

$$[CF_3CONH_2]^+ \rightsquigarrow CF_3CO\dot{N}H + H^+ \qquad \dot{C}F_3 + O{=}C{=}NH \qquad (6)$$

In addition, the anion could decay via equation (5) to give $\cdot CF_2CONH_2$.

However a recent detailed study of a transient radical which was observed at 77 K to occur in trifluoroacetamide single crystals irradiated at 77 K showed that mechanisms (5) and (6) were inadequate to explain the complete radical formation and decay scheme in trifluoroacetamide (19). In fact, the concentration of $CF_3CO\dot{N}H$ at 77 K is approximately 6 times that of $\cdot CF_2CONH_2$ while the concentration of $\cdot CF_3$ is twice that of $\cdot CF_2CONH_2$ after correcting for saturation and esr line width differences found for the three radicals. What is really surprising is that as the CF_3-$CO\dot{N}H$ radical decays by one unit of intensity the concentration of the $\cdot CF_3$ radical remains constant while the concentration of $\cdot CF_2CONH_2$ increases by a half unit of intensity. Why this occurs is not clear. However recent X-ray crystallography studies of some perfluoro crystals have shown some interesting molecular packing arrangement which may upon further study explain some of the apparent inconsistences.

Crystal Structures. One of the first uses of a crystal structure to explain radical formation in fluorine-substituted acetamide or carboxylic acid was carried out by Iwasaki and Toriyama (43). They observed an esr spectrum due to the trapping of a radical pair between $\cdot CH_2CONH_2$ and $\cdot CFHCONH_2$ in addition to a spectrum of an isolated $\cdot CH_2CONH_2$ radical in irradiated CFH_2CONH_2 crystals at 77 K. The formation of the radical pair can be readily explained by taking note of the CFH_2CONH_2 crystal structure given in Figure 10. CFH_2CONH_2 crystallizes as a triclinic crystal made up of layers of CFH_2CONH_2 molecules. This layering of the CFH_2CONH_2 molecules places (Figure 10) the fluorine atom directly above the methyl hydrogen of the adjacent molecule along the c axis. The appearance of

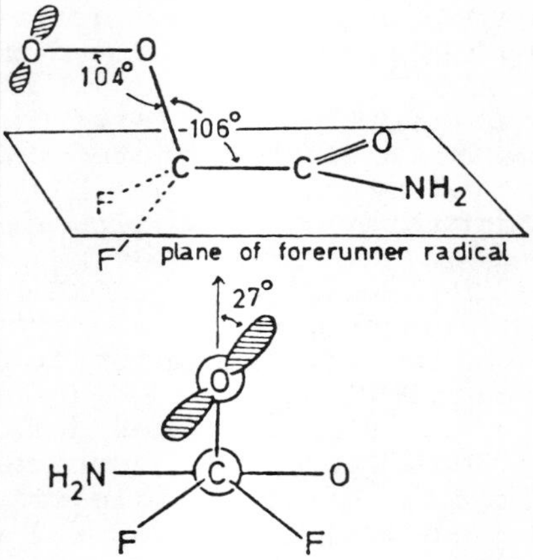

Journal of Physical Chemistry

Figure 9. $\dot{O}OCF_2CONH_2$ radical showing the position of the half-filled 2P orbital and the OOC and OCC angles. Adapted from Reference 25.

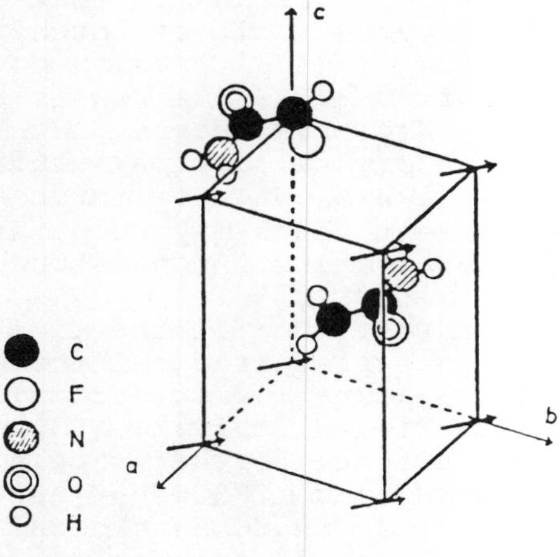

Journal of Chemical Physics

Figure 10. Location and structure of the two radicals $\dot{C}FHCONH_2$ and $\dot{C}H_2CONH_2$ which make up a radical pair in the triclinic unit cell. The location of the undamaged molecules is indicated by the arrows on the corners of the unit cell. Adapted from Reference 43.

$\cdot CH_2CONH_2$ at 77 K suggests the loss of fluorine atoms as a primary process. Thus to form the radical pair would require only the translation of a hot fluorine atom along the c axis followed by an abstraction of H by the F atom according to equation (7).

$$CH_2FCONH_2 \xrightarrow{\gamma} \left[\cdot CH_2CONH_2 \right] + \cdot F$$
$$\cdot F + CH_2F\overset{\curvearrowright}{C}ONH_2 \longrightarrow \left[\cdot CFHCONH_2 \right] + HF \qquad (7)$$

Alternatively, assuming the initial formation of a cation and anion in adjacent molecular layers gives rise to a similar but preferred (<u>42</u>) scheme depicted by equation (8).

$$CH_2FCONH_2 \xrightarrow{\gamma} [CFH_2CONH_2]^+ + [CH_2FCONH_2]^{\underline{-}} \rightarrow \left[\dot{C}H_2CONH_2 \right] + F^-$$
$$F^- + [CH_2FCONH_2]^+ \longrightarrow HF + \left[\dot{C}FHCONH_2 \right]$$
$$(8)$$

As the crystal is warmed the $\cdot CH_2CONH_2$ radical decays via a hydrogen abstraction with a neighboring molecule to form CH_3CONH_2 and $\cdot CFHCONH_2$.

<u>Crystal Disorder</u>. Similar radical reactions do not occur in CF_3CONH_2 as this compound does not crystallize as a layered structure (Figure 11). Therefore the elimination of HF between $[CF_3CONH_2]^-$ and $[CF_3CONH_2]^+$ is not possible via the loss of F^- and the amide proton (H^+) respectively. However the reactions are still influenced by the molecular packing scheme which occurs for CF_3CONH_2 and its chloro or bromo analogs. In fact, recent crystal structures of CF_3-CONH_2, $CBrF_2CONH_2$ and $CClF_2CONH_2$ show (<u>44</u>) that they all exhibit two rotationally disordered conformations of the CXF_2 group. When X = F or Cl, the two conformations are nearly equally populated. From an esr study of the irradiated crystals (<u>44</u>), it appears that the $\cdot CF_2CONH_2$ radical is formed via the loss of X from one or both of the conformations depending on the orientation of the C-X bond relative to the CON plane. Which of these cases actually occurs for a crystal is easily detected in an esr spectrum.

In general esr studies of irradiated crystals show that radicals which are formed retain the spatial arrangement of the parent molecule, except for a small change due to the rehybridization necessary for the formation of a radical center. Because of this, radicals formed in irradiated crystals of monoclinic symmetry are located in two magnetically nonequivalent sites in the a^*b and bc planes where a, b and c are the monoclinic crystal axes and a^* is an axis normal to the

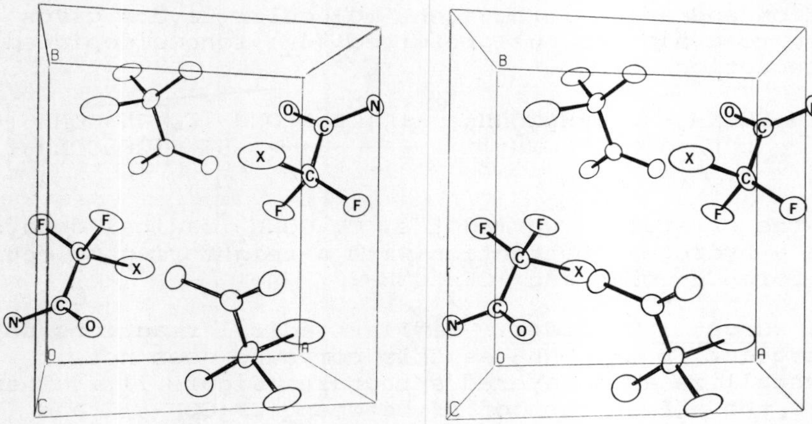

Figure 11. The packing diagram for the isostructural series XCF_2CONH_2 viewed down the [001] direction where X can equal F, Cl, or Br. The intermolecular N– – –O distance suggests possible hydrogen bonding.

bc plane. Along the a*, b, and c axes, only one magnetic site is observed. On the other hand, when disorder occurs for a crystal of monoclinic symmetry, then multiple magnetic sites occur along the a*, b or c axes due to the magnetic nonequivalence of the disordered sites.

In Figure 12 is given the esr spectrum obtained at 90 K for a crystal of irradiated chlorodifluoroacetamide when the magnetic field is parallel to the a* axis of a monoclinic crystal exhibiting $P2_1/c$ symmetry. Lines A and B represent the high- and low-field esr lines of two magnetically nonequivalent crystal sites of the radical $\cdot CF_2CONH_2$ (a four-line spectrum of intensity 1:1:1:1 due to two nonequivalent fluorines is expected for each site) when the crystal initially irradiated at 77 K is warmed to 90 K. The separation between the low-field or high-field lines A and B does not represent a small hyperfine splitting as the low-field separation equals 9.0 Gauss while the high-field separation equals 10.25 Gauss. An unequal "splitting" such as this can only be due to two magnetically nonequivalent sites each with a slightly different g value. The lines labelled C are spectral lines due to a precursor radical stable at 77 K but not at 90 K. As the temperature is raised to 170 K, the spectrum in Figure 12a changes to that in Figure 12b where only a three-line spectrum of intensity 1:2:1 is observed. In addition the spectral width has increased from 266 to 291 Gauss. This spectral change is due to a reorientation of the radical $\cdot CF_2CONH_2$ as well as a torsional oscillation of the CF_2 group (5) about the C-C bond. At 170 K, any rotational disorder observed below 170 K is averaged out by the torsional motion. Upon recooling to 90 K (which stops the torsional oscillation on an esr time scale) the two magnetically nonequivalent sites A and B (Figure 12c) reappear. However, the separation between the low-field lines is 21.5 Gauss while the separation between the high-field lines is 24.5 Gauss. This change is a result of an irreversible reorientation of the radical as the temperature is raised.

Further evidence that the small separation between the low-field lines A and B is due to magnetically nonequivalent sites is given in Figure 13a. The esr spectrum was obtained with the magnetic field parallel to the a*b plane, 35.5° from a* crystal axis. In this direction two magnetically nonequivalent sites are expected. These are observed as sites I and II. In addition, the nonequivalent sites due to disorder are also observed as unequal splittings of the high- and

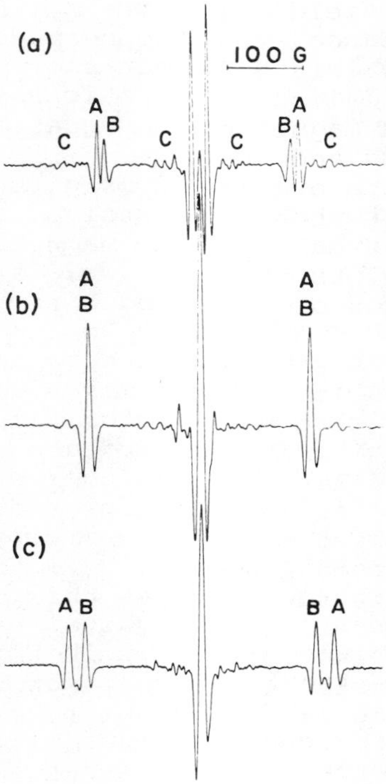

Figure 12. ESR spectra of irradiated chlorodifluoroacetamide crystals with the magnetic field parallel to the ESR a axis when (a.)* T = *90°K and no previous warming, (b.) crystal warmed to* T = *170°K and (c.) crystal cooled to* T = *90°K. Lines A and B indicate the high and low field lines of two magnetically nonequivalent sites of* $\dot{C}F_2CONH_2$ *caused by rotational disorder. Lines C are caused by a partially decayed precursor radical.*

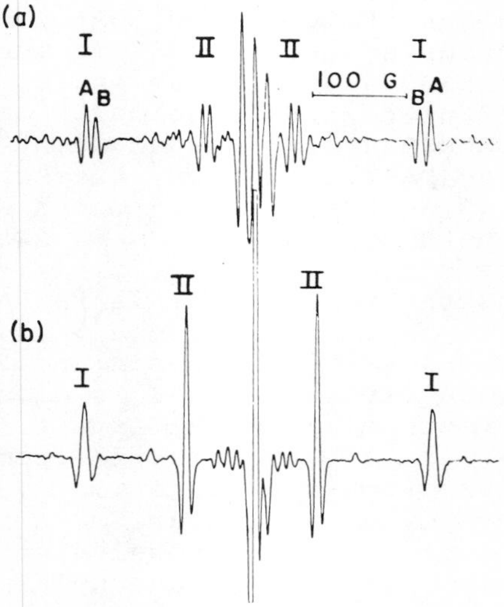

Figure 13. ESR spectra of irradiated chlorodifluoroacetamide crystals with the magnetic field in the direction (cos 35°, O, sin 35°). (a.) Expected monoclinic nonequivalent sites I and II are split further into lines A and B because of the appearance of rotational disorder for the $\cdot CF_2CONH_2$ *radical at 90°K. (b.) At 170°K, lines A and B for monoclinic sites I and II coalesce because of the known torsional oscillation of the* CF_2 *group about the C–C bond. The difference in ESR line width between sites I and II is caused by the appearance of unresolved hyperfine splitting from N and H at certain crystal angles.*

low-field lines of the $\cdot CF_2CONH_2$ radical. For instance, the separation A-B of the low-field line of site I is 9.75 Gauss, while the separation of the high field line A-B is 10.75 Gauss. For splittings of this magnitude, second-order effects would be $\leq$ 0.1 Gauss and thus the unequal separation can only be due to the overlap of two magnetic nonequivalent radical sites whose g values are not quite equal for the observed direction. The onset of torsional oscillation, as the temperature is raised, averages out the effect of the disorder at 170 K and thus only magnetic sites I and II are observed (Figure 13b).

A similar esr study for irradiated trifluoroacetamide showed no evidence for rotational disorder. In Figure 14 is given the esr spectrum taken along the esr a^* axis. No unequal splitting of the high- and low-field lines of $\cdot CF_2CONH_2$ (denoted as D) nor of the high- and low-field lines of $\cdot CF_3$ (denoted by C) are observed at 90 K. In addition the precursor radicals observed at 77 K did not show any evidence of rotational disorder. Also, in contrast to the irreversible orientational change of $\cdot CF_2CONH_2$ in irradiated chlorodifluoroacetamide on initial warming, no such irreversible change in orientation for $\cdot CF_2CONH_2$ was observed in irradiated trifluoroacetamide. An interpretation of these esr results in light of the crystal structure will be given below.

The appearance of two magnetically nonequivalent radical sites along the esr a^* axis in $CClF_2CONH_2$ rather than the one expected is presumably due to the equal probability of chlorine loss from both conformations of $CClF_2CONH_2$ as the C-Cl bond in the two disordered sites makes an approximately equal angle (23.0° and 27.3°) with respect to a direction perpendicular to the CON plane (44). The formation of $\cdot CF_2CONH_2$ is favored by the presence of the C-X bond located in a direction perpendicular to the CON plane as loss of X placed in such a direction would require a minimum shift in the remaining atoms to produce $\cdot CF_2CONH_2$. Numerous studies of polymerization reactions carried out in single crystals have demonstrated that reactions in the solid state proceed along a path where the shift in atom positions is minimized. Based on these observations, it must be assumed that only the disordered site where the C-F bond in CF_3CONH_2 makes an angle of 17° to the perpendicular direction to the CON plane gives rise to $\cdot CF_2CONH_2$. The C-F bonds in the other disordered site all form large acute angles (45.0° and 38.2°) to this direction. According to this assumption only one magnetic nonequivalent site

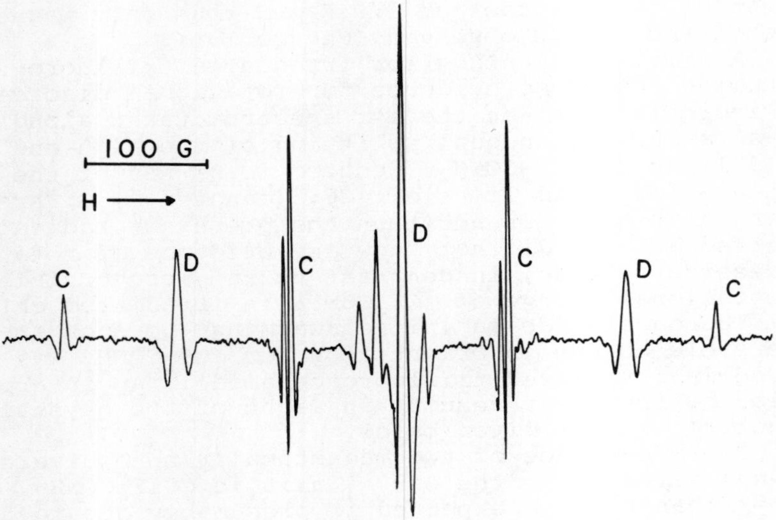

Figure 14. The ESR spectrum of irradiated trifluoroacetamide at 90°K caused $\dot{C}F_3$ (C) and $\dot{C}F_2CONH_2$ (D). No unequal splitting of the high and low field lines of either radical was observed at any crystal angle between 77° and 200°K, indicating the absence of rotational disorder.

would be observed along the esr a* axes as was verified experimentally, and no irreversible reorientation of the CF_2 group upon warming would occur as only one site is possible.

Thus the location of the atoms apparently plays an important role in determining the particular reaction mechanism operative for each crystal system. Therefore in general, crystal structures are needed for a given problem before a reaction mechanism can be assigned.

Prognosis for Future Studies

It is clear from the previous section that the mechanism by which fluorine-containing radicals are formed or decay in irradiated crystals is relatively unknown. Complicating the study is the recent discovery by X-ray crystallography measurements that some of the fluorine-containing crystals possess regularly disordered substituents. This causes more than one possible route along which the radical reaction can proceed. In addition relatively little work has been carried out to determine the stable products produced by the X- or γ-irradiation of a crystal. Without the knowledge of the end products, numerous reaction mechanisms can be proposed which are consistent with the radicals formed. Generally this procedure leads to more confusion rather than clarification. It is obvious that all future work will require not only the crystal structure be determined for each compound, but also a product analysis of the irradiated crystals.

Acknowledgement

The author is indebted to Mr. B. Kalyanaraman for use of some of his esr spectra and Dr. T. C. S. Chen for discussions. This work was supported in part through a contract with the United States Energy Research and Development Administration.

Literature Cited

1. Lontz, R. J. and Gordy, W., J. Chem. Phys. (1962), 37, 1357.
2. Cook, R. J., Rowlands, J. R. and Whiffen, D. H., Mol. Phys. (1963), 7, 31.
3. McConnell, H. M., Heller, C., Cole, T. and Fessenden, R. W., J. Amer. Chem. Soc. (1960), 82, 766.
4. Rogers, M. T. and Kispert, L. D., J. Chem. Phys. (1971), 55, 2360.

5. Bogan, C. M. and Kispert, L. D., J. Chem. Phys. (1972), 57, 3109.
6. Herring, F. G., Lin, W. C., and Mustafa, M. R., J. Mag. Resonance (1970), 2, 9.
7. Iwasaki, M., Noda, S., and Toriyama, K., Mol. Phys. (1970), 18, 201.
8. Cook, R. J., Rowlands, J. R., and Whiffen, D. H., Mol. Phys. (1963), 7, 31.
9. Kispert, L. D. and Myers, F., Jr., J. Chem. Phys. (1972), 56, 2623.
10. Lontz, R. J., J. Chem. Phys. (1966), 45, 1339.
11. Rogers, M. T. and Whiffen, D. H., J. Chem. Phys. (1964), 40, 2662.
12. Kispert, L. D. and Rogers, M. T., J. Chem. Phys. (1971), 54, 3326.
13. Rogers, M. T. and Kispert, L. D., J. Chem. Phys. (1967), 46, 3193.
14. Maruani, J., McDowell, C. A., Nakajima, H., and Raghunathan, P., Mol. Phys. (1968), 4, 349; (1970), 18, 165.
15. Srygley, F. D. and Gordy, W., J. Chem. Phys. (1967), 46, 2245.
16. Herring, F. G., Lin, W. C., and Mustafa, M. R., Can. J. Chem. (1970), 48, 447.
17. Fessenden, R. W. and Schuler, R. H., J. Chem. Phys. (1965), 43, 2704.
18. Jingiyi, M., Lin, K. C., McDowell, C. A., and Raghunathan, P., J. Chem. Phys. (1976), 65, 3910.
19. Chen, T. C. S. and Kispert, L. D., J. Chem. Phys. (1976), 65, 2763.
20. Rossi, A. R. and Wood, D. E., J. Amer. Chem. Soc. (1976), 98, 3452.
21. Yim, M. B. and Wood, D. E., J. Amer. Chem. Soc. (1976), 98, 3457.
22. Iwasaki, M., Int. Rev. Sci., Phys. Chem. Ser. (1972), 4, 317, and refs. cited therein.
23. Kevan, L. and Kispert, L. D., "Electron Spin Double Resonance Spectroscopy", John Wiley and Sons, New York (1976).
24. Bogan, C. M. and Kispert, L. D., J. Phys. Chem. (1973), 77, 1491.
25. Toriyama, K. and Iwasaki, M., J. Phys. Chem. (1969), 73 , 2663.
26. Fraenkel, G., K., J. Phys. Chem. (1967), 71, 139, and references within.
27. Iwasaki, M., Toriyama, K. and Eda, B., J. Chem. Phys. (1965), 42, 63.
28. Ichikawa, T., Iwasaki, M., and Kuwata, K., J. Chem. Phys. (1966), 44, 2979.

29. Iwasaki, M., Toriyama, K., Muto, H., and Nunome, K., Chem. Phys. Letters (1976), 39, 90.
30. Nunone, K., Muto, H., Toriyama, K., and Iwasaki, M., Chem. Phys. Letters (1976), 39, 542.
31. Ayscough, P. B., and Mach, K., J. C. S. Faraday I (1974), 70, 118.
32. Toriyama, K. and Iwasaki, M., J. Chem. Phys. (1976), 65, 2883.
33. Rogers, M. T. and Kispert, L. D., Adv. Chem. Ser. (1968), 2, 327.
34. Roberts, H. C. and Eachus, R. S., J. Chem. Phys. (1973), 59, 5251.
35. Eachus, R. S. and Herring, F. G., Can. J. Chem. (1971), 49, 2866.
36. Eachus, R. S. and Herring, F. G., Can. J. Chem. (1971), 49, 562.
37. Booth, R. J., Starkie, H. C., and Symons, M. C. R., J. Chem. Soc. Faraday II (1972), 68, 638.
38. Booth, R. J., Starkie, H. C., and Symons, M. C. R., J. Phys. Chem. (1972), 76, 1876.
39. Rogers, M. T., and Schoening, R. C., Abstract 46, physical section, First Chemical Congress of the North American Continent, 1975.
40. Ayscough, P. B., Machová, J. and Mach, K., J. C. S. Faraday II (1973), 69, 750.
41. Betts, J. and Cheriniak, E. A., Canad. J. Chem. (1971), 49, 3389.
42. Toriyama, K. and Iwasaki, M., J. Phys. Chem. (1972), 76, 1824.
43. Iwasaki, M. and Toriyama, K., J. Chem. Phys. (1967), 46, 4693.
44. Kalyanaraman, B., Kispert, L. D. and Atwood, J. L., unpublished results.

14

ESR Spectra and Structure of Inorganic Fluorine-Containing Free Radicals *(1)*

J. R. MORTON and K. F. PRESTON

Division of Chemistry, National Research Council of Canada, Ottawa, Canada

In the past few years there has been a sufficient proliferation of data on the ESR spectra of inorganic fluorine-containing free radicals that a review seems in order. The bell-wether of recent progress was the discovery in 1966 by Fessenden and Schuler (2) of the radical PF_4 and a number of other species in irradiated SF_6 doped with various additives. In the intervening period, a variety of paramagnetic fluorides of the elements of Groups III-VII have been detected by ESR (3-21). The present article will be concerned solely with such species, i.e. paramagnetic fluorides of the main group elements, and will be restricted to spectra in the condensed phases.

Of the two interactions which govern the appearance of an ESR spectrum, the electronic Zeeman term and the hyperfine interaction term, it is the latter which is most informative in establishing the geometry and electronic structure of a free radical. This term, which represents the interaction between the unpaired electron and a nuclear magnetic moment, gives rise to the hyperfine splittings characteristic of most ESR spectra. A knowledge of the hyperfine tensors for all nuclei in a radical provides an excellent basis for a description of the semi-occupied orbital. In many instances, unfortunately, such detailed information is not available; in fact for many of the paramagnetic species to be discussed here only isotropic spectral parameters are available. However, even such limited data often enable one to assign a structure and a ground-state electronic symmetry to the free radical in question.

Hyperfine splittings are only resolved if the product of the nuclear magnetic moment and the unpaired electron spin density at the nucleus is sufficiently large. From this point of view, studies of fluorine-containing free radicals are particularly fruitful because of the large nuclear magnetic moment of fluorine. A second factor which favours the presence of large hyperfine interactions in paramagnetic inorganic fluorides lies in the make-up of the semi-occupied orbital in such species. Whereas the unpaired electron in many organic free radicals occupies an orbital which is antisymmetric (π) with respect to an atomic plane, in inorganic free radicals derived from the main-group elements it is more usual for the unpaired electron to occupy a totally symmetric (σ) orbital. In such molecules both atomic p and s orbitals may contribute directly to the half-filled orbital, resulting in large hyperfine interactions for the constituent magnetic nuclei. A striking example is found in the comparison (3) of the ^{13}C hyperfine interactions of the σ-radical CF_3 (271 G) and the π-radical CH_3 (38.5 G).

Although the large hyperfine interactions commonly encountered in inorganic fluorine radicals usually result in well-resolved spectra, they pose certain analytical problems for the spectroscopist. A large separation of hyperfine components means that (a) each line must be measured individually, and (b) the hyperfine interactions cannot be determined from field differences, nor the g-values from field averages. A compensating feature of such spectra is that a thorough and rigorous analysis often leads to information regarding the relative signs of different hyperfine interactions in the same radical (3, 22).

Analysis of ESR Spectra

As mentioned above, the analysis of ESR spectra showing large hyperfine splittings presents a number of difficulties. The underlying problem in cases of a strong interaction between the electron and nuclear spins is that of an incomplete Paschen-Back effect: The Hamiltonian is not even approximately diagonal in the high-field, uncoupled representation and the usual "first-order" solution, which ignores off-diagonal terms, is incorrect. We will restrict further discussion here to the isotropic case; the analysis of anisotropic spectra is similar, although more complicated, and has in any case been adequately treated elsewhere (23).

As an example of the effect of complete neglect of off-diagonal elements, we consider the energy matrix for a single nucleus $I = \frac{1}{2}$ coupled to the electron spin $S = \frac{1}{2}$ in a magnetic field of intensity H. From the Hamiltonian

$$\mathcal{H} = g\beta H\cdot S + aI\cdot S - \gamma H\cdot I \quad , \tag{1}$$

where a is the hyperfine interaction in MHz, g is the electronic g-factor, β is the Bohr magneton (1.399611 MHz/G) and γ is the magnetogyric ratio of the nucleus, one obtains the spin-matrix:

m_s, m_I	$\vert\frac{1}{2}, \frac{1}{2}\rangle$	$\vert\frac{1}{2}, -\frac{1}{2}\rangle$	$\vert-\frac{1}{2}, \frac{1}{2}\rangle$	$\vert-\frac{1}{2}, -\frac{1}{2}\rangle$
$\langle\frac{1}{2}, \frac{1}{2}\vert$	$\frac{1}{2}g\beta H - \frac{1}{2}\gamma H + \frac{1}{4}a$	0	0	0
$\langle\frac{1}{2}, -\frac{1}{2}\vert$	0	$\frac{1}{2}g\beta H + \frac{1}{2}\gamma H - \frac{1}{4}a$	$\frac{1}{2}a$	0
$\langle-\frac{1}{2}, \frac{1}{2}\vert$	0	$\frac{1}{2}a$	$-\frac{1}{2}g\beta H - \frac{1}{2}\gamma H - \frac{1}{4}a$	0
$\langle-\frac{1}{2}, -\frac{1}{2}\vert$	0	0	0	$-\frac{1}{2}g\beta H + \frac{1}{2}\gamma H + \frac{1}{4}a$

Since this matrix factorizes into submatrices of order no greater than two (as is invariably the case for a single nuclear spin), it may be diagonalized algebraically to yield the following expressions for the eigenvalues:

$$|\tfrac{1}{2}, \tfrac{1}{2}\rangle = \tfrac{1}{2}g\beta H - \tfrac{1}{2}\gamma H + \tfrac{1}{4}a \qquad (2)$$

$$|\tfrac{1}{2}, -\tfrac{1}{2}\rangle = \tfrac{1}{2}g\beta H + \tfrac{1}{2}\gamma H - \tfrac{1}{4}a + \tfrac{1}{2}(g\beta H + \gamma H)\left\{\sqrt{1 + \frac{a^2}{(g\beta H + \gamma H)^2}} - 1\right\} \qquad (3)$$

$$|-\tfrac{1}{2}, -\tfrac{1}{2}\rangle = -\tfrac{1}{2}g\beta H + \tfrac{1}{2}\gamma H + \tfrac{1}{4}a \qquad (4)$$

$$|-\tfrac{1}{2}, \tfrac{1}{2}\rangle = -\tfrac{1}{2}g\beta H - \tfrac{1}{2}\gamma H - \tfrac{1}{4}a - \tfrac{1}{2}(g\beta H + \gamma H)\left\{\sqrt{1 + \frac{a^2}{(g\beta H + \gamma H)^2}} - 1\right\} \qquad (5)$$

Cast in slightly different form, (2) - (5) become the Breit-Rabi (24) expressions for the energies of the magnetic sublevels. The normal ESR transitions $\langle\frac{1}{2}, \frac{1}{2}| \leftarrow \langle-\frac{1}{2}, \frac{1}{2}|$ and $\langle\frac{1}{2}, -\frac{1}{2}| \leftarrow \langle-\frac{1}{2}, -\frac{1}{2}|$ are obtained by subtracting (5) from (2) and (4) from (3) respectively, and equating the differences to the microwave frequency (ν, MHz). Assuming a has a positive sign this gives for the low-field line:

$$\nu = g\beta H_l + \tfrac{1}{2}a + \tfrac{1}{2}(g\beta H_l + \gamma H_l)\left\{\sqrt{1 + \frac{a^2}{(g\beta H_l + \gamma H_l)^2}} - 1\right\} \qquad (6)$$

and for the high-field line:

$$\nu = g\beta H_h - \tfrac{1}{2}a + \tfrac{1}{2}(g\beta H_h + \gamma H_h)\left\{\sqrt{1 + \frac{a^2}{(g\beta H_h + \gamma H_h)^2}} - 1\right\} \qquad (7)$$

A "first-order" solution is obtained by the complete neglect of the terms $a^2/(g\beta+\gamma)^2H^2$ in (6) and (7). This will usually only correspond within experimental error to the exact solution when a is very small (less than 30 G). An exact solution is achieved by a process of iteration in which successive approximations to g and a are substituted into the square root terms of (6) and (7) and the resulting linear simultaneous equations solved. The first-order and exact solutions differ markedly for large values of $\underline{a}$. This is shown in Table I for the case of FCO where the low- and high-field resonances were observed at 3097.76 G and 3430.19 G with an observing frequency of 9164.78 MHz.

Table I First-order, second-order and exact solutions of equations (6) and (7) as applied to FCO in liquid Freon 13 at 110°K.

Solution	$\underline{g}$	$\underline{a}_{19}$(G)
First-order	2.00617	332.43
Second-order	2.00097	331.57
Exact	2.00102	331.58

A good approximation to the exact solution can be obtained in many instances by applying second-order shifts (25) to the measured line positions. This amounts to retaining the first two terms only in the binomial expansion of the square root in (6) and (7). Shifts of $\underline{a}^2/4\underline{g}^2\beta^2H$ (neglecting γ) are added to H_l and H_h before computing the ESR parameters (Table I), H_h-H_l being used as the initial estimate of $\underline{a}/\underline{g}\beta$.

An algebraic solution is always possible when the unpaired electron interacts with a single nucleus or with an effective single nucleus. The latter situation arises when magnetically equivalent nuclei are present. Thus, with four equivalent ^{19}F ($\underline{I} = \frac{1}{2}$) nuclei, as in the radical SF_5, the observed spectrum (Fig. 1) is most simply treated by considering the interactions of the unpaired electron with all the nuclear states which arise through combination of four nuclear spins $\frac{1}{2}$; i.e. $\underline{I}$ = 2, 1 (triply degenerate) and 0 (doubly degenerate). One then expects the spectrum to consist of three superposed sub-spectra: a quintet (from $\underline{I}$ = 2) of unit intensity, a triplet (from $\underline{I}$ = 1) of intensity 3 and a singlet (from $\underline{I}$ = 0) of intensity 2. The entire spectrum has the appearance 1 : 1, 3 : 1, 3, 2 : 1, 3 : 1 with 1, 3 and 3, 2 second-order splittings of approximately $2\underline{a}^2/\underline{g}^2\beta^2H$ and $\underline{a}^2/\underline{g}^2\beta^2H$, respectively (25). Only for a small hyperfine interaction (or large line-width) do the second-order components merge to give the more familiar 1 : 4 : 6 : 4 : 1 pattern associated with four equivalent spins $\frac{1}{2}$.

The presence of resolved second-order components in an ESR spectrum is of great value in establishing firstly the equivalence or otherwise of magnetic nuclei in the free radical, and secondly a statistical best-fit set of parameters to the observed spectrum.

In a radical with more than one large hyperfine interaction it is sometimes possible to establish the relative signs of the hyperfine interactions on the basis of such statistical analyses (3). When the second-order components are unresolved there is, of course, always some doubt as to the significance of the measured "centers" of the lines and a consequent uncertainty in the parameters derived therefrom.

In cases where two or more large hyperfine interactions are present it is not always possible to obtain algebraic expressions for the magnetic sublevels. Good approximations to the spectral parameters may usually be obtained by the procedure, outlined above, of applying second-order shifts. This method breaks down for very large hyperfine interactions or for cases where the hyperfine interactions differ from each other by the order of a second-order shift (22, 25). An exact solution can always be obtained, however, by computerized diagonalization of the Hamiltonian (26). When excess data are available, i.e. when the number of measured lines exceeds the number of spectral parameters, the best-fit parameters should be determined by a least-squares method for all possible relative signs of the hyperfine interactions (3). A comparison of the standard deviations obtained for the various sign combinations using the F-test may then permit a judgment as to which combination is most probably the correct one.

Very large hyperfine interactions. A number of paramagnetic fluorides exhibit extremely large central-atom hyperfine interactions, e.g. 1576 G for ^{75}As in AsF_4(12), 6200 G for ^{127}I in IF_6 (13). In such instances, where $\underline{a}$ approaches or exceeds $\nu(\underline{I} + \frac{1}{2})^{-1}$ MHz, ESR spectra bear no resemblance to the hyperfine patterns predicted by first-order theory. The resonances are not symmetrically disposed about $\underline{g} = 2$ and when $\underline{a} > \nu(\underline{I} + \frac{1}{2})^{-1}$ only one ESR transition, the highest-field transition, is detectable (22).

It is of great help in such cases to express both the magnetic field and the hyperfine interaction in dimensionless form: i.e. $\underline{A} = \underline{a}/\nu$, $\underline{H} = H\underline{g}\beta/\nu$. If one neglects γ, the magnetogyric ratio, solutions to (1) take on a remarkably simple form when expressed in terms of these dimensionless variables (22). Thus, the highest field ESR transition is given by

$$\underline{A} = (2\underline{H} - 2)/(2\underline{I} + 1 - \underline{H}) \quad , \qquad (8)$$

an expression which may be used to calculate an approximate value of the hyperfine interaction from an assumed $\underline{g}$ value and the position of the transition. Plots of $\underline{A}$ versus $\underline{H}$ for a given spin are invaluable in assigning the observed transitions. Fig. 2,

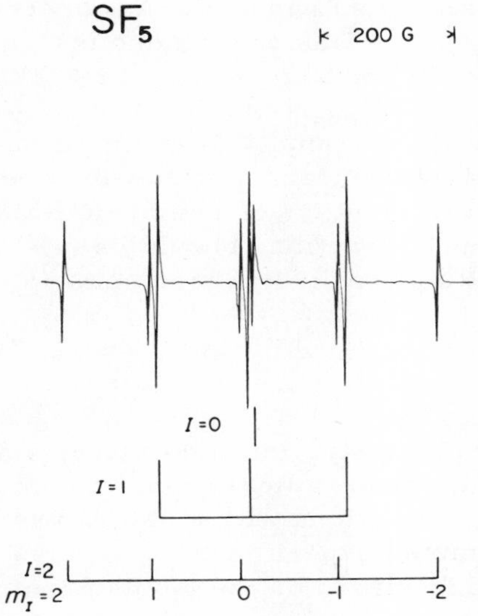

Figure 1. ESR spectrum of SF_5 at 100°K, obtained by γ-irradiation of SF_6

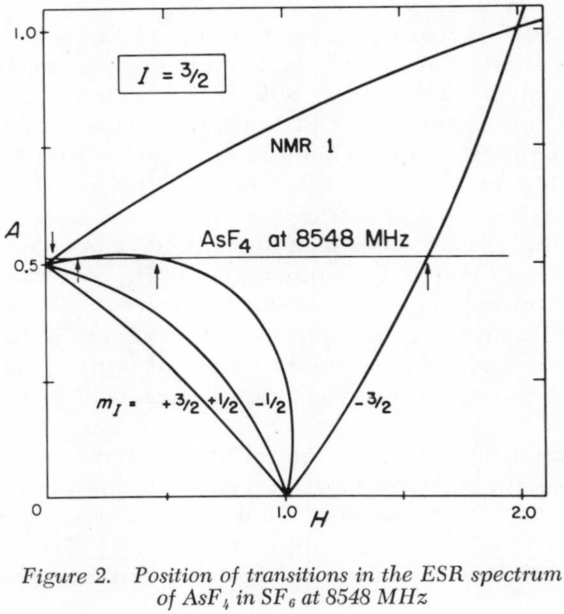

Figure 2. Position of transitions in the ESR spectrum of AsF_4 in SF_6 at 8548 MHz

for example, shows that the spectrum of AsF_4 in SF_6 at an observing frequency of 8548 MHz consists of the ESR transitions $\underline{m}_{I(As)} = -3/2$, $\underline{m}_{I(As)} = -\frac{1}{2}$ (appearing twice) and a forbidden, or NMR, transition.

Provided the unshifted centres of the transitions can be located the central-atom hyperfine interaction and g value may be determined exactly from an iterative solution of the Breit-Rabi equations. Ligand fluorine hyperfine interactions should be determined from the highest-field transition available where spin-spin decoupling is essentially complete.

Predicting Hyperfine Interactions

The process of establishing a geometry for and assigning a ground-state electronic symmetry to a free radical is one of comparison of the measured hyperfine interactions with those (a) predicted theoretically, and (b) determined experimentally for isoelectronic species. Great reliance is placed on the isoelectronic principle: free radicals having the same numbers of electrons and nuclei will have similar structures and unpaired electron spin density distributions. In cases where the isoelectronic principle cannot be invoked for lack of examples one must resort to the theoretical estimation of hyperfine interactions in order to ascertain that a proposed electronic structure is consistent with the experimental parameters.

The most satisfactory theoretical approach, ab initio calculation, is costly and to date has only been used for small molecules (27). An alternative approach, which has proved to be of great value, is to estimate atomic orbital spin densities by semi-empirical molecular orbital methods such as the Huckel or INDO methods (28). The choice of factors for converting the unpaired a.o. spin densities so obtained to hyperfine interactions is a subject of some contention. Empirical conversion factors have been suggested for use with INDO - derived spin densities (28). Alternatively, one may employ atomic coupling constants derived ab initio from a particular set of atomic wave functions. This is the approach which we favour, and, if only for the sake of consistency, we have used the atomic constants derived from Froese's (29) wave function (Table II).

We have found that VSEPR theory (30) is remarkably successful in predicting the structures of inorganic free radicals providing one assumes that the unpaired electron behaves very much as would a lone pair. This is tantamount to saying that free-radical cations (and radicals isoelectronic with such species) have the

Table II

One-electron Hyperfine Constants

$$A_0 = \left(\frac{8\pi}{3}\right) \gamma_e \gamma_M \psi_{ns}^2(0) \text{ Gauss}$$

Nucleus	A_0	Nucleus	A_0
^{7}Li	103.2	^{75}As	3393
^{9}Be	-32.8	^{77}Se	4792
^{11}B	720.8	^{79}Br	7737
^{13}C	1110	^{83}Kr	-1434
^{14}N	549.6	^{85}Rb	199.4
^{17}O	-1649	^{87}Sr	-165.0
^{19}F	17100	^{107}Ag	-417.3
^{21}Ne	-2041	^{111}Cd	-2317
^{23}Na	223.4	^{115}In	3412
^{25}Mg	-118.6	^{119}Sn	-7554
^{27}Al	980.0	^{121}Sb	6032
^{29}Si	-1209	^{125}Te	-9690
^{31}P	3638	^{127}I	7295
^{33}S	969.5	^{129}Xe	-11786
^{35}Cl	1666	^{133}Cs	349.2
^{39}K	51.7	^{137}Ba	525.7
^{43}Ca	-148.9	^{197}Au	311.2
^{63}Cu	1764	^{199}Hg	3401
^{67}Zn	452.5	^{205}Tl	14984
^{69}Ga	2658	^{207}Pb	6824
^{73}Ge	-530.9	^{209}Bi	6336

same structure as the parent molecule from which they are derived. Thus, NF_3^+, CF_3 and BF_3^- are expected to have the same geometry (pyramidal) as NF_3 itself; SF_4^+ (if and when found) and PF_4 should, like SF_4, have trigonal bipyramidal structures; ClF_3^+ and SF_3 should be T-shaped like ClF_3.

A second generalisation concerns the ^{19}F hyperfine interactions in isoelectronic pairs of radicals, one of which is an anion radical, the other neutral: the ^{19}F hyperfine interaction in the anion is larger than it is in the neutral species. Many examples may be quoted: PF_5^-, SF_5; SiF_4^-, PF_4; SF_6^-, ClF_6; OPF_3^-, OSF_3. In fact, we know of no exceptions to this rule and believe that it may be a consequence of weaker, more polarizable bonds in the negatively charged species. There is as yet insufficient data to establish a similar generalisation for cation radicals.

Semi-occupied Orbital of Inorganic Fluorine Radicals

The free radicals we shall discuss in this article have in common the basic structure of their semi-occupied orbitals. In its simplest form this semi-occupied molecular orbital is the linear combination F(2p) - M(ns) - F(2p), in which the fluorine 2p orbitals point towards the central atom (M) and are antibonding with respect to the M(ns) contribution (Fig. 3, top left). Also illustrated in Figure 3 are the prototypes of some of the fluorine-containing free radicals we shall be discussing. These examples illustrate the different ways linear F-M-F arrays can be combined to form molecular orbitals of various symmetries. In the case of the radicals SF_3 and PF_4 a single F-M-F array is used, but their low symmetries permit central-atom $3p_z$ and apical fluorine $2p_z$ orbitals (z is the two-fold axis) also to be incorporated.

For radicals of higher symmetry, such as ClF_4 (D_{4h}) and SF_5 (C_{4v}) two mutually perpendicular F-M-F arrays are utilized in the formulation of the semi-occupied orbital. Again, the lower symmetry of SF_5 permits a contribution from sulfur $3p_z$ and fluorine $2p_z$ atomic orbitals.

Finally, in the case of the halogen hexafluorides the high symmetry (O_h) permits a description of the semi-occupied orbital in terms of a three-dimensional F-M-F array. In each case, the ligand (F) 2p orbitals point towards the central atom and the orbital is antibonding in each M-F distance.

The contribution of central-atom ns character to the semi-occupied orbital gives rise to a distinctive feature in the ESR spectra of many of these radicals, namely a very large central atom hyperfine interaction, often several kiloGauss. On the other hand, the isotropic ^{19}F hyperfine interactions are surprisingly small (∿200 G). A pure fluorine 2p contribution to the semi-occupied molecular orbital would generate no isotropic hyperfine interaction with the unpaired electron. The small, though finite, ^{19}F hyperfine interactions in these radicals are thought to be due entirely to M-F bond polarisation effects, and not to any significant direct contribution from fluorine 2s atomic orbitals.

We shall frequently have occasion to discuss the effect on the various spectral parameters of changes in the electronegativity of the central atom, or of an RO ligand, in a series of radicals. In this connection it is important to remember that the semi-occupied orbital is anti-bonding (i. e. has a node) between the central atom and its ligands. An increase in the central atom (M) electronegativity will therefore result in a decrease in the contribution of M(ns, np) orbitals to the semi-occupied orbital (31). Conversely, an increase in the electronegativity of a ligand (e. g. an RO group) will result in an increase in the central-atom contribution to the semi-occupied molecular orbital.

Preparation of Inorganic Fluorine-Containing Radicals

Three methods of preparation are commonly used:

a) Photolysis of hypofluorites. If a hypofluorite such as CF_3OF is photolysed with UV light the radicals CF_3O and F are generated in the liquid. These species cannot be detected by ESR, and in the absence of any third entity, they merely recombine. If, however, the CF_3OF contains a solute, such as SF_4, then both CF_3O radicals and F atoms attack the SF_4, resulting in the formation of the free-radicals CF_3OSF_4 and SF_5 respectively (6). (Fig. 4). A variation of this method is the use of the peroxide ROOR, rather than the hypofluorite. In this case only a single adduct ($ROSF_4$) is formed. This technique has been used to study the ESR spectra of a series of adducts to SF_4 and to PF_3, by varying the nature of the peroxide ROOR (32).

b) Gamma-irradiation in neo-pentane. A solid solution of the inorganic fluorine compound in neo-pentane often reveals, after γ-irradiation at 77°K, the ESR spectrum of a negative ion. For example, the spectrum of SiF_4^- may be observed after neo-pentane containing dissolved SiF_4 is γ-irradiated at 77°K (19). This technique is not restricted to inorganic fluorides: the spectrum of $C_4F_8^-$ has recently been observed in irradiated C_4F_8:neo-pentane solutions (33).

c) Gamma-irradiation in SF_6. By far the most fruitful technique of the three has been that of γ-irradiation of a solid solution

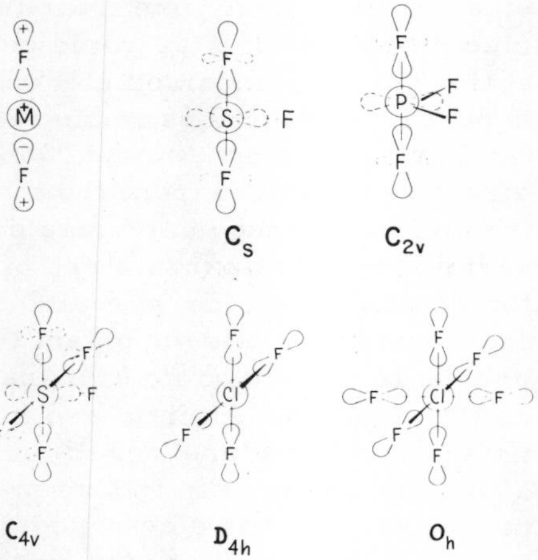

Figure 3. Fluorine-containing inorganic radicals of various symmmetries

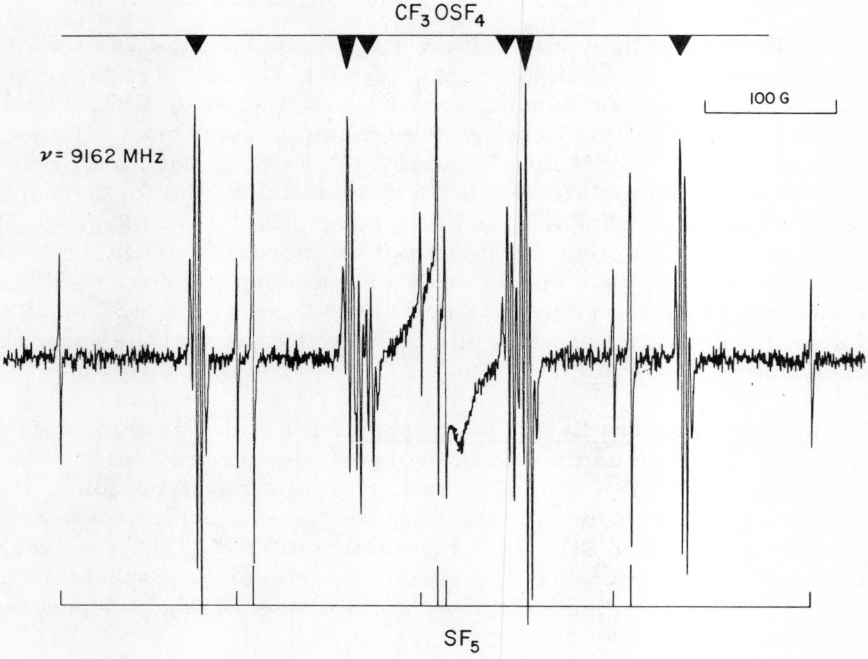

Figure 4. ESR spectra of CF_3OSF_4 and SF_5 obtained by photolysis of CF_3OF:SF_4 at 180°K

of a solute in SF_6, introduced by Fessenden and Schuler in 1966 (2). Irradiation of the pure solvent (SF_6) yields ESR spectra of the radicals SF_3, SF_5 and SF_6^-. If a solute is present, one usually obtains, in addition, the spectrum of a fluorine atom adduct. For example, irradiation of PF_3:SF_6 yields the spectrum of PF_4 (2). This spectrum, and many others which may be generated by this technique, will be discussed later.

From the spectroscopic point of view, the advantage of the SF_6 matrix technique is that isotropic spectra are obtained, in contrast to the neo-pentane matrix, in which the spectra are usually anisotropic. This means that in SF_6 the spectroscopist is dealing with a spectrum of sharp ($\Delta H \approx 1$ G), well-resolved lines susceptible to accurate measurement. Interpretation of the spectrum and identification of its carrier is thus facilitated.

Tri-coordinated Fluoride Radicals

The trivalent fluoro-radicals which we shall discuss are SF_3 and its derivatives. The ESR spectra of CF_3 and other perfluoro-alkyl radicals have been adequately treated elsewhere (34, 35). The radical SF_3 was first detected in electron-irradiated SF_6(9). Its spectrum indicates the presence of two, rather than three equivalent ^{19}F nuclei: $a_{19}(2) = 54.3$ G, $a_{19}(1) = 40.4$ G. This observation is consistent with the structure of SF_3 shown in Fig. 3, in which the semi-occupied orbital is described as an antibonding combination of $F_{ap}(2p)$ orbitals and the S(3s, $3p_z$) orbitals. The bonds to the apical fluorines are weaker, longer, and more polarizable, resulting in the somewhat larger hyperfine interaction of the apical ^{19}F nuclei.

Table III Spectral data for SF_3 and derivatives in solution (36).

Radical	g	a_{19}(apical)	a_{19}(equat)	
SF_3	2.0050	48.7	41.0	Gauss
$(CF_3O)_2SF$	2.0074	1.93	6.96	
$(SF_5O)_2SF$	2.0076	2.28	6.16	
$(CF_3O)_3S$	2.0060	1.38	0.27	
$(SF_5O)_3S$	2.0059	1.49	(not resolved)	

Certain derivatives of SF_3 have been prepared. These are $(R_fO)_2SF$ and $(R_fO)_3S$, where $R_f = CF_3$ or SF_5 (36). In the case of the $(R_fO)_2SF$ radicals, the R_fO ligands occupy equivalent apical positions. As in the case of the corresponding derivatives of PF_4, the apical position seems to be the preferred site for R_fO substitution. As expected, the trisubstituted derivatives have larger ^{19}F hyperfine interactions for the R_fO ligands in the apical positions than for those in the equatorial positions (Table III).

A species closely related to SF_3, AsF_3^-, has been detected in γ-irradiated polycrystalline AsF_3 (37). The ESR parameters for this radical are consistent with a planar T-shaped structure; we feel, however, that the unpaired electron occupies a totally symmetric orbital, rather than an orbital antisymmetric with respect to the molecular plane as suggested by Subramanian and Rogers (37).

Tetra-coordinated Fluoride Radicals

The prototype of this group of radicals is PF_4, discovered in 1966 by Fessenden and Schuler (2) who generated its spectrum by irradiation of a solid solution of PF_3 in SF_6 with 2.8 MeV electrons. This radical, and its derivatives, have a "trigonal" bipyramidal structure. The apical and equatorial ^{19}F nuclei of PF_4 have isotropic hyperfine interactions of 282 G and 59 G respectively, the ^{31}P hyperfine interaction being 1330 G. A more recent statistical treatment of the PF_4 spectrum showed that all three hyperfine interactions in PF_4 had the same (positive) sign (3).

The radical AsF_4 has also been detected by this technique, being observed at 110°K in samples of AsF_3:SF_6 γ-irradiated at 77°K (12). Its spectrum has several features of interest, but for the moment we note that the hyperfine interaction of the ^{75}As ($I = 3/2$) nucleus is so large that the appearance of the spectrum (Fig. 2) bears little similarity to that of the analogous PF_4 radical. Its spectral parameters can, however, readily be obtained with the aid of the Breit-Rabi equations (24).

Although SF_6 is a useful matrix in which to prepare neutral fluorine containing radicals, it is not usually as effective as neopentane for the preparation of negative ions. Thus, in order to compare PF_4 with SiF_4^-, its negatively charged analogue, a solid solution of SiF_4 in neo-pentane was irradiated at 77°K (19).

In Table IV the isotropic hyperfine parameters of SiF_4^-, PF_4 and AsF_4 are listed, together with an estimate of the central-atom ns contribution to the semi-occupied orbital. For this purpose, we have used Froese's UHF wavefunction (29) to obtain estimates of $\psi^2(0)$ and hence A_0 (Table II).

Table IV Hyperfine interactions for tetravalent radicals of Group V.

Radical	Hyperfine Interactions (Gauss) M ($\underline{a}_M/A_0$)	$F_{(ap)}$	$F_{(eq)}$
$^{29}SiF_4^-$	-415.0 (0.34)	309.6	80.7
$^{31}PF_4$	1321.5 (0.36)	293.0	59.8
$^{75}AsF_4$	1576.4 (0.46)	238.2	49.5
PF_3OEt_{eq}	1267	285	63.0
$PF_2(OEt_{eq})_2$	1201	280	-
$PF_{ap}(OEt)_3$	1033	306	-

As discussed in an earlier section, the semi-occupied orbital of the radicals we consider in this article are antibonding between ligand F(2p) atomic orbitals and central atom M(ns) atomic orbitals. In the present case the semi-occupied orbital is primarily an anti-bonding $F_{ap}(2p)$-M(ns)-$F_{ap}(2p)$ combination.

An increase in the electronegativity of M will result in a decrease in the M(ns) contribution to all anti-bonding orbitals, including the semi-occupied orbital. This effect is noted in Table IV for AsF_4 and PF_4, where it is seen that both the ^{19}F and the central-atom hyperfine constants indicate higher spin-density on the central-atom in AsF_4 than in PF_4.

Comparing PF_4 and SiF_4^-, however, there is a much smaller difference in the central atom character. The ^{19}F hyperfine interactions are, nevertheless, significantly larger in SiF_4^- than in PF_4, an effect we attribute to the weaker, more polarizable bonds of the anionic species. We regard the ^{19}F hyperfine interactions in these radicals, although positive in sign, as arising from polarization effects and not from any direct F(2s) contribution to the semi-occupied molecular orbital.

It is also possible to examine the effects of ligand electronegativity-change in a series of radicals such as $(RO)_{ap}PF_3$.

These species, prepared by photolysis of a peroxide ROOR containing dissolved PF_3 (10), show a linear correlation between $\underline{a}_{31}$ and $\underline{a}_{19(ap)}$ (Fig. 5). The more electronegative apical ligands (FSO_3, FCO_2) are associated with higher ^{31}P hyperfine interactions than the less electronegative CF_3O and $(CF_3)_3CO$ ligands. The $^{19}F_{(eq)}$ hyperfine interactions are not significantly affected by changes in electronegativity at the apical position. Increase in the electronegativity of the apical RO ligand results in greater P(3s, 3p) contributions to the semi-occupied orbital, because the latter is anti-bonding in the P-O distance. Increased P(3s) character is measured directly as an increased ^{31}P hyperfine interaction. The increased P(3p) contribution, however, is registered only by its effect on the apical fluorine, causing increased $P\text{-}F_{(ap)}$ polarization and a reduction in the $^{19}F_{(ap)}$ hyperfine interaction. An exactly analogous effect has been observed in a series of $ROSF_4$ radicals (32).

The dominance of the apical ligands in the LCAO description of the semi-occupied orbital of these radicals is emphasised not only by the invariance of $\underline{a}_{19(eq)}$ to apical substitution, but also by the relatively minor changes in $\underline{a}_{31}$ and $\underline{a}_{19(ap)}$ on equatorial substitution. Alkoxy groups, as opposed to fluoro alkoxy groups, appear preferentially in equatorial positions in phosphoranyl radicals. A comparison of the data for PF_4 with those for its ethoxy-substituted derivatives (Table IV) shows that apical substitution of OEt for F has a far greater effect on the ^{31}P and $^{19}F_{(ap)}$ hyperfine interactions than does equatorial substitution (38).

Mention should perhaps be made here of another kind of tetravalent fluoride radical, ClF_4, obtained by UV photolysis of CF_3OF containing dissolved Cl_2 (7), or by γ-irradiation of a solid solution of ClF_3 in SF_6 (14). The spectral parameters of ClF_4 are $\underline{g}$ = 2.0118, $\underline{a}_{35}$ = 288.2 G, $\underline{a}_{19}(4)$ = 78.8 G. The value of the ^{35}Cl hyperfine interaction indicates a chlorine 3s contribution to the semi-occupied molecular orbital of approximately 0.17, which seemed reasonable for a square-planar configuration. The alternative bisphenoidal structure was ruled out on the grounds that (a) the spectrum of ClF_4 indicated it contained four equivalent fluorines and (b) such a structure would imply a considerable chlorine 4s contribution to the semi-occupied molecular orbital, and a correspondingly small ^{35}Cl hyperfine interaction. Our conclusion regarding the planarity of ClF_4 has generated some debate among theoretical chemists (39), but now appears to have been accepted (40).

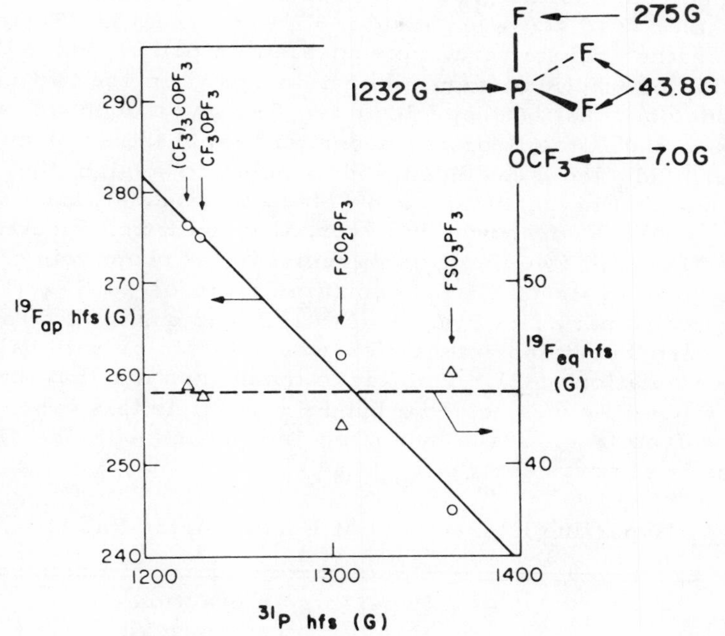

Figure 5. Correlation between a_{31} *and* $a_{19(ap)}$ *in radicals of the type* R_fOPF_3

Penta-coordinated Fluoride Radicals

In 1963 one of the present authors obtained at 300°K the ESR spectrum in the top half of Fig. 6 by γ-irradiation of powdered NH_4PF_6 (41). It is a spectrum showing hyperfine interactions with a ^{31}P nucleus (1354 G) and four equivalent ^{19}F nuclei (198 G). This spectrum was assigned to the radical PF_4. Later, Fessenden and Schuler (2) discovered another PF_4 spectrum in SF_6 at 140°K in which the ^{19}F nuclei were equivalent in pairs. Since the ^{31}P hyperfine interactions were almost identical in the two cases it was concluded that both species were PF_4 and that in an NH_4PF_6 matrix at 300°K the fluorines underwent rapid intramolecular exchange. This led Fessenden and Schuler to assign the lower-spectrum in Fig. 6, obtained by electron-irradiation of SF_6, to SF_4^+. It was later shown, however, that the latter spectrum was that of SF_5 (6). Two methods were used: (a) photolysis of SF_5Cl and (b) photolysis of CF_3OF containing dissolved SF_4 (Fig. 4). The extraordinary fact that the fifth ^{19}F nucleus in SF_5 had a virtually zero hyperfine interaction was later to be rationalized by INDO calculations (42). This led to the suggestion that the spectrum in NH_4PF_6 was not PF_4 but PF_5^- (43); in this case, as can be seen from Fig. 6, the hyperfine interaction with the fifth ^{19}F nucleus is almost resolved.

Table V Hyperfine interactions of Pentafluoride Radicals.

Radical	Hyperfine Interaction (G) M (a_M/A_0)	F
PF_5^-	1354.4 (0.37)	197.6
SF_5	306.8 (0.32)	143.0
AsF_5^-	1762 (0.52)	187.0
SeF_5	1877 (0.39)	118.0

The spectrum of AsF_5^- has been obtained in SF_6 (21), and in Table V its parameters are compared with those of PF_5^-, SF_5, and SeF_5. We see that the radicals centred on the more electro-

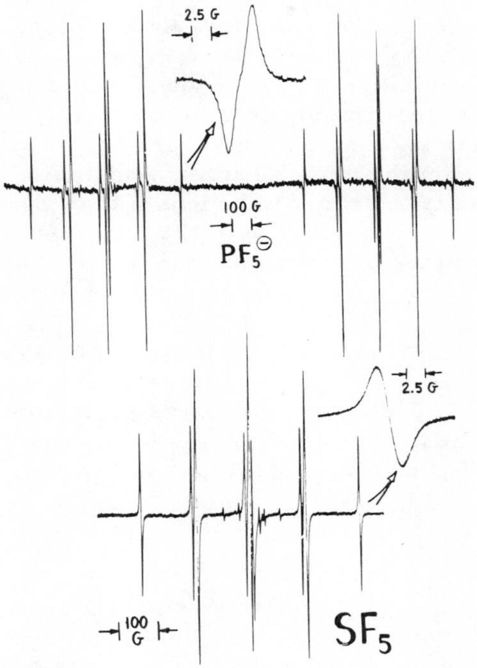

Figure 6. ESR spectra of PF_5^- (top) and SF_5 (bottom) in γ-irradiated NH_4PF_6 and SF_6, respectively

negative third row elements (PF_5^-, SF_5) have larger ^{19}F hyperfine interactions than the corresponding species centered on the fourth row elements (AsF_5^-, SeF_5). In addition, the central atom character, measured as a fraction of A_0 (Table II), is lower for PF_5^- and SF_5 than for AsF_5^- and SeF_5 respectively. These facts tend to confirm the hypothesis that increasing the electronegativity of M (e.g. As→P) will result in a lower M(ns) contribution to the semi-occupied molecular orbital.

As between the isoelectronic radicals PF_5^-, SF_5 and AsF_5^-, SeF_5 we conclude that the difference in the ^{19}F hyperfine interactions is essentially Coulombic in origin. The anionic species have less charge on the central atom, and their bonds are therefore slightly weaker and more polarizable than those of the neutral species. The dominant contribution to the ^{19}F hyperfine interaction being polarization by spin-density in the F(2p) atomic orbitals, the net result is larger isotropic ^{19}F hyperfine interactions.

It is possible to prepare a series of pentavalent derivatives of SF_5 by photolysis of various peroxides containing dissolved SF_4 (32). These $ROSF_4$ radicals have C_s symmetry, and their dominant hyperfine interactions are: $\underline{a}_{19}(1)$, that of the ^{19}F nucleus trans to the RO ligand and $\underline{a}_{19}(2)$ those of the two ^{19}F nuclei cis to the RO ligand:

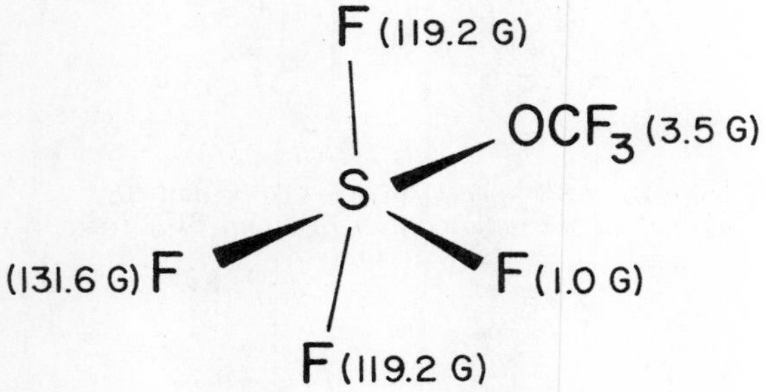

The data in Table VI illustrate the effect on $\underline{a}_{19}(1)$ and $\underline{a}_{19}(2)$ of changing the electronegativity of the ligand RO. It will be seen that there is a dramatic effect on $\underline{a}_{19}(1)$, which decreases with increasing electronegativity in R. The value of $\underline{a}_{19}(2)$, however, is virtually unaffected by this change.

Table VI Hyperfine interactions of various $ROSF_4$ radicals at 228°K.

R	$a_F(1)$	$a_F(2)$	Other hfs
$(CH_3)_3Si$	184.3	121.5	
CH_3	171.6	122.3	$a_1(3) = 2.4$
CF_3	131.6	119.2	$a_{19}(3) = 3.5$; $a_{19}(1) = 1.0$
SF_5	127.1	118.9	$a_{19}(4) = 5.0$
$(CF_3)_3C$	117.6	117.6	$a_{19}(6) = 6.6$
FSO_2	111.4	120.2	

This effect is probably due to an increased sulfur $3p_z$ contribution to the semi-occupied molecular orbital as the electronegativity of RO increases. An increased sulfur $3p_z$ contribution will cause increased polarization of the S-F bond, resulting in a decrease in the ^{19}F hyperfine interaction. It is not clear, however, why the effect is transmitted to the trans fluorine nucleus only, and it is a pity that no data are yet available on the ^{33}S hyperfine interaction and hence on changes in the sulphur 3s contribution to the semi-occupied molecular orbital.

Hexafluoride Radicals

The last group of radicals we shall discuss are the hexafluoride radicals of Groups VI and VII. The radical SF_6^-, first discovered by Fessenden and Schuler (2), is obtained by irradiation of SF_6. Its analogues SeF_6 and TeF_6 are obtained by irradiation of SF_6 containing dissolved SeF_6 and TeF_6 respectively (11). Their ESR data are given in Table VII. The halogen hexafluorides ClF_6 and BrF_6 were prepared by radiolysis of ClF_5:SF_6 (13, 14) and BrF_5:SF_6 (13, 15) respectively, whereas IF_6 was obtained by UV photolysis of IF_7:SF_6 (13).

The most abundant isotopes of sulfur, selenium and tellurium have zero spin, and the spectra of SF_6, SeF_6 and TeF_6 containing these isotopes show interactions of six equivalent ^{19}F nuclei only. In addition, weak satellite spectra have been detected from which it was possible to obtain the hyperfine interactions of the isotopes ^{33}S, ^{77}Se and ^{125}Te in SF_6, SeF_6 and TeF_6 respectively (11).

The halogen hexafluorides ClF_6, BrF_6 and IF_6, on the other hand, have central atoms whose nuclei have non-zero nuclear spin. The appearance of their spectra is therefore dominated by the central-atom hyperfine interactions, which (particularly for BrF_6 and IF_6) are extremely large and necessitated unusual methods of analysis (22).

Table VII ESR Parameters for Hexafluoride Radicals.

Radical	g	Hyperfine Interactions (Gauss)	
		a_M	$a_{19}(6)$
$^{33}SF_6^-$	2.0078	643	195.4
$^{77}SeF_6^-$	2.0098	3634	173.3
$^{125}TeF_6^-$	2.0070	-10081	212.0
$^{35}ClF_6$	2.0181	775.5	89.6
$^{79}BrF_6$	2.0158	4160	88.5
$^{127}IF_6$	2.0098	6140	150.2

These radicals are thought to have octahedral symmetry, and their semi-occupied orbital is thought to be a three-dimensional version of the F(2p) - M(ns) - F(2p) array illustrated in Figure 1 (44). The central atom ns character has been estimated for SF_6^-, SeF_6^- and TeF_6^- with the aid of Froese's wavefunction (Table II) to be 0.66, 0.76 and 1.04 respectively. The latter value being in excess of unity leads one to suspect that values obtained by incorporating the Mackey-Wood (45) correction would be more appropriate: 0.65, 0.66, 0.68. The corresponding figures for ClF_6, BrF_6 and IF_6 are: 0.46, 0.46, 0.54.

An attempt was made to verify the latter figures by an analysis of the anisotropic ^{19}F hyperfine pattern in ClF_6, BrF_6 and IF_6 at 27°K (44). This analysis proved conclusively that the halogen hexafluoride radicals are octahedral, since the anisotropic hyperfine patterns were those of pairs of equivalent ^{19}F nuclei. The individual F(2p) spin-densities were found to be $\sim$0.2. The overall estimated spin-density in all three halogen hexafluorides is thus considerably greater than unity, and one is forced to conclude that polarization phenomena have generated considerable negative spin-density in various atomic orbitals (INDO calculations suggest primarily the M(np) orbitals). These arguments tend to reinforce our conclusion that in many of these σ- or $^2A_1$ radicals polarization effects are of great importance and are, in fact, primarily responsible for the non-zero ^{19}F hyperfine interactions.

Literature Cited

1. Issued as NRCC No. 15631.
2. Fessenden, R.W. and Schuler, R.H., J. Chem. Phys. (1966) 45, 1845.
3. Fessenden, R.W., J. Magnetic Resonance (1969) 1, 277.
4. Nelson, W. and Gordy, W., J. Chem. Phys. (1969) 51, 4710.
5. Patten, F.W., Chem. Phys. Lett. (1973) 18, 112.
6. Morton, J.R. and Preston, K.F., Chem. Phys. Lett. (1973) 18, 98.
7. Morton, J.R. and Preston, K.F., J. Chem. Phys. (1973) 58, 3112.
8. Morton, J.R. and Preston, K.F., J. Chem. Phys. (1973) 58, 2657.
9. Colussi, A.J., Morton, J.R., Preston, K.F. and Fessenden, R.W., J. Chem. Phys. (1974) 61, 1247.
10. Colussi, A.J., Morton, J.R. and Preston, K.F., J. Phys. Chem. (1975) 79, 651.
11. Morton, J.R., Preston, K.F. and Tait, J.C., J. Chem. Phys. (1975) 62, 2029.
12. Colussi, A.J., Morton, J.R. and Preston, K.F., Chem. Phys. Lett. (1975) 30, 317.
13. Boate, A.R., Morton, J.R. and Preston, K.F., Inorg. Chem. (1975) 14, 3127.
14. Nishikida, K., Williams, F., Mamantov, G. and Smyrl, N., J. Amer. Chem. Soc. (1975) 97, 3526.
15. Nishikida, K., Williams, F., Mamantov, G. and Smyrl, N., J. Chem. Phys. (1975) 63, 1693.
16. Nishikida, K. and Williams, F., J. Amer. Chem. Soc. (1975) 97, 7166.
17. Hasegawa, A., Sogabe, K. and Miura, M., Mol. Phys. (1975) 30, 1889.
18. Mishra, S.P., Symons, M.C.R., Christe, K.O., Wilson, R.D. and Wagner, R.I., Inorg. Chem. (1975) 14, 1103.
19. Morton, J.R. and Preston, K.F., Mol. Phys. (1975) 30, 1213.
20. Vanderkooi, N., Mackenzie, J.S. and Fox, W.B., J. Fluorine Chem. (1976) 7, 415.

21. Boate, A.R., Colussi, A.J., Morton, J.R. and Preston, K.F., Chem. Phys. Lett. (1976) 37, 135.
22. Boate, A.R., Morton, J.R. and Preston, K.F., J. Magnetic Resonance (1976) 24, 259.
23. Farach, H.A. and Poole, C.P., Adv. Magnetic Resonance (1971) 5, 229.
24. Breit, G. and Rabi, I.I., Phys. Rev. (1931) 38, 2082.
25. Fessenden, R.W., J. Chem. Phys. (1962) 37, 747.
26. Morton, J.R. and Preston, K.F., Program 311, Quantum Chemistry Program Exchange, Indiana University, Indiana.
27. Hudson, A. and Treweek, R.F., Chem. Phys. Lett. (1976) 39, 248.
28. Pople, J.A. and Beveridge, D.L., "Approximate Molecular Orbital Theory", McGraw-Hill Inc., 1970.
29. Froese, C., J. Chem. Phys. (1966) 45, 1417.
30. Gillespie, R.J., "Molecular Geometry" van Nostrand-Reinhold Co., London, (1972).
31. Atkins, P.W. and Symons, M.C.R., "The Structure of Inorganic Radicals," p. 103, Elsevier Publishing Company, Amsterdam (1967).
32. Gregory, A.R., Karavelas, S.E., Morton, J.R., and Preston, K.F., J. Amer. Chem. Soc. (1975) 97, 2206.
33. Shiotani, M. and Williams, F., J. Amer. Chem. Soc. (1976) 98, 4006.
34. Fessenden, R.W. and Schuler, R.H., J. Chem. Phys. (1965) 43, 2704.
35. Lloyd, R.V. and Rogers, M.T., J. Amer. Chem. Soc. (1973) 95, 1512.
36. Morton, J.R. and Preston, K.F., J. Phys. Chem. (1973) 77, 2645.
37. Subramanian, S. and Rogers, M.T., J. Chem. Phys. (1972) 57, 4582.
38. Elson, I.H., Parrott, M.J. and Roberts, B.P., J. Chem. Soc. Chem. Comm. (1975) 586.
39. Gregory, A.R., J. Chem. Phys. (1974) 60, 3713.
40. Ungemach, S.R. and Schaefer, H.F., J. Amer. Chem. Soc. (1976) 98, 1658.
41. Morton, J.R., Can. J. Phys. (1963) 41, 706.
42. Gregory, A.R., Chem. Phys. Lett. (1974) 28, 552.
43. Mishra, S.P., Symons, M.C.R., J. Chem. Soc. Chem. Comm. (1974) 279.
44. Boate, A.R., Morton, J.R. and Preston, K.F., J. Phys. Chem. to be published.
45. Mackey, J.H. and Wood, D.E., J. Chem. Phys. (1970) 52, 4914.

Acknowledgments

Figure 4 is reproduced from (6) by permission of the North-Holland Publishing Company, Amsterdam; Figure 5 is reproduced from (10) by permission of the American Chemical Society.

INDEX

D

E

G

H

J

K

L

M

N

R

U

V